T0328263

ENERGY DEMOCRACIES FOR SUSTAINABLE FUTURES

ENERGY DEMOCRACIES FOR SUSTAINABLE FUTURES

Foreword by

BENJAMIN SOVACOOL

Edited by

MAJIA NADESAN

*School of Social and Behavioral Sciences, Arizona State University,
Glendale, AZ, United States*

MARTIN J. PASQUALETTI

*School of Geographical Sciences and Urban Planning, Arizona State University, Tempe, AZ, United States;
Senior Global Futures Scientist, Julie Ann Wrigley Global Future Laboratory,
Arizona State University, Tempe, AZ, United States*

JENNIFER KEAHEY

School of Social and Behavioral Sciences, Arizona State University, Glendale, AZ, United States

ELSEVIER

ACADEMIC PRESS

An imprint of Elsevier

Academic Press is an imprint of Elsevier
125 London Wall, London EC2Y 5AS, United Kingdom
525 B Street, Suite 1650, San Diego, CA 92101, United States
50 Hampshire Street, 5th Floor, Cambridge, MA 02139, United States
The Boulevard, Langford Lane, Kidlington, Oxford OX5 1GB, United Kingdom

Notices
Knowledge and best practice in this field are constantly changing. As new research and experience broaden
our understanding, changes in research methods, professional practices, or medical treatment may become
necessary.

Practitioners and researchers must always rely on their own experience and knowledge in evaluating and
using any information, methods, compounds, or experiments described herein. In using such information or
methods they should be mindful of their own safety and the safety of others, including parties for whom they
have a professional responsibility.

To the fullest extent of the law, neither the Publisher nor the authors, contributors, or editors, assume any
liability for any injury and/or damage to persons or property as a matter of products liability, negligence or
otherwise, or from any use or operation of any methods, products, instructions, or ideas contained in the
material herein.

ISBN 978-0-12-822796-1

For information on all Academic Press publications
visit our website at https://www.elsevier.com/books-and-journals

Publisher: Charlotte Cockle
Acquisitions Editor: Graham Nisbet
Editorial Project Manager: Andrae Akeh
Production Project Manager: Kiruthika Govindaraju
Cover Designer: Vicky Pearson Esser

Typeset by STRAIVE, India

Contents

Theme 2

Futures

12. The future of energy ownership

Clark A. Miller

II

Transitions

Introduction to Part II: Energy Futures

Martin J. Pasqualetti

Theme 1

Organizing

13. Energies of resistance? Conceptualizing resistance in and through energy democratization

Joshua K. McEvoy

14. The role of ownership and governance in democratizing energy: Comparing public, private, and civil society initiatives in England

M. Lacey-Barnacle and J. Nicholls

15. Lessons from electric cooperatives: Evolving participatory governance practices

Stephanie Lenhart

16. Bringing democratic transparency to Karachi's electric sector

Ijlal Naqvi

Theme 2

Communities

III

Risks

Theme 1

Assemblages

Theme 2
Security

List of figures

List of tables

Contributors

Lourdes Alonso-Serna Universidad del Mar, Huatulco, Oaxaca, México

Peta Ashworth School of Chemical Engineering, The University of Queensland, Brisbane, QLD, Australia

Bidtah Becker CalEPA, Sacramento, CA, United States

Heather Plumridge Bedi Environmental Studies Department, Dickinson College, Carlisle, PA, United States

Melissa Bollman Graduate School of Geography, Clark University, Worcester, MA, United States

Johanna Bozuwa Climate+Community Project, San Francisco, CA, United States

Christian Brannstrom Department of Geography, Texas A&M University, College Station, TX, United States; Department of Geography, Federal University of Ceará, Fortaleza, Brazil

Ry Brennan Department of Sociology, University of California, Santa Barbara, CA, United States

Marie Claire Brisbois Science Policy Research Unit, University of Sussex, Brighton, United Kingdom

Matthew Burke Leadership for the Ecozoic, University of Vermont, Burlington, VT, United States

John Byrne Center for Energy & Environmental Policy, University of Delaware, Newark, DE; Foundation for Renewable Energy & Environment, New York, NY, United States

Zoé Chateau Department of Geography, College of Life and Environmental Sciences, University of Exeter, Exeter, United Kingdom

Araceli Clavijo Research Institute on Renewable Energy (INENCO), National Research Council of Argentina (CONICET), National University of Salta (UNSa), Salta, Argentina

Stan Cox Ecosphere Studies, The Land Institute, Salina, KS, United States

Jessica Craigg Energy & Resources Group, University of California—Berkeley, Berkeley, CA, United States

Patrick Devine-Wright Department of Geography, College of Life and Environmental Sciences, University of Exeter, Exeter, United Kingdom

Walter F. Diaz Paz Research Institute on Renewable Energy (INENCO), National Research Council of Argentina (CONICET), National University of Salta (UNSa), Salta, Argentina

Alexander Dunlap The Centre for Development and the Environment, University of Oslo, Oslo, Norway

Niall P. Dunphy Cleaner Production Promotion Unit, School of Engineering and Architecture; Environmental Research Institute, University College Cork, Cork, Ireland

Danielle Endres Communication, University of Utah, Salt Lake City, UT, United States

Melisa Escosteguy Research Institute on Renewable Energy (INENCO), National Research Council of Argentina (CONICET), National University of Salta (UNSa), Salta, Argentina

Andrea M. Feldpausch-Parker Environmental Studies, State University of New York College of Environmental Science and Forestry, Syracuse, NY, United States

Adryane Gorayeb Department of Geography, Federal University of Ceará, Fortaleza, Brazil

Matthew J. Herington Centre for Communication and Social Change, School of Communication and Arts, The University of Queensland, Brisbane, QLD, Australia

Anthony P. Heynen Energy & Poverty Research Group, School of Chemical Engineering, The University of Queensland, Brisbane, QLD, Australia

Karen Hudlet-Vazquez Graduate School of Geography, Clark University, Worcester, MA, United States

Marc Hufty Graduate Institute of International and Development Studies, Geneva, Switzerland

Martín A. Iribarnegaray Research Institute on Renewable Energy (INENCO), National Research Council of Argentina (CONICET), National University of Salta (UNSa), Salta, Argentina

Craig B. Jacobson Energy & Poverty Research Group, School of Chemical Engineering, The University of Queensland, Brisbane, QLD, Australia

Rudy Kahsar Environmental Studies, University of Colorado Boulder, Boulder, CO, United States

Jennifer Keahey Social and Behavioral Sciences, Arizona State University, Glendale, AZ, United States

Jörg Kemmerzell Institute of Political Science, Technical University of Darmstadt, Darmstadt, Germany

M. Lacey-Barnacle Science Policy Research Unit, University of Sussex, Brighton, United Kingdom

Paul A. Lant Energy & Poverty Research Group, School of Chemical Engineering, The University of Queensland, Brisbane, QLD, Australia

Joohee Lee Center for Energy & Environmental Policy, University of Delaware, Newark, DE; Foundation for Renewable Energy & Environment, New York, NY, United States

Stephanie Lenhart Energy Policy Institute – Center for Advanced Energy Studies and School of Public Service, Boise State University, Boise, ID, United States

Breffní Lennon Cleaner Production Promotion Unit, School of Engineering and Architecture; Environmental Research Institute, University College Cork, Cork, Ireland

J. Macgregor Wise School of Social and Behavioral Sciences, Arizona State University, Glendale, AZ, United States

Arthur Mason Department of Social Anthropology, Norwegian University of Science and Technology, Trondheim, Norway

Setsuko Matsuzawa Sociology and Anthropology, The College of Wooster, Wooster, OH, United States

James McCarthy Graduate School of Geography, Clark University, Worcester, MA, United States

Joshua K. McEvoy Department of Political Studies, Queen's University, Kingston, ON, Canada

Clark A. Miller School for the Future of Innovation in Society, Arizona State University, Tempe, AZ, United States

Majia Nadesan Arizona State University, Glendale, AZ, United States

Ijlal Naqvi Singapore Management University, Singapore

Jatin Nathwani Waterloo Institute for Sustainable Energy, University of Waterloo, Waterloo, ON, Canada

J. Nicholls Bristol Law School; School of Sociology, Politics and International Studies, University of Bristol, Bristol, United Kingdom

Avino Niphi Indian Institute of Technology, Madras, Chennai, Tamil Nadu, India

Ambika Opal Waterloo Institute for Sustainable Energy, University of Waterloo, Waterloo, ON, Canada

Carlos Ortega Insaurralde Research Institute on Renewable Energy (INENCO), National Research Council of Argentina (CONICET), National University of Salta (UNSa), Salta, Argentina

Martin J. Pasqualetti School of Geographical Sciences and Urban Planning, Arizona State University, Tempe, AZ, United States

Tarla Rai Peterson Communication, University of Texas at El Paso, El Paso, TX, United States

Dana E. Powell Department of Anthropology, Appalachian State University, Boone, NC, United States; College of Indigenous Studies, National Dong Hwa University, Hualien, Taiwan

Thomas Ptak Department of Geography and Environmental Studies, Texas State University, San Marcos, TX, United States

Steven M. Radil Department of Economics and Geosciences, United States Air Force Academy, Colorado Springs, CO, United States

M.V. Ramana University of British Columbia, Vancouver, BC, Canada

Harald Rohracher Linköping University, Linköping, Sweden

Jean-Pierre Roux Department of Geography, College of Life and Environmental Sciences, University of Exeter, Exeter, United Kingdom

Stacia Ryder Department of Geography, College of Life and Environmental Sciences, University of Exeter, Exeter, United Kingdom

Lilly P. Sar Centre for Social and Creative Media, University of Goroka, Goroka, Papua New Guinea

Lucas Seghezzo Research Institute on Renewable Energy (INENCO), National Research Council of Argentina (CONICET), National University of Salta (UNSa), Salta, Argentina

Veith Selk Institute of Political Science, Technical University of Darmstadt, Darmstadt, Germany

Jeongseok Seo National Assembly, Seoul, Republic of Korea

Carla Santos Skandier Climate and Energy Program, The Democracy Collaborative, Washington, DC, United States

Nicholas J. Sokol Found Spatial, Knoxville, TN, United States

Alevgül H. Şorman Basque Centre for Climate Change (BC3), Leioa; IKERBASQUE, Basque Foundation for Science, Bilbao, Spain

Helen Stern Research Institute on Renewable Energy (INENCO), National Research Council of Argentina (CONICET), National University of Salta (UNSa), Salta, Argentina

Kacper Szulecki Climate and Energy Research Group, Norwegian Institute of International Affairs, Oslo, Norway

Edgar Talledos-Sánchez CONACYT/EL Colegio de San Luis, A.C., San Luis Potosí, Mexico

Ekaterina Tarasova Linköping University, Linköping; Södertörn University, Huddinge, Sweden

Abraham Tidwell Pacific Northwest National Laboratory, Richland, WA, United States

Jacqueline Hettel Tidwell Knowlicy Group, Richland, WA, United States

Ethemcan Turhan Department of Spatial Planning and Environment, University of Groningen, Groningen, the Netherlands

Cristian D. Venencia Research Institute on Renewable Energy (INENCO), National Research Council of Argentina (CONICET), National University of Salta (UNSa), Salta, Argentina

Chad Walker Department of Geography, College of Life and Environmental Sciences, University of Exeter, Exeter, United Kingdom

Barbara Wejnert Department of Environment and Sustainability, University at Buffalo, Buffalo, NY, United States

Cam Wejnert-Depue Environmental Sciences and Policy, Krieger School of Arts and Sciences, John Hopkins University, Baltimore, MD; Post-Masters Research Associate, Pacific Northwest National Laboratory and University of Maryland's Joint Global Change Research Initiative, Washington, DC, United States

Kathy Witt Centre for Natural Gas, The University of Queensland, Brisbane, QLD, Australia

Caroline G. Wright Communication Studies, Arizona State University, Glendale, AZ, United States

Thomaz Xavier Department of Geography, Federal University of Ceará, Fortaleza, Brazil

Editors biographies

Majia H. Nadesan is a professor at Arizona State University's New College. Majia's research addresses risk, security, and biopolitics. Relevant publications include *Fukushima and the Privatization of Risk* (2013, Palgrave); *Crisis Communication, Liberal Democracy, and Ecological Sustainability* (2016, Lexington), and "Nuclear governmentality" (*Security Dialogue*, 2019).

Martin J. Pasqualetti is a professor at the School of Geographical Sciences and Urban Planning at Arizona State University and Senior Global Futures Scientist in the Julie Ann Wrigley Global Futures Laboratory. He is an elected fellow of the American Association of Geographers, a recipient of the Alexander and Ilse Melamid Medal (American Geographical Society), and the 2018 Distinguished Alumnus of the Year at the University of California (Riverside). His general research interests encompass three areas of emphasis: energy and society, energy and land use, and renewable energy development. His current research concentrates on the social acceptance of renewable energy landscapes and recycling of energy landscapes. He has served two Arizona governors as chair of the Arizona Solar Energy Advisory Council and was a founding member of the Arizona Solar Center. He serves on 10 editorial boards, including *Energy Research & Social Science*.

Jennifer Keahey is an assistant professor in the School of Social and Behavioral Sciences at Arizona State University and a senior sustainability scholar in the Global Institute of Sustainability. As a development sociologist with action research expertise, Keahey is interested in questions pertaining to sustainability, participation, and social change. She has conducted extensive fieldwork on Fairtrade and sustainable agriculture movements in post-authoritarian societies. In her previous work with South African Rooibos tea farmers, she generated a participatory commodity networking approach to Fairtrade research and support as well as a multi-paradigmatic foundation for sociopolitical analysis. Her current scholarship is focused on the ethical and political dilemmas of development, with particular focus on the ethics of development research and the role of cultural worldviews in supporting transitions to democracy and sustainability.

Contributors biography

Lourdes Alonso-Serna is an associate lecturer at Universidad del Mar, Oaxaca. She gained her PhD in human geography from the University of Manchester. Her research lines include the political ecology of wind energy in Mexico, land rent and wind energy, and geographies of energy transition.

Peta Ashworth is a professor and chair in Sustainable Energy Futures at the University of Queensland, Brisbane, Australia. Peta's research interests span the integration of science and technology in society with a focus on low-carbon energy technologies and climate change solutions. Peta has recently been focusing on public attitudes to future fuels, hydrogen and biomethane, as low-carbon technology solutions.

Heather Plumridge Bedi is an associate professor of environmental studies at Dickinson College. Her research focuses on environmental and social justice, agricultural dispossession, and just energy transition geographies. She received a Fulbright Nehru Academic and Professional Excellence Fellowship in India and the American Association of Geographers award for undergraduate teaching.

Bidtah N. Becker has dedicated her career to the Navajo Nation and its natural resources. Bidtah Becker is Deputy Secretary for Environmental Justice, Tribal Affairs and Border Relations at California's Environmental Protection Agency (CalEPA). Prior to this position, she had the honor of serving as the director of the Navajo Nation Division of Natural Resources from May 2013 to January 2019, as an appointee of President Begaye and Vice President Nez, after serving 11 years as an attorney for the Navajo Nation, focusing on water rights and natural resources issues. Continuing her deep interest and passion for water, she serves on the leadership team for the Water and Tribes Initiative in the Colorado River Basin and is honored to serve as a commissioner on the New Mexico Interstate Stream Commission, as an appointee of Governor Lujan Grisham, and on the Navajo Nation Water Rights Commission, as an appointee of Speaker Damon. Ms. Becker is equally passionate about supporting artists and serves as a trustee for the Institute of American Indian Arts and Culture (IAIA), as an appointee of President Obama. Ms. Becker is a member of the Nation and lives on the Navajo Nation in Fort Defiance with her husband and two school-age children.

Melissa Bollman is a doctoral candidate at Clark University's Graduate School of Geography in Worcester, Massachusetts, United States. Her research focuses on comparing the different low-carbon energy transition trajectories of US states, with an emphasis on renewable energy resource policy, political rationality, and subjectivity production. Melissa holds a Master's in Energy and Environmental Policy from the University of Delaware.

Johanna Bozuwa is the co-manager of the Climate and Energy Program at the Democracy Collaborative. Her research focuses on transitioning from the extractive, fossil fuel economy and building toward resilient and equitable communities based on energy democracy. She has organized around climate justice both in the United States and in the Netherlands.

Christian Brannstrom is professor of geography and associate dean for academic affairs at Texas A&M University. He studies geographical dimensions of wind-power expansion in Brazil, where he has partnered with geographers at the Federal University of Ceará. He also studies social and political aspects of renewable energy and unconventional fossil fuels in Texas.

Ry Brennan is a doctoral researcher in sociology at the University of California, Santa Barbara. They study the nexus of ecology, technology, and democracy with a focus on energy infrastructure decentralization. Their empirical work currently explores three sites of energy transformation in Santa Barbara County, California: community choice aggregation, decentralized energy resources, and environment-building trades relations.

Marie Claire Brisbois is a lecturer in energy policy in the Science Policy Research Unit at the University of Sussex. Her work centers on issues of power and politics in sustainability transitions with a particular focus on the potential of energy decentralization to shift larger political power dynamics and catalyze widespread social change.

Matthew Burke is a postdoctoral associate at the Rubenstein School of Environment and Natural Resources and Gund Institute for Environment at the University of Vermont, and a postdoctoral fellow with the leadership for the Ecozoic project. He explores sociopolitical and ecological dimensions of renewable energy transitions. Matthew earned his PhD in renewable resources—environment from McGill University, examining energy democracy.

John Byrne is the director of the Center for Energy and Environmental Policy and distinguished professor in the Biden School of Public Policy, University of Delaware. He is also the president of the Foundation for Renewable Energy and Environment. A contributor since 1992 to the Intergovernmental Panel on Climate Change, his research focuses on political economy, climate-energy-environmental justice conflicts, and energy transformation.

Zoé Chateau is a PhD student in human geography at the University of Exeter, United Kingdom, with broad interests in the interrelations between low-carbon transitions and sociospatial transformations. Her PhD addresses the controversial deployment of wind energy in France and focuses on how competing actors use particular discourses on the "local" to legitimate and negotiate specific imaginaries of wind energy.

Stan Cox, formerly a geneticist with the US Department of Agriculture, is a research scholar in Ecosphere Studies at The Land Institute in Salina, Kansas. He is the author of *The Green New Deal and Beyond* (2020) and the upcoming *The Path to a Livable Future: Forging a New Politics to Fight Climate Change, Racism, and the Next Pandemic.*

Jessica Craigg is a PhD student working at the intersection of renewable energy and gender in Tanzania. A geographer at heart, she is currently working with advisor Dr. Youjin Chung at the Energy and Resources Group at UC Berkeley. Outside of academia, she is a pet lover, novel reader, and dystopian video game player.

Patrick Devine-Wright holds a chair in geography at the University of Exeter. With expertise spanning human geography and environmental psychology, he conducts theoretically driven research on a range of environmental challenges including sustainable energy transitions and climate change. He is an IPCC lead author and has been ranked in the world's top 1% of social science scholars in 2019 and 2020.

Alexander Dunlap is a postdoctoral fellow at the Centre for Development and the Environment, University of Oslo. His work has critically examined police-military transformations, market-based conservation, wind energy development, and extractive projects more generally in Latin America and Europe. He has published two books: *Renewing Destruction: Wind Energy Development, Conflict and Resistance in an American Context* (Rowman and Littlefield, 2019) and co-authored *The Violent Technologies of Extraction* (Palgrave, 2020).

Niall Dunphy is the director of the Cleaner Production Promotion Unit, fellow of the School of Engineering and Architecture, and principal investigator of the Environmental Research Institute at University College Cork, Ireland. He leads an interdisciplinary team, operating at the intersection of social sciences with science and engineering, conducting engaged research on the theme of society, sustainability, and energy.

Melisa Escosteguy is a doctoral student at the University of Buenos Aires, with a scholarship from the National Research Council of Argentina (CONICET). She holds a Bachelor's in Anthropology from the National University of Salta (UNSa), Argentina. She is currently doing research on political ecology, energy transitions, environmental justice, and the lithium global production network.

Danielle Endres is a professor of communication and environmental humanities at the University of Utah. Her research focuses on rhetoric of science/environmental controversies including nuclear waste siting, climate justice, and energy democracy. Her research seeks to contribute to developing more just and equitable systems of environmental decision-making. Endres enjoys walking, live music, reading, and being with her partner and kids.

Andrea M. Feldpausch-Parker is an associate professor of environmental communication at the State University of New York College of Environmental Science and Forestry (SUNY-ESF). Her research is interdisciplinary in nature, focusing on science, environmental, and energy communication with interests in natural resources conservation through communication and collaboration. Much of her research addresses public participation in environmental decision-making and social movements.

Adryane Gorayeb is a professor of geography and coordinator of the graduate program in geography at the Federal University of Ceará, Brazil. She also coordinates the GIS Laboratory of the Department of Geography and a network on impacts of wind farms in Brazil, the Observatório da Energia Eólica. She studies participatory cartography and social and environmental impacts of renewable energy implementation.

Matthew Herington is an adjunct fellow with the Centre for Communication and Social Change at the University of Queensland. He is also a senior consultant with Ndevr Environmental in Perth, Australia, and has more than 10 years of experience in energy and climate change spanning academia, government, and private industry. Matt's principal research interests are transdisciplinary by nature and lie in building an understanding of the political, social, and economic challenges to sustainable energy transitions globally, with a particular focus on South Asia and the Pacific.

Tony Heynen is program coordinator of Sustainable Energy and a senior lecturer in the School of Chemical Engineering at The University of Queensland, Australia. Following a career in the energy and resources sector as an environmental engineer, Tony's research lies at the nexus of the energy and development. His research projects in Papua New Guinea, India, and Timor-Leste explore electrification in Base of the Pyramid communities.

Craig Jacobson leverages his broad background in system dynamics, process engineering, and energy modeling to characterize wicked problems. Craig received a bachelor of science in systems engineering and design from University of Illinois—Urbana Champaign (2006), as well as a master of engineering and master of science in energy, environmental, and chemical engineering from Washington University in St. Louis (2010,2012).

Jörg Kemmerzell is a lecturer in political science at the Technical University of Darmstadt, Germany. He is also a senior researcher in the Kopernikus Project "ARIADNE—Evidence-Based Assessment for the Design of the German Energy System Transformation" funded by the German Federal Government. His research focuses on energy transition, climate policy, and democracy research.

Max Lacey-Barnacle is a research fellow at the Science Policy Research Unit, University of Sussex. His PhD explored the energy justice implications of energy decentralization, with a focus on the impact of community and municipal energy organizations, intermediaries, and local energy networks. Max previously worked in policy for the Energy

Saving Trust, one of the UK's leading sustainable energy organizations.

Paul Lant is a professor of chemical engineering at The University of Queensland (UQ). Paul co-founded the Energy and Poverty Research Group at UQ. The mission is to support positive social, environmental, and health outcomes that are vital for sustainable and productive livelihoods in energy impoverished communities in the developed and developing world.

Joohee Lee is a PhD candidate at the Center for Energy and Environmental Policy, University of Delaware, and research fellow at the Foundation for Renewable and Environment. She researches issues of energy (in)justice, energy transitions, climate-energy policy, and capability approaches. Her dissertation proposes a conceptual and analytical framework to understand systemic energy injustice associated with large-scale, centralized, and high-risk energy systems.

Breffní Lennon is a research fellow and co-investigator at the Cleaner Production Promotion Unit, Environmental Research Institute at University College Cork, Ireland. He is a human geographer researching the social and economic dimensions to the energy transition, along with wider sustainability issues concerning long-term community engagement and the role energy citizenship can play in transforming our energy system.

Stephanie Lenhart is a senior research associate at the CAES Energy Policy Institute at Boise State University. Her research focuses on the governance of electricity systems and energy transitions. Her work examines how institutional designs, stakeholder participation, and the negotiation of authority are changing in response to a changing climate, new technologies, and increasingly diverse policy goals.

Rudy Kahsar holds a PhD in chemical engineering and has worked in energy consulting, in the US Department of Energy, Rocky Mountain Institute, and as a lecturer in the Masters of Environment program at CU Boulder where he taught classes in energy policy, energy systems and technologies, data science and visualization in energy, the energy-water nexus, and international energy and sustainability.

Arthur Mason is an associate professor in social anthropology at the Norwegian University of Science and Technology. He holds a PhD in cultural anthropology from University of California at Berkeley. He has published an edited volume titled *Subterranean Estates: Life Worlds of Oil and Gas* with co-editors Michael Watts and Hannah Appel. He studies oil and gas development, esthetics, and energy consultant futures.

Setsuko Matsuzawa is an associate professor of sociology at the College of Wooster. Her research interests include Development, Environment, Transnational Activism, and East Asian Societies (China and Japan). She has examined environmental activism, transnational activism, and the effects of donor projects. Her publications include *Activating China: Local Actors, Foreign Influence, and State Response* and her articles have appeared in *Mobilization, Sociology of Development,* and *AsiaNetwork Exchange,* among others.

James McCarthy is the director of the Graduate School of Geography and the Leo L. and Joan Kraft Laskoff Professor of Economics, Technology, and Environment at Clark University. He earned a BA from Dartmouth College and an MA and PhD from the University of California, Berkeley. His areas of research include energy geographies, environmental politics, political ecology, and political economy.

Joshua McEvoy is a PhD candidate and teaching fellow in political studies at Queen's University. His dissertation focuses on the potential for just transition movements, including community energy, transit, and green labor movements, to contribute to transformative change. Joshua is also interested in the role of energy in (de)colonization. He is active in various social and environmental justice advocacy organizations.

Clark Miller is the director of the Center for Energy and Society and an associate director of the Consortium for Science, Policy, and Outcomes at Arizona State University. His research focuses on technology and globalization. He writes about the design and critical analysis of knowledge systems in support of United States and global policymaking, the governance challenges posed by new and emerging technologies, and about the social sustainability of transitions in complex, large scale, sociotechnological systems.

Jatin Nathwani is a professor at the University of Waterloo and founding executive director of the Waterloo Institute for Sustainable Energy. Dr. Nathwani's research focuses include sustainable energy policy and global access to energy. Prior to his appointment at the University of Waterloo, Dr. Nathwani worked in a leadership capacity in the Canadian energy sector for over 30 years.

Ijlal Naqvi is an associate professor of sociology at Singapore Management University. He studies governance and development in the global south, using infrastructure as a lens on state building and the citizen's engagement with the state on an everyday basis. He is currently working on a book titled *Access to Power: Electricity and the Infrastructural State in Pakistan*.

Jack Nicholls is a senior research associate based in the School of law at the University of Bristol. Jack's research centers on the democratization of the low-carbon transition, with a particular focus on deliberative politics and exploring alternative socioeconomic organizational structures. He is also an academic supervisor to students working on research projects that advance a civic university agenda.

Avino Niphi holds a Master's in Development Studies with a specialization in science and technology studies from the Indian Institute of Technology, Madras. Her research interests lie in exploring the intersections between climate change, energy politics, and security.

Ambika Opal is the manager of Global Programs at the Waterloo Institute for Sustainable Energy, a research center at the University of Waterloo. In her role Ambika manages the Affordable Energy for Humanity initiative—a global consortium of more than 150 energy access experts and practitioners. Ambika's research focus includes energy sovereignty and universal energy access through the deployment of smart energy systems.

Dana E. Powell studies the cultural politics of energy infrastructure and how to foster decolonial ecologies. Her first book, *Landscapes of Power: Politics of Energy in the Navajo Nation*, traces the work of a Diné-led movement against a coal-fired power plant slated for tribal land. Powell's current multi-sited ethnography partners with water protectors in Navajo Nation and North Carolina, and she is developing a comparative project in Taiwan.

Tarla Rai Peterson is a professor of communication and affiliated faculty of Environmental Science and Engineering at the University of Texas El Paso. Her research examines how intersections between communication and democracy enable and constrain science and environmental policy. She seeks to facilitate the emergence of an inclusive and nurturing community for Earth's citizens, while critiquing the normativity of that goal.

Thomas Ptak is a broadly trained geographer and social scientist in the Department of Geography, Texas State University, San Marcos. Ptak's scholarship critically investigates human-environment centered phenomena, policies, and practices shaped by contemporary energy transitions. His current research projects include small hydropower development in China, micro-grid creation through irrigation modernization, wildfire forced grid disruptions in California, and the rapidly emerging Community Solar landscape in the United States.

Steven Radil is an assistant professor of geospatial science in the Department of Economics and Geosciences at the US Air Force Academy. He is a political geographer, and his research interests include the politics of governance and the use of geospatial tools for applications of participatory governance, including energy.

M.V. Ramana is the Simons Chair in Disarmament, Global and Human Security and director of the Liu Institute for Global Issues at the School of Public Policy and Global Affairs, University of British Columbia. He is the author of *The Power of Promise: Examining Nuclear Energy in India* and contributes regularly to the annual *World Nuclear Industry Status Report*.

Harald Rohracher is a professor of technology and social change at Linköping University, Sweden. His research focuses on the governance of sociotechnical change particularly in the fields of energy and digital infrastructures, urban low-carbon transitions, and transformative innovation policies. He is an associate editor of the journal *Environmental Innovation and Societal Transitions*.

Jean-Pierre Roux is a PhD candidate at the University of Exeter. He is writing a comparative historical case study on political agenda setting in support of offshore wind power in Ireland, the United Kingdom, and France since the commercialization of the industry in the late 1990s. His PhD research is funded through a Marie Skłodowska-Curie Action (MSCA) research fellowship on the MISTRAL Innovation Training Network.

Stacia Ryder is a postdoctoral researcher in geography at the University of Exeter and co-founder of the Center for Environmental Justice at Colorado State University (CSU). Her research focuses on power, place, and participation in environmental, energy, and climate justice contexts. She is the lead editor of *Environmental Justice in the Anthropocene: From (Un)Just Presents to Just Futures*.

Lilly Sar is the director of Center for Social and Creative Media, University of Goroka Papua New Guinea. She has worked extensively with rural households addressing global issues such as poverty alleviation, food security, environmental sustainability, and gender imbalance. She specializes in using adult learning and participatory practices in building local capacity of communities.

Jeong-Seok Seo is a senior policy secretary to an elected Member of the Korea National Assembly and was the research director of FREE's East Asia office. His research focuses on sustainable energy policy and financing. He earned an MA in climate and society from Columbia University and a PhD from the Center for Energy and Environmental Policy, University of Delaware.

Lucas Seghezzo is a researcher of the National Research Council of Argentina (CONICET). He is also a professor of environmental sociology at the National University of Salta (Argentina). He holds a PhD degree in environmental sciences from Wageningen University, the Netherlands. His current research focuses on political ecology, environmental justice, the sustainability of social-ecological systems, and social perception of environmental issues.

Veith Selk is a lecturer in political science at the Technical University of Darmstadt, Germany. His research focuses on political theory, the history of political thought, populism, and democracy. Recent publications cover topics such as right-wing populism and environmentalism, critical transformations of democracy, and pragmatist political thought (Rorty and Dewey).

Carla Skandier is the co-manager of the Climate and Energy program at the Democracy Collaborative. She earned her master of law in energy from Vermont Law School in 2015. She has a background in international environmental law, climate change, renewable energy, and sustainable development, particularly in developing countries including Brazil and China.

Nicholas Sokol is a geographer with a focus on climatology, geocomputation, and renewable energy. His current research explores climate impacts on renewable energy efficiency and adaptation. He currently works as the CEO of Found Spatial Institute.

Alevgül H. Şorman is an Ikerbasque Research Fellow at the Basque Centre for Climate Change (BC3). Her research focuses on energy transformations and metabolism, low-carbon energy, gender justice, co-creation with stakeholders, and dialog on changing energy frontiers.

Kacper Szulecki is a research professor in the Department of Political Science, University of Oslo, and the Center for Energy Studies, Norwegian Institute of International Affairs. He is also affiliated with the Center for Socially Inclusive Energy Transitions "Include" at the University of Oslo. He works on the politics of energy transitions, European energy and climate policy, and energy security.

Ekaterina Tarasova is a postdoctoral researcher in the Department of Thematic Studies, Theme Technology and Social Change, Linköping University, Sweden. She currently works in the research project that explores matters of exclusion and marginalization in smart grids.

Edgar Talledos-Sánchez is a researcher at CONACYT/El Colegio de San Luis A.C. He earned a PhD in geography from UNAM. He is a member of the National System of Researchers (SNI), level 1. His research lines are political geography; studies on the politics, culture, and territory of water; political conflicts over natural resources (water, forests, land, beaches); and megaprojects in Mexico and Latin America.

Abraham Tidwell is an energy social scientist at Pacific Northwest National Laboratory. His research focuses on the social and behavioral dimensions of energy behaviors and uses in buildings, with research spanning commercial and residential construction. Abraham earned a PhD in the Human and Social Dimensions of Science and Technology from Arizona State University.

Jacqueline Hettel Tidwell is the founder of the Knowlicy Group and an expert in energy industry communications. Jacqueline bridges research in data science, linguistics, and science and technology studies to understand how actors within energy systems make sense of the role(s) of energy in daily life. Jacqueline holds a PhD in English from the University of Georgia.

Ethemcan Turhan is an assistant professor of environmental planning in the Department of Spatial Planning and Environment, University of Groningen (RUG). His research interests are at the intersection of climate justice and energy democracy.

Chad Walker is an environmental social scientist with broad interests around just and inclusive low-carbon transitions. He is currently a postdoctoral research fellow at the University of Exeter and part of a consortium called EnergyREV, looking at engagement and public participation in smart local energy systems. Other published research has focused on Indigenous-led renewable energy and opposition toward wind turbines.

Barbara Wejnert is a professor in the Department of Environment and Sustainability and Department of Global Gender Studies at the University at Buffalo. She is an author or editor of 14 books published in academic presses and 60 peer-reviewed journal articles on the worldwide diffusion of democracy and globalization and the effects of these changes on environmental sustainability and gender equity.

Cam Wejnert-Depue is a graduate student at John Hopkins University, in the Environmental Science and Policy Program and a fellow at the Environmental Protection Agency in Washington, DC. He is a recipient of the William K. Reilly Scholar award from the Center for Environmental Policy, American University, and was awarded an internship for Environmental Governance and Leadership by the American Lung Association. He specializes in the protection of clean air and renewable energy.

Katherine Witt is senior research fellow in sociology at the University of Queensland's Centre for Natural Seam Gas. She studies the cumulative social and economic effects of unconventional gas development for regional communities, community acceptance, and non-technical risks of a range of energy technologies, including hydrogen. She is interested in the social and economic effects of energy transition and the contribution of new energy projects to sustainable regional development. Katherine is also co-chair of the Social Impact Assessment section of the International Association for Impact Assessment.

Karen Hudlet Vazquez is a doctoral candidate at Clark University's Graduate School of Geography. She earned an MSc in international development from Utrecht University. Her research focuses on energy justice and energy imaginaries in Latin America. Her interests include legal geographies, feminist political ecology, and animal geographies.

J. Macgregor Wise is a professor of communication studies and social technologies at Arizona State University, where he studies cultural studies approaches to technology and media, especially questions of globalization and surveillance. He is the author of several books, including *Surveillance and Film* (2016, Bloomsbury) and, with Jennifer Daryl Slack, *Culture and Technology: A Primer* (2nd edition, 2014, Peter Lang).

Caroline Wright is a graduate from the MACS program at Arizona State University. She has presented at several conferences and is a published academic author. Her interests include ecology, methods of resistance, and avenues of social change.

Thomaz Xavier is a PhD student in geography at the Federal University of Ceará with an emphasis on the potential socioenvironmental impacts of offshore wind farms in Ceará, Brazil. He is interested in research and education in the areas of cartography and social cartography, participatory GIS, GIS analysis, impact assessment, socioenvironmental planning, and ocean environmental policies.

Foreword

Benjamin Sovacool

**Director of the Institute for Sustainable Energy, Boston University, Boston, MA, United States;
Center on Innovation and Energy Demand, University of Sussex, Brighton, United Kingdom;
Danish Center for Energy Technology, Aarhus University, Aarhus, Denmark**

The sociotechnical systems behind energy provision and climate change are so materially intensive and expensive that we could easily consider infrastructure and investment as the two most salient aspects of any energy transition that might include decarbonizing our atmosphere and democratizing our energy.

For instance, Fig. 1 illustrates how human population growth matches up closely with primary energy and economic development. Indeed, the world now boasts almost 8 billion people (7.674 billion) and an average GDP per capita of almost US$15,500. As such, complex global energy systems have arisen to connect energy resources and fuels, prime movers, and delivery infrastructure. Humanity as a whole used about 580 exajoules (EJ) of energy last year. Despite such increases, the demand for energy will grow 45% between now and 2030, and this amount will almost triple by the end of the century.

The technology behind this ubiquitous and growing demand for energy can be breathtakingly complex. For instance, BP's oil platform *Deepwater Horizon*—which famously caused the largest oil spill in US history in the Gulf of Mexico in 2010—had drilled to a measured depth of 35,000 ft—a depth equal to the altitude commonly

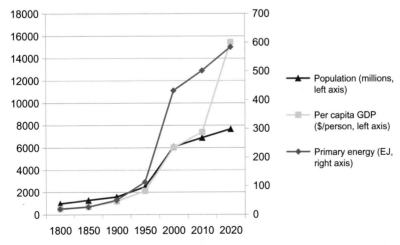

FIG. 1 World population growth *(left axis)*, per capita GDP *(left axis)*, and primary energy use *(right axis)*, 1800–2020 [1].

flown by commercial airlines. Also impressive, the shale gas wells being exploited by ExxonMobil are greater in length than the height of *10* Empire State Buildings. Coal mines, like oil and gas infrastructure, can become quite extensive. One of the world's largest coal mines in the world, the Black Thunder Mine in Wyoming, mines annually millions of tons of coal from its 6 draglines, 22 power shovels, and 148 haul trucks—enough to fill 25 miles of railroad cars per day. A single 1000-MW nuclear reactor, the standard size, will need upwards of 179,000 tons of concrete, 36,000 tons of steel, and 729 tons of copper, among other items [2].

The world's electricity is currently supplied by approximately 200,000 generators at almost 100,000 commercial power plants (about 440 of them nuclear-powered), 3500 underground natural gas storage fields, 1000 nuclear waste storage facilities, and millions of kilometers of transmission and distribution lines. The system behind automobility and industry relies on producing and consuming almost 90 million barrels of oil per day worldwide, down from 100 million barrels per day (bpd) in 2019. Whether or not consumption is 90 or 100 million bpd, it requires a massive system of oil extraction, production, about 700 refineries, and about 1 million gasoline stations to supply the world's roughly 1 billion automobiles that drive on 11.1 million miles of paved roads—enough to drive to the moon and back 46 times [3]. If we could speed up time, the growth of the global economy and its energy infrastructure would appear to be literally crashing into the earth like an asteroid [4].

Accompanying these features of modern energy production, delivery and use are admittedly shocking levels of financial commitment and investment. One report suggested that a future, decarbonized global energy sector is expected to need cumulative investments of at least $110 trillion between now and 2050, representing an average of 2% of global GDP per year, in perpetuity [5].

Similarly, the International Finance Corporation examined the national commitments submitted by 21 emerging market countries as part of the Paris Agreement and found they need $23 trillion in missing investment if they were to achieve their targets by 2030; the total amount of infrastructure investment required across all countries surpasses $90 trillion in present and future investment [6]. Post-2050 carbon abatement options could be even more expensive. For example, drawing down 50 ppm of atmospheric CO_2 with enhanced or accelerated weathering (using artificially created chemicals and minerals to store carbon) could cost $60–600 trillion for mining, grinding, and transporting rock, with further similar costs for distributing it [7].

However, despite these alarming figures, the energy and climate conundrum is *not* only a techno-economic affair. Because so much is at stake, and because almost every human being on the planet depends on or aspires to benefit from modern energy services, energy has also become an important social, cultural, political, and behavioral phenomenon. There are broader changes and drivers afoot that could make carbon and fossil fuels as obsolete as the horse stagecoach or telephone booth.

Thus, as humanity enters what may be the most compelling, and far reaching, sociotechnical challenge of this century—how to phase out carbon and achieve a net-zero, socially just, and economically stable energy system—it will need guidance from thoughtful, reflective scholarship. How will the growing demand for energy comport with the ideals and practicalities of democracy?

And it is here that I wholeheartedly recommend and endorse *Democratizing Energy* for providing us with an extensive, interdisciplinary, but nevertheless fascinating and well-supported tome of insight. Never before have I seen so many excellent colleagues teaming up to pen such erudite chapters on an important array of topics, from how democratic transitions are envisioned to how we grapple with power dynamics and relations, to who owns future energy systems (or who ought to own them). Chapters discuss at length how democracy and justice intersect with energy issues—a topic dear to my own heart and research agenda—as well as the promise and pitfalls of community energy. Other chapters apply a much-needed spatial approach to consider geographies of disaster, risk, and security, or sound the alarm bell over emerging threats to democratic or egalitarian principles and procedures.

Essentially, all of the chapters in this masterful volume come to articulate a better way forward—a roadmap for slowing, and even reversing, the metaphorical asteroid strike by pursuing energy options that are more sustainable, democratic, and thoughtful. Despite how radical such thinking may sound to some, especially those that profit from the present system of fossil fuel abundance and the ecological (and social) catastrophes that accompany it, the book would nevertheless ensure our future economies come to be built on ecological realities and human-centered priorities. This serves as an essential antidote to harmful discourses and falsely assuring platitudes that humanity is somehow exempt from the laws of ecology, atmospheric science, and physics, or the social justice imperative of fairness and equality.

With a blend of bluntness and sensitivity, *Democratizing Energy* shines a hard light on the ways that we have limited our own thinking with regard to the global energy challenge. The book opens the chance for us to go beyond our past and into a future that we consciously make better together. Collectively, we owe these contributors a debt of gratitude for showing us the limits of our thoughts so that we can transcend such limits in our future.

References

[1] Data for 1800 to 2010 from Table 20.1 from C.J. Cleveland, C.G. Morris, Handbook of Energy Volume i: Diagrams, Charts, and Tables, Elsevier Science, London, 2013. 2020 data for primary energy consumption comes from BP's Statistical Review of World Energy 2020, p. 8; 2020 per capita GDP and population data from the World Bank.

[2] B.K. Sovacool, Exploring the hypothetical limits to a nuclear and renewable electricity future, Int. J. Energy Res. 34 (2010) 1183–1194.

[3] A. Goldthau, B.K. Sovacool, The uniqueness of the energy security, justice, and governance problem, Energy Policy 41 (2012) 232–240.

[4] J.G. Speth, The Bridge at the End of the World: Capitalism, the Environment, and Crossing From Crisis to Sustainability, Yale University Press, New Haven, 2008.

[5] International Renewable Energy Agency (IRENA), Global Energy Transformation: A Roadmap to 2050, IRENA, Abu Dhabi, 2018.

[6] IFC, Climate Investment Opportunities in South Asia an IFC Analysis, 2017, World Bank Group, Washington, DC, 2017.

[7] H.J. Buck, Rapid scale-up of negative emissions technologies: social barriers and social implications, Clim. Chang. 139 (2016) 155–167.

Acknowledgment

We would like to acknowledge Caroline G. Wright's important contributions to the success of this collection.

Introduction to collection

Majia Nadesan[a], Martin J. Pasqualetti[b], and Jennifer Keahey[c]

[a]Arizona State University, Glendale, AZ, United States

[b]School of Geographical Sciences and Urban Planning, Arizona State University, Tempe, AZ, United States

[c]Social and Behavioral Sciences, Arizona State University, Glendale, AZ, United States

1 Introduction

Energy is integral to life, animating all we do, crave, cherish, envision, desire, loathe, buy, need, steal, sell, appropriate, move, love, and detest. Given its core importance, it is baffling that decision-making regarding energy production has been, since the 19th century, dominated mainly by technological and political elites. Only in the context of the looming climate crisis has the general population awakened to energy's vital role in determining our welfare, health, security, happiness, and the future livability of the planet.

Yet, even with new sensitivities toward energy's social importance, the idea of energy democracy is often posited as a simple negation of our vast and entrenched energy inequities and externalities (i.e., offloaded costs). The concept of energy democracy is idealistically articulated with sustainable and socially just energy assemblages that are capable of providing green jobs, operating according to fair labor standards, and ensuring universal energy access [1]. However, this utopic aggregation of social and technological outcomes is prescriptive, with no clear road map for navigating technological, political, and social barriers.

Energy is inescapably political. Energy infrastructures are among the most determinate in shaping conditions for economic and social life; hence, control over energy production, distribution, and utilization has been highly coveted and largely monopolized. Such a grip on energy resources will not be readily relinquished and can sometimes even lead to regional and global wars. Given these constraints, visions of greater energy democracy through decentralized energy production might not necessarily yield utopian aspirations.

History holds numerous illustrations of energy's vital role in shaping the development of world politics. Examples of how energy politics affect world affairs are endless, but here are a few examples drawn from the history of oil [2]:

- 1872: Rockefeller's Standard Oil is incorporated in Ohio, and by 1878 Standard controlled 90% of US refining capacity.
- 1909: Anglo Persian Oil Company (APOC) was founded and in 1911 British navy shifts from coal to oil and the government begins investing in the company.
- 1914–1918: WWI driven in part by control over oil, as illustrated by the 1916 *Sykes-Pikot Agreement*.
- 1930s: Beginning of resistance from the developing world with the nationalization of the Mexico oil industry in the 1930s.
- 1939: Hitler invaded Poland and headed to the oil fields of Romania and the Caucasus while Japan overran the oil fields of the Dutch East Indies.

- 1973: Arab oil embargo by OPEC followed by the response of the US Nixon administration to call for energy independence followed by the "moral equivalent of war" declarations of the Carter administration.
- 1980s: Attempt by the Reagan administration to thwart the construction of oil and gas pipelines from Russia to Europe.
- 1990s: US-led coalition of countries determined to stop Saddam Hussein from marching into Saudi Arabia and taking control of the largest oil field in the world.
- 2000s: China's construction of military outposts in the South China Sea to secure critical shipping lanes for oil and gas tankers.

What does this all mean? Simply put, energy—in all its forms and stages—remains as political as it has always been.

By focusing on the political, this collection unpacks the prospects facing energy democracies. While recent history suggests that the democratization of energy is an inchoate trend, we take a longer view, arguing that energy equity is more possible than conventional politics assumes. In the following sections, we consider the evolution of energy over the longue durée, showing how the rise of modern energy assemblages has caused energy ownership and decision-making to become concentrated through corporatized monopolies. In contrast to the modern energy sector, which has become a key vector for reproducing social inequity and deepening systemic risk, alternative energy assemblages are responding to social and environmental crises by investing in renewable technologies and participatory governance. After framing energy democracy challenges and potentials, we briefly outline the scope of this collection, a richly woven body of knowledge that joins the intellectually diverse expertise of established and new energy scholars from around the world.

2 From prehistoric energy consumption to modern energy politics

For most of human history, access to energy has been relatively equitable. In gatherer-hunting, horticultural-husbandry, and pastoral societies, everyone produced energy within households and communities as needed and when available. Early energy assemblages were non-commercial in form. Like other species, prehistoric humans foraged for primary energy resources, including various forms of food, such as nuts, berries, fruits, and occasional meats. The rapid expansion of anatomically modern humans across the globe suggests that prehistoric populations were highly mobile. There was little opportunity for hoarding and no need to establish an energy mercantile.

The rate of population growth accelerated substantially after the agricultural revolution. As sedentary life became feasible, the world witnessed the creation of the pre-industrial city state. Food surplus allowed for the specialization of skills and facilitated excess energy accumulation, sometimes in the form of food, other times in the form of biomass for heating and cooking. These developments spawned growing energy inequalities.

Agriculture, especially in arid areas, required irrigation, and this need eventually led to the invention of various devices to harness the power of moving water in the form of water wheels for lifting and later for grinding and many other tasks. The waterpower of streams was largely localized, and therefore the machines and their utility took on a value that was defended and commercialized.

The use of wind—first for sailing—came later. It, too, was commercialized, both for draining wetlands and grinding grain. Wind and water were site-specific resources, making it feasible for ancient elites to centralize control over surplus energy.

Access to energy became more inequitable with growing social complexity. Wood resources were commercialized as stands of forest were bought and sold and more easily gathered within human jurisdictions, thereby limiting equal access. Sometimes it was a landowner who monopolized this critical energy resource. Other times it was a sovereign government. Wood was harvested, transported, and sold as building materials, heating fuel, and eventually for early industrialization in the form of charcoal. In premodern Western societies, the control, provision, sale, and use of energy were controlled by those with authority, not by the general public. This stage in the evolution of energy development became globalized as European kingdoms began colonizing societies worldwide. The globalization of food and energy production under imperial rule further centralized ownership, as global society became increasingly hierarchically institutionalized and dependent on slave labor.

This trend intensified at the dawn of the Industrial Revolution around the mid-18th century, when reliance on renewable energy resources such as wind and rivers gave way to increasing dependence on energy-dense fossil fuels. The exploration, recovery, transportation, and sale of these commodities favored commercialization on a grand scale. Monopolies were common, most notoriously with Rockefeller's Standard Oil (1870–1911).

Timothy Mitchell argues that growing dependence on oil as an energy resource gave rise to "dematerialized and de-natured" politics [2, p. 235]. This shift disempowered workers as technical, natural, and human entanglements reinforced distinctions between experts and laypersons, not just in energy production but also in attendant areas such as science, warfare, industrial management, public health, law, and economic planning (p. 241). For example, petroleum engendered imperialist aspirations in the oil-rich Middle East, causing European ideas regarding national security to become linked to the control of oil. At the turn of the 20th century, the declining power of the Ottoman Empire in Mesopotamia rendered the region ripe for intervention.

3 Energy security and concentrated energy ownership and decision-making

The Anglo-Persian Oil Company (APOC) was founded in 1909 after discovering a large oil field in Iran (eventually renamed the British Petroleum Company, BP, in 1954). In 1911, Winston Churchill, then First Lord of the Admiralty, declared oil was of strategic importance to the Imperial Navy. Once oil was linked to security, the British government took a more active role in securing the resource by purchasing controlling stock in the APOC in 1914 [3]. Although many forces contributed to World War I (1914–1918), oil was an indisputable motive.

In 1916, France and Great Britain entered into a secret agreement, the *Sykes-Pikot Agreement* (1916), which divided the (dissolving) Ottoman Empire into a patchwork of states including Syria, Lebanon, Transjordan, Iraq, and Palestine that would be ruled by the French and British [4]. This document shaped the post-WWI geography, leaving Britain in control of much of the oil in the Mesopotamian area because it had taken the Deutsche Bank shares of the TPC [4]. The British engineered the election of an Arab leader, Faisal, as King in the area now known as Iraq and created a separate area called "Transjordan," where they installed Abdullah, the new Iraqi leader's brother emir (leader). Iraq remained under British control for several decades, with Iraq granted nominal independence in 1930 and Transjordan achieving independence in 1946. However, Transjordan was so poor that its existence as a state depended on direct support from Britain and the United States (of which the latter supported Transjordan on condition that the country would be friendly to the newly established state of Israel).

World War II (1939–1945) was fueled by oil; thus, control over oilfields was strategically important. If the Nazis sought control over the oil-producing Caucuses, Japan sought to control Indonesian oil fields. In return, allied forces sought to maintain their tenuous control over Middle Eastern oil resources. Against this oil-drenched backdrop, warring global powers raced to unleash the power of the atom through nuclear testing in an effort to win the war and dominate the post-war environment. Despite the promises of atomic energy, however, the old Western oil powers emerged from the war having secured yet more centralized control of the world's oil reserves. By 1948, a small number of Western companies controlled Middle Eastern oil, including Anglo Persian Oil Company (today's British Petroleum), Gulf Oil (most of which became part of British Petroleum and the other parts which joined Chevron), Standard Oil of California or SoCal (today's Chevron), Texaco (later a part of Chevron in a merger), London headquartered Royal Dutch Shell, Standard Oil Company of New Jersey (Esso which became Exxon), and Standard Oil Company of New York or Socony (Mobil, which merged with Exxon to become ExxonMobil).

Wars of colonial independence led to the nationalization of critical resources, such as rubber and bauxite, and newly freed nations struggled to create non-aligned movements that would operate outside of the shadow of the Cold War. The establishment of the Organization of Petroleum Exporting Countries (OPEC) in 1960 illustrated such efforts to seize control over a key resource held by colonizers, followed in the 1970s by the nationalization of oil concessions in Saudi Arabia, Kuwait, and Venezuela.

The industrial world's commitments to democratic self-determination were challenged as state authorities in the developing world wrestled control over national resources from colonizers. In fact, global energy assemblages often demarcated the limits of democratic practices, as observed by Michael Watts [5]. Efforts to renew control over key resources took innovative forms in the post-WWII environment as new financial agreements, instruments, and organizations were instituted after the Bretton Woods Agreement in 1944. The United States benefited from its currency acting as the global reserve currency used for international trade and the purchase of oil from OPEC, leading to creating "petro-dollar" surpluses that circulated the world looking for returns.

Oil was still desirable but also inescapably limited. New energy forms were required, and atomic energy promised to deliver energy too cheap to meter. However, the global public was appalled by the concentrated power and destructiveness of the atomic bombs dropped on Japanese civilians. In 1947, David Lilienthal, chairman of the Atomic Energy Commission, promised: "Atomic energy and scientific discoveries have not and need not change the fundamental principles of democracy, which rest upon faith in the ultimate wisdom of the people when they have been truthfully and clearly informed of the essential facts" [6]. Challenging such platitudes, Richard Monastersky notes that the first atomic blast represented the human-engineered mass extinction event termed the Anthropocene [7].

The knowledge and organizing nexus that produced the atomic explosions in Trinity, New Mexico, United States, and the Japanese cities of Hiroshima and Nagasaki were subsequently institutionalized in Cold War military technology and civilian energy development. Gabrielle Hecht describes this nexus as *nuclearity*, a social-technical *assemblage* of actors, knowledge, technology, and linked organizations that govern, produce, distribute, and consume energy [8]. If oil, as an energy resource, is highly centralized, nuclear energy assemblages demand yet more concentrated control of energy and decision-making, and, like oil, have been historically expansionary in their extractive aims [9].

The history of atomic energy is marked by catastrophe, with inextricable institutional linkages across nuclear energy, medicine, and weapons [10–12]:

- 1945: Atomic bombs dropped on the Japanese cities of Hiroshima and Nagasaki, ending WWII, and heralding the Atomic Era.
- 1946–1963: Atomic testing of at least 504 nuclear devices above ground by the United States, Soviet Union, and the United Kingdom [13].
- 1953: US President Eisenhower's Atoms for Peace created the conditions for the international proliferation of nuclear reactors.
- 1963: Partial Test Ban Treaty signed by the United Kingdom, the United States, and the Soviet Union limiting nuclear detonations.
- 1986: Chernobyl nuclear disaster in Ukraine.
- 2011: Fukushima Daiichi nuclear meltdown.
- 2019: Shift away from nuclear power announced by Chancellor Merkel in Germany in favor of renewable energy resources.

Nuclear power was attractive because it promised endless energy and provided the materials needed to develop atomic weapons. Yet the discourse of endless energy and scientifically engineered safety was always contested by scientists, engineers, doctors, and everyday citizens who understood that the nuclear mythos was wracked with contradictions and strategic elisions. Nuclear power is expensive, and plant construction historically has been characterized by delays and over-expenditures, the costs of which are usually absorbed by governments and ratepayers [14]. Even proponents of nuclear power acknowledge unique risks from its operations because of the toxicity and long-lasting nature of nuclear waste, the inherent dangers of controlled fission, and potential targeting by terrorists.

By the turn of the 21st century, ownership of these energy forms, supply chains, and board membership had become highly concentrated, as all are the largest global energy corporations. According to an important study by Vitali, Glattfelder, and Battison, energy assemblages of oil, gas, and nuclear not only institutionally concentrate ownership over resources and decisions, but also are closely connected with global finance and Western governments' core conceptions of national security [15]. Moreover, innovative financial instruments such as exchange-traded funds (ETFs) and futures markets have functioned to swallow ever more of the world's commodity supply chains, allowing localized activities to be drawn into global commodities markets. Resulting expropriations of value have caused prices to rise in nations worldwide, engendering market volatility, especially in emerging markets.

This energy industrial complex aims to gather ever more global resources into its highly centralized networks of control through expanded commodity markets and financial instruments, such as exchange-traded funds and exotic futures contracts [9]. In tandem with the broader politics of neoliberal globalization, which have exacerbated systemic inequities in countries around the world, the deregulation of energy has given corporate monopolies free rein to global energy sources while failing to hold these entities politically or economically accountable for the catastrophic energy crises that they cause, such as oil spills and nuclear disasters. This global process of regulatory capture derives from revolving door relationships between power elites situated in energy, finance, and government, as well as from mercantile and war-like formulations of national security interests who purchase massive amounts of energy to extend territorial control.

4 Rise of renewables and challenges of energy democratization

Over the last few decades, renewable energy resources have begun to loosen the stranglehold of giant energy companies. Here are 10 the most significant "push and the pull" factors leading us in this new direction:

- Global climate change from the burning of carbon-rich fuels has created an existential threat to the welfare of humanity, and multiple waves of climate refugees are gathering on the horizons of many countries.
- Water demand for thermal power plant cooling now exceeds water used in agriculture.
- Competition for secure fossil fuel supplies is creating multiple "flash points" for future military conflict.

- Safe and secure strategies for permanent storage of long-lived nuclear waste have eluded all of us.
- Dissatisfaction with dismal standards of living resulting from corrupt and greedy exploitations in energy-rich countries is stoking many fires of discontent and rebellion.
- Continuous forays by oil and gas companies searching for riches in deeper waters and the last remaining wild places threaten entire ecosystems with extermination.
- Conventional 20th-century energy infrastructures are crumbling in many places.
- Environmental damage to air, land, and water has reached intolerable levels in thousands of places where fossil and nuclear fuels are being developed.
- Costs of generating electricity from wind and solar are now on par with—or cheaper than—that from fossil or nuclear fuels.
- Solar energy is everywhere people want to live. It is not just in scattered deposits begging for someone to control them for the profit of the few.

Despite acting as drivers for change, it must be remembered that renewables are neither universal remedies nor guaranteed strategies for realizing more sustainable energy futures as they face key resource, network, and social challenges, including the cooptation of energy democracy discourse and practices by hegemonic energy.

The most pressing resource challenge for renewables derives from dependencies on rare earths, whose contemporary supply chains are fraught with environmental, social, and political externalities [16]. Additionally, networked forms of energy ownership and distribution currently favor centralized decision-making by large and publicly traded utilities seeking to extend their span of control over renewable development, policies, and distribution-utilization networks. This centralizing network model is inconsistent with the aims and means of energy democratization, which favors decentralization and loose couplings over centralized network control. Using electricity near the production source is the most efficient option. Still, decentralization is challenging to achieve financially. It produces its own risks, including single point of failure threats and the attendant necessity for converting from unidirectional power distribution to bidirectional systems.

In some cases, this conversion will turn a grid into a smart grid. Smart grids would allow communities and/or utilities to measure energy production and handle demand fluctuations, among other functions. The language around smart grid utilization often invokes the idea of the energy citizen who responsibly monitors and modulates their consumption, but this formulation is troubled with contradictions and may not deliver the promised energy democratization.

Sustaining a status quo built on the crumbling pillars of energy oligarchies is untenable. Conventional governance of renewable energy sources can only reintroduce the same or similar systems of power, likely leading to the acceleration of environmental and social risks, as has occurred in the past. Energy democracy scholars are problematizing these challenges, focusing on issues about scale, ownership, and governance. Given the critical importance of energy in world affairs—and the health of the planet and all its inhabitants—this volume addresses two primary questions:

(1) What approaches should we adopt to improve energy governance—including ownership, technological mode of production, distribution, and end consumption—in order to achieve ecologically and socially sustainable energy futures?

(2) Are energy democracies achievable and capable of resolving the wave of energy-related issues moving forward?

5 The social relations of energy governance

The questions posed above demand consideration of the social relations of energy production. According to Kacper Szulecki, a physical framing of energy has been built into technoscientific approaches [17]. This orientation is highly problematic as *energy development should be envisioned as a social issue with a technical component, not the other way around.* By delimiting energy as the domain of engineers (who develop infrastructures) and scientists (who develop energy technologies), traditional energy approaches *exclude* broad political participation in energy planning and only narrowly incorporate economy as the source of demand. As a result, little focus is given to social matters such as energy ownership, energy access, and energy futures. Similarly, a technoscientific approach to energy relations hinders societies from effectively addressing energy insecurities and risks. These are typically formulated in state-centric ways that preclude social and environmental justice concerns and prioritize professional risk management that has demonstrated spectacular failures internationally with catastrophic nuclear accidents and oil spills.

Related to the issue of technoscientific engagement, Goldthau and Sovacool find that decision-makers in top-down energy approaches tend to treat energy as just another commodity, thereby obscuring inextricable linkages across energy forms and other sectoral and policy contexts [18]. As explained by Pasqualetti and Stremke, the tendency to reify energy as a distinct commodity disassociates it from the material and social conditions of production, erasing "energy landscapes" that encompass "marks, structures, excavations, creations, and supplements that energy developments produce" [19]. According to Tidwell and Smith, however, the logic of security organizing many energy landscapes across nation-states, natural resources, energy technoscience, and markets is antithetical to participatory governance [20].

While the growing sense of climate crisis adds urgency to the topic of the energy transition, it does so in a context of established institutional interests and organizations that are technocratic and, too often, undemocratic. In fact, energy often demarcates the limits of democratic practices, as observed by Michael Watts [5]. The sociologist Saskia Sassen states that national and international energy landscapes erode support for participative governance and give rise to a global economy characterized by gross inequities in energy access with "expulsions" growing in frequency. More broadly, her work demonstrates how "predatory operations" of technological assemblages found in sectors such as banking and energy services concentrate resources and capital while brutally expelling de-valued people and places [21]. Indeed, several studies have documented the landscapes of energy expulsions and injustices [22–25].

National attempts to affect energy transition without redressing energy expulsions have led to problematic outcomes. As a recent case in point, the former Bolivian President Evo Morales has alleged that Bolivia's 2019 political crisis was driven by the efforts of international capital to access and control the nation's rich but difficult to extract lithium resources, which are vital for alternative energy storage technologies that are currently gaining purchase in the renewable energy space.

The crisis language used to encourage renewables can deflect attention away from proposed alternatives' political and bio-ecological sustainability. This deflection has been illustrated by problematic language in the 2018 US Green New Deal, a populist approach to the energy transition. The civil rights scholar and activist Harry Boyte notes that the proposal's language raises questions regarding stakeholder involvement in social change by invoking the federal government's mobilization for World War II as an exemplar, potentially reifying an ideology of wartime state authoritarianism predicated upon practices of censorship and crackdowns on labor activism [26]. By conflating WWII mobilization with the 1930s-era New Deal, the proposal's language obfuscates the differing relational arrangements of multi-stakeholder participation and creates room for authoritarian control.

The prioritization of rights and participatory governance may seem excessive to some, positioned as we are in the "fulcrum of the Anthropocene" described by Jedediah Purdy in *After Nature: A Politics for the Anthropocene* [27]. Indeed, contemporary, so-called "crisis" thinking erodes support for inclusion and democratic deliberation. Still, Purdy argues that appeals to expertise and efficiency, coupled with "economics-style neutrality" approaches to pricing energy alternatives, ultimately diminish our social capacity to reflect on and debate fundamental values and disagreements.

In addition to investing in renewable means of energy production, a profound realignment in social relations is integral to avoid reproducing the grave injustices of today's energy infrastructures. According to Frank Fischer in his *Climate Crisis and the Democratic Prospect*, the argument for democracy more generally is *especially* relevant in the fulcrum of the Anthropocene [28]. Fischer similarly identifies the limitations of expert-dominated techno-optimism and the prospects for ecological citizenship and environmental democracy in ecological cities and transition towns through participatory governance.

These issues suggest the importance of inverting top-down energy systems that relate only to the techno-rational knowledge systems produced by scientific experts and neoliberal commercial interests. Rather than replicating technocratic relations in renewable energy assemblages, energy democracies seek to operate according to a social-technological model that foregrounds issues about energy equity, justice, and rights.

Despite facing significant challenges, energy democracy remains a potential pathway to more socially just and environmentally sustainable futures. In an essay titled "Making Technology Democratic," Sclove argues that "if citizens ought to be empowered to participate in determining their society's basic structure, and technologies are an important species of social structure, it follows that *technological design and practice should be democratized*" [emphasis added] [29]. Like Szulecki, Sassen, and Boyte, Sclove's work on energy democracy work assumes that a democratic environmentalist approach characterized by participatory energy governance can capture the imagination for building more sustainable energy futures.

So far, however, most experiments in energy democracy have involved local energy cooperatives [30], with success explained by the extent of broader infrastructural support. Researchers also note that financing such cooperatives can be challenged with availability and forms driving local ownership models [31]. Thus, a key question remains: Is energy democracy capable of delivering energy transitions in countries worldwide?

Whereas utopic visions of sustainable energy landscapes often prioritize a politics of localism that views city and regional governments as central to processes of social change, national governments arguably play a more significant role in enabling and disabling future energy landscapes. This is because states play a key role in developing energy policies and financing infrastructural development through taxation. In addition, states are better positioned to work with international governments and energy markets to develop global policies and practices. Yet national energy efforts historically have lacked democratic proceduralism, with policy often shaped by the interests and objectives of private corporations in both liberal democratic and conservative or traditional societies.

6 Directions

The book you hold in your hands (or are reading on your computer screen) is divided into three interrelated parts, each of which commences with an introduction to the section and its contributors. Bringing established and new energy scholars worldwide into dialog, the collection begins with theoretical and utopian scholarship on energy imaginaries, moves to pragmatic and empirical studies on energy transitions, and culminates with critical coverage of institutional and social barriers to change. The contributors to this collection are an amazingly transdisciplinary group who have found a common interest in the possibilities and challenges of energy democracy. Their contributions are divided into three parts:

Part I: Imaginaries identifies utopian solutions to the energy crisis, with contributors envisioning energy assemblages that support sustainable and just futures. This book section begins with a brief introductory chapter that contextualizes the relationship between social utopias and energy production by covering key contributions to the section. Part I is further divided into two themes. The first of these, *Knowledges*, pushes the boundaries of energy discourse by giving voice to diverse perspectives. Considering the phenomenology of closeness, Kashar emphasizes the need for communities to establish relational ties with distant energy grids through democratic governance systems that enable passive consumers to become involved in governance and eventually as energy co-producers. Employing a postcolonial and feminist lens, Hudlet-Vazquez and colleagues call for the formation of radically egalitarian energy assemblages that are grounded in Indigenous and feminist sciences. Brennan applies decentralist anarchist reasoning to argue in favor of regional energy insurrections that employ technology to subvert big energy's control over local communities. The second theme, *Futures*, extends these debates by considering how today's energy alternatives may become tomorrow's energy systems. While pragmatic contributors such as Lee and colleagues and Brisbois examine the potential of working within existing political orders to enact incremental change, critical scholars such as Ptak and Radil note that even the most radical of alternative energy assemblages are threatened by technocratic cooptation given the demands of capital in a neoliberal world order.

Part II: Transitions considers the application of energy alternatives, focusing on efforts to transition to energy democracy. The contributors to this section challenge the discourse on energy utopias by noting that energy transitions cannot be guided by ideas alone. Empirical analysis is needed to develop a pragmatic understanding of matters such as the scope and breadth of investments occurring worldwide, the challenges of decentralizing ownership, the capacities needed to democratize governance, and so forth. This section engages the themes of *Organizing* and *Communities* to provide multilayered insight into cross-cutting questions. If Lenhart problematizes the organizational dynamics of participatory governance by

examining the infrastructural evolution of energy cooperatives, Wright draws from research on worker and energy cooperatives to assess community and worker benefits. Whereas Naqvi engages the case of the Karachi electric sector to examine the challenges facing activists who are demanding democratic transparency from electrical utilities, Sokol examines Puerto Rico's rapid transition to energy microgrids as a case of community adaptation to alternative energy. Like Xavier and colleagues, who argue in favor of instituting participatory research to prevent potential conflicts arising between fishers and offshore wind farms in Brazil, Heynen and colleagues draw from an energy literacy training initiative in Papua New Guinea to argue the importance of capacity building and the potential role that community youth leadership may play in fostering energy transitions.

Part III: Risks grapples with the gap between normative energy ideals and entrenched energy interests in chapters led by Walker, Lennon, and Selk. Contributors to the section themes *Assemblages* and *Security* address the riskiest and most entrenched energy systems. Deconstructing the "logic of security," Szulecki shows how this framing serves elite interests and prevents needed change. In a somewhat different vein, chapters by Sorman and Turhan, Wejnert and Depue, and Niphi and Ramana focus on the catastrophic risk and concentrated political power defining the nuclear industry, whose geopolitical approach to security threatens life on Earth. Highlighting factors driving concentrations of influence over decision-making, several chapters seek to identify avenues for decentralized energy as well as problematize institutional challenges to funding change in a world controlled by powerful energy actors. For example, Alonso-Serna and Talledos-Sánchez focus on the contradictions between the promise of decentralized democratization and the reality of these entrenched interests. Examining new technologies, such as smart grids, Tarasova and Rohrbacher come to a similar conclusion, arguing that these are failing to deliver their promise of empowering citizens. Although renewables are often marketed as pro-social and "green," contributors such as Escosteguy and colleagues find that many renewable energy development projects are monopolized by powerful utilities and/or wealthy entities able to afford energy innovations. In this context of concentrated ownership, local sovereignty over energy production tends to be the exception rather than the norm. Finally, several scholars focus on cultural issues, with Mason arguing that cultural change is needed if we are to effectively combat the worst energy abuses and demystify energy supply chains in ways that enable end consumers to become involved as agents of change. While Ashworth and Witt describe the "psychic numbing" that erodes energy transitions, Becker and Powell suggest that one antidote to alienation resides in socially just storytelling, a qualitative mechanism for developing democratic, sustainable, and socially just values that spark change.

Concluding the volume, Dunlap provides a call to action that asks energy democracy scholars and practitioners to squarely confront intractable energy assemblages by committing ourselves more fully to socially transformative research and practice.

References

[1] Energy Democracy Map, International Energy Democracy Alliance. https://energy-democracy.net/. Retrieved in May 2021.

[2] T. Mitchell, Carbon Democracy: Political Power in the Age of Oil, Verso Press, London, 2011.

[3] The international petroleum cartel, in: Staff Report to the Federal Trade Commission, Released through Subcommittee on Monopoly of Select Committee on Small Business, U.S. Senate, 83d Cong., 2nd Session, Washington, DC, 1952, pp. 47–112. http://www.mtholyoke.edu/acad/intrel/Petroleum/ftc4.htm.

[4] P. Mansfield, A History of the Middle East, second ed., Penguin, London, 2003.

[5] M. Watts, Accumulating insecurity and manufacturing risk along the energy frontier, in: S. Soederberg (Ed.), Risking Capitalism (Research in Political Economy, Volume 31), Emerald Group Publishing Limited, 2016.

[6] D. McClure, Social-studies textbooks and atomic energy, School Rev. 57 (10) (1949) 540–546.

[7] R. Monastersky, First atomic blast proposed as start of anthropocene, Nature (2015) 16739, https://doi.org/10.1038/nature.2015.

[8] G. Hecht, Being Nuclear: Africans and the Global Uranium Trade, MIT Press, Cambridge, 2012.

[9] M. Nadesan, Crisis Communication, Liberal Democracy and Ecological Sustainability, Lexington, Lantham, 2016.

[10] S. Lindee, Suffering Made Real: American Science and the Survivors at Hiroshima, University of Chicago, Chicago, 1994.

[11] A. Petryna, Life Exposed: Biological Citizens after Chernobyl, Princeton University Press, Princeton, NJ, 2002.

[12] K. Brown, Manual for Survival: A Chernobyl Guide to the Future, WW Norton, New York, 2019.

[13] S.L. Simon, A. Bouville, C. Land, Fallout from nuclear weapons tests and cancer risk, Am. Sci. 94 (2006) 48, https://doi.org/10.1511/2006.1.48.

[14] R. Smith, Nuclear power firms feel squeeze, Wall Street J. B3 (2015).

[15] S. Vitali, J.B. Glattfelder, S. Battiston, The Network of Global Corporate Control, Swiss Federal Institute of Technology, 2011. arXiv:1107.

[16] J. Shih, J. Linn, T. Brennan, J. Darmstadter, M. Macauley, L. Preonas, The supply chain and industrial organization of rare earth materials: Implications for the U.S. wind energy sector, in: Resources for the Future, 2018. https://media.rff.org/documents/RFF-Rpt-Shih20etal20RareEarthsUSWind.pdf?_ga=2.167331454.112967261.1556308170-719866651.1556308170.

[17] K. Szulecki, Conceptualizing energy democracy, Environ. Polit. 27 (1) (2018) 21–41, https://doi.org/10.1080/09644016.2017.1387294.

[18] A. Goldthau, B.K. Sovacool, The uniqueness of the energy security, justice, and governance problem, Energy Policy 41 (2012) 232–240.

[19] M. Pasqualetti, S. Stremke, The renewable energy landscape: preserving scenic values in our sustainable future, Energy Res. Soc. Sci. 36 (2018) 94–105, https://doi.org/10.1016/j.erss.2017.09.030.

[20] S.D. Abraham, J.M. Smith, Morals, materials, and technoscience: the energy security imaginary in the United States, Sci. Technol. Hum. Values 40 (5) (2015) 687–711, https://doi.org/10.1177/0162243915577632.

[21] S. Sassen, Expulsions, Harvard University Press, Cambridge, MA, 2016.

[22] H. Appel, A. Mason, M. Watts, Subterranean Estates: Lifeworlds of Oil and Gas, Cornell University Press, New York, 2015.

[23] A. Blowers, Nuclear Wastelands Part I: Landscapes of the Legacy of Nuclear Power, Town & Country Planning, 2017, pp. 303–308.

[24] M.T. Klare, Blood and Oil: The Dangers and Consequences of America's Growing Dependency on Imported Petroleum, Metropolitan Books, New York, 2004.

[25] B.K. Sovacool, M.H. Dworkin, Global Energy Justice: Problems, Principles, and Practices, Cambridge University Press, Cambridge, 2014.

[26] H. Boyte, Populism or socialism: The Divided Heart of the Green New Deal. Minnesota Post, 2019. https://www.minnpost.com/community-voices/2019/03/populism-or-socialism-the-divided-heart-of-the-green-new-deal/.

[27] J. Purdy, After Nature: A Politics for the Anthropocene, Harvard University Press, Cambridge, 2015, pp. 256, 265.

[28] F. Fischer, Climate Crisis and the Democratic Prospect, Oxford University Press, Oxford, UK, 2017.

[29] R.E. Sclove, Making technology democratic, in: J. Brooke, I. Boal (Eds.), Resisting the Virtual Life: The Culture and Politics of Information, City Lights, San Francisco, 1995, pp. 85–101.

[30] J.A.M. Hufen, J.F.M. Koppenjan, Local renewable energy cooperatives: revolution in disguise? Energy Sustain. Soc. 5 (2015) 18, https://doi.org/10.1186/s13705-015-0046-8.

[31] S. Hall, T.J. Foxon, R. Bolton, Financing the civic energy sector: how financial institutions affect ownership models in Germany and the United Kingdom, Energy Res. Soc. Sci. 12 (2015) 5–15, https://doi.org/10.1016/j.erss.2015.11.004.

Further reading

D. Yergin, J. Stanislaw, Commanding Heights, Simon and Schuster, New York, 1998.

Imaginaries

Introduction to Part I: Energy imaginaries

Jennifer Keahey

Social and Behavioral Sciences, Arizona State University, Glendale, AZ, United States

1 Introduction

Energy has many meanings. In simple terms, it denotes the ability to work, move, and to change. As a physical force, it is kinetic or potential, manifesting through movement or stored in reserve. Yet energy can neither be created nor can it be destroyed, for according to the law of conservation, the total amount of energy in the universe is finite, with currents simply changing in form [1]. There would be no existence without energy conversion. If songbirds know how to change the potential chemical energy stored within fruit into the energy of flight, humans have long since learned how to transform wood into thermal and radiant energies that heat and light a homestead at night. While one may reasonably argue that energy is the driving force behind the order, it also represents disorder, as isolated energy systems devolve into a state of entropy, or chaos over time. Yet perfectly isolated systems do not exist within the known universe, and it is through the passage of energy between systems that new orders emerge. Energy and knowledge systems share an important feature in common: they combine what already exists to bring futures into being.

The contributors to this section interrogate critical energy junctures to rethink energy in development and identify more egalitarian futures. The first theme—*Knowledges*—is situated at the fulcrum of debates on energy and society. It moves from Tidwell's discussion of a centralized energy imaginary that has become the hegemonic norm to Wise's radical vision of posthuman energy assemblages that relationally intersect across dimensions and scales. The second theme—*Futures*—identifies alternative practices that may be harnessed to democratize energy and then concludes with two accounts of possible energy futures, both of which emphasize ownership. Brannstrom cautions that renewable technologies may enable capitalism to overcome the climate change crisis by opening new possibilities for capital accumulation and territorial control. While Miller is somewhat more optimistic, stating that energy transitions tend to reconfigure social and market formations, this final chapter also recognizes that the redistribution of energy ownership is key to our planetary future.

The concept of social imaginaries has its roots in classical sociology, and more specifically, in Marx's vision of a post-capitalist and post-socialist utopia. When writing on the alienation of labor, Marx and Engels imagined a radically egalitarian society wherein people would be free to "hunt in the morning, fish in the afternoon, rear cattle in the evening, criticize after dinner … without ever becoming hunter, fisherman [sic], shepherd or critic" [2]. Structural interpretations have focused on the material means for realizing this liberatory vision; however, this passage may also be read in a post-structural light, for it identifies enforced categorizations as alienating prisons that structure the social relations of capitalism. Marx's differentiation between the free act of fishing and the subjugated fisher brings to mind the more recent work of feminist post-structural scholars, such as Butler who articulates social categories as a paradoxical affirmation of social existence and tool for oppression [3].

Taken together, the chapters in *Imaginaries* suggest that the construction of energy democracies cannot be accomplished without a dual focus on structural and relational change. In structural terms, Marx's articulation of the internal contradictions of capital has been demonstrated by the rise of treadmills of production and accumulation that are fatally designed to require infinite growth in a world of finite resources [4]. Like other classical sociologists, however, Marx's ability to predict the future was less certain. Failing to recognize the relational dynamics of social change, Marx presents us with a universal and linear supposition of utopian liberation from a dystopian past. Such an abstract reading ignores mixed lived realities and sociocultural differences. It also tragically fails to problematize the interrelated effects of process and outcome. Consider, for example, the impact of Stalinism on the Soviet Union. Under the modernizing cloak of science and objectivity, Stalin's regime reenacted the traditional values of patriarchy and empire, using educated professionals to subordinate people, other species, and ecologies to the yoke of the authoritarian master. Far from engendering socialism, the culture of violence unleashed by Stalin replicated the social relations of the Tsarist regime, betraying utopian revolutionaries who had envisioned a more collaborative and egalitarian approach to social change [5].

2 Knowledges

Knowledges begins with Tidwell, who shows how a past utopian vision became the hegemonic norm through the charismatic leadership of Samuel Insull. During the early 20th century, Insull played a key role in establishing an integrated energy infrastructure in the United States by fostering a shared culture of public responsibility. This transition ushered in a state-regulated public utility model that delivered initial benefits but since has become corrupted by monopolization and neoliberal policymaking. If Tidwell uses the case of Insull to emphasize the importance of motivating the public to challenge political leaders who are in bed with big energy, Kashar posits that individuals are more likely to care about distributed energy systems when emotionally invested in their governance. Drawing upon the phenomenology of closeness, Kashar argues that it is not the geographic proximity of energy grids to consumers that ensures public responsibility, but rather the ability of communities to build relational ties with energy.

Feldpausch and colleagues inject an energy justice voice, arguing that energy democracies must be attentive to the concerns of powerless social groups, including members of other species and interspecies ecologies. While this justice vision perceives power in terms of empowering people to act for themselves and on behalf of silent stakeholders, Hudlet-Vazquez and colleagues offer a more critical assessment. Engaging a postcolonial lens, these scholars argue that hegemonic renewable energy visions are not in alignment with the interests of energy justice. Alternative imaginaries also run the risk of reproducing traditional power dynamics, particularly when founded upon expert worldviews that retain Western, white, and patriarchal assumptions about the nature of reality. By infusing social imaginaries with Indigenous, feminist, and global South knowledge systems that are capable of navigating pluriversality, or complex worlds of difference, we may begin to root out the cognitive impulses that produce either-or solutions and reify social inequity.

Calling for the formation of "technoregions of insurrection," Brennan envisions energy democracies that are rooted in decentralist politics yet capable of enacting change at scale. By harnessing the liberatory potential of renewable technologies within the spatial utility of the region, Brennan argues that energy anarchists can more effectively undermine centralized energy management systems. Finally, Wise advocates for the creation of posthuman energy assemblages that incorporate awareness of relationality and affect. Employing Illich's concept of conviviality, Wise argues that we need to radically expand the notion of community to include the human/non-human and social/ecological assemblages of which we are all a part. Energy democracies are not simply about building physical systems, but also about investing in emotional maturity through governance technologies that support inter-relational growth. In *Futures*, several scholars identify tools for engagement.

3 Futures

Lee, Seo, and Byrne use the case of an alternative energy initiative in Seoul to develop social fundamentals for energy democracy that center democratic and commonwealth practices. By replacing cornucopian energy economics with a commonwealth approach, Seoul has begun to address the uneven distribution of privatized risks, pointing to the potential of commonwealth energy to ameliorate systemic injustices within city infrastructures. Bozuwa and colleagues expand the discussion by reviewing eight participatory and deliberative governance structures that may be implemented to wind down global dependency upon fossil fuels. At the local level, participatory budgeting initiatives can bring civic interest groups, such as Black Lives Matter, into the governance of public utilities. Local worker-owned energy cooperatives may develop regional cooperative networks through the formation of a regional cooperative congress. In a similar vein, transnational Indigenous sovereignty networks may coordinate efforts to reclaim energy sovereignty in settler colonial and postcolonial territories. At the global level, the Fossil Fuel Non-proliferation Treaty initiative is pressuring international governance organizations to place moratoria on and eventually ban the use of fossil fuels.

Brisbois likewise discusses the potential of working within the system to enact change. Although centralized energy systems wield a great deal of power, the introduction of new political actors into electricity systems has impacted energy decision making in states such as the Netherlands, Germany, and Australia. While new energy actors are increasing their effectiveness in lobbying, cities are also taking a more active role in transitioning to renewable energy, and if it is unclear whether decentralized infrastructures will go so far as to become decentrally owned, the rise of non-profit energy initiatives nevertheless hold promise for the realization of energy democracies. Ptak and Radil shift focus by looking at the disjuncture between theoretical visions and empirical realities. Interrogating community solar development in the United States, these authors argue that the interests of local stakeholders are all too often ignored due to overreliance upon technical expertise. The demands of capital cause the egalitarian impulses of community-based energy initiatives to fall victim to technocratic governance practices and a return to business as usual.

Brannstrom assumes a yet more pessimistic view. As scientists and engineers have imagined the conversion of wind and solar into commodity chemicals, big energy may harness these technologies to intensify capital accumulation and neocolonial territorial control. It is entirely possible that electrochemical interventions will enable oil and gas to embrace renewables without changing the social relations of production. In such a world, affluent communities would continue to overconsume energy produced in impoverished territories, poor countries in the global South would host the multinational firms that patent and own renewable technologies, and the global poor would be left in the dark. Miller argues that the stakes could not be higher, for the shape that renewable energy structures take today will determine the hegemonic political and economic order in the 21st century and beyond.

4 Conclusions

According to the antiracist feminist poet, Audre Lorde, "the master's tools will never dismantle the master's house" [6]. In terms of energy imaginaries, the contributions to this book section suggest that the master's tools are expert knowledge systems that embed socially alienating assumptions about the nature of reality. Feminist, Indigenous, and global South sciences provide critical insight into the prospects of establishing democratic and sustainable energy systems in a complex and divided world.

While the democratization of energy governance is important, life on Earth cannot thrive without decentralizing power. Apart from prioritizing equitable distributions of ownership, we also must invest in inter-relational transformations that enable us to transcend the limitations of our past ideologies. Pathways to socially just and environmentally sustainable futures are structurally diverse, deceptively simple, and relationally challenging. Given the dichotomies of public/private ownership; centralized/decentralized governance; and equal/ unequal economies, we must figure out how to reconcile categorical divisions in ways that challenge the underlying social relations of oppression. Without combining structural change with inter-relational transformation, even the most equitable imaginaries can become corrupt tomorrow.

References

[1] P. Atkins, The Laws of Thermodynamics: A Very Short Introduction, Oxford University Press, Oxford, 2010.
[2] K. Marx, F. Engels, The German Ideology, Electric Book Company, 1998 (1932).
[3] J. Butler, The Psychic Life of Power: Theories in Subjection, Stanford University Press, Stanford, 1997.

[4] J.B. Foster, The treadmill of accumulation: Schnaiberg's "environment" and Marxian political economy, Organ. Environ. 18 (1) (2005) 7–18.

[5] D.L. Hoffmann, Stalinist Values: The Cultural Norms of Soviet Modernity, 1917–1941, Cornell University Press, Ithaca, 2003.

[6] A. Lorde, The master's tools will never dismantle the master's house, in: C. Moraga, G. Anzaldua (Eds.), This Bridge Called My Back: Writings by Radical Women of Color, Kitchen Table Press, New York, 1983.

Knowledges

1

Serving in the public interest: Samuel Insull and the public service utility imaginary

Abraham Tidwell[a] *and Jacqueline Hettel Tidwell*[b]

[a]Pacific Northwest National Laboratory, Richland, WA, United States [b]Knowlicy Group, Richland, WA, United States

1 Introduction

The public interest is an oft-repeated term within the context of the regulation of electric utilities in the United States, used to justify any number of arguments both in favor of and against significant transitions in the composition and function of energy systems. Nevertheless, what do we mean when we talk about the public interest in the context of charting the future of our electrification systems? The public interest as an organizing concept for understanding the relationship between tangible policies and the values embedded in a society has a complicated intellectual past. Bozeman [1] notes that during the mid-20th century, mainstream public administration scholars distanced themselves from the concept of "public interest theory," deeming it unsuitable for scientific study. Referring to such work as the embodiment of individualistic values and rhetoric, these midcentury scholars argued in favor of objective assessment of a state's policies via preordained value systems.

By contrast, public interest theorists of the period, such as Richard Flathman, argued that a study of the public interest was necessary and should begin within the context (historical, political, ethical, moral) from which understandings of "the good" for a society emerge [2]. Flathman's normative point provides valuable guidance for the study of values and the public interest in the context of electrification systems. It is instructive to explicate what we mean by the regulation of energy *in the public interest* to understand the values underpinning energy transitions and the purpose behind why they are regulated. Energy transitions involve complex rearrangements of people, technology, and, importantly, the values underpinning energy production and its use. Miller et al. [3] call these constellations "socio-energy systems," and it is helpful to think about energy through the lens of a sociotechnical systems framework, especially when considering how transitions between energy system configurations form, are adjudicated, and become the "rules of the game" that govern a sector.

Orienteering during transitions is a complicated task. The pathways originating from our current energy system configurations are in-flux. They are also highly negotiable at varying scales of intervention: from the individual/community through the country and between countries. In the spirit of learning from examples of both what we should do and what we should not do, finding applicable models through a genealogical approach can elucidate the formations of power and knowledge as arrayed through varying instruments of utility operation.

To that end, this chapter provides a close reading of one crucial moment in the construction of the public interest within the context of a critical energy transition in the United States: the shift to the sizeable state-regulated power utilities common in the United States today. We focus our attention on one important actor: Samuel Insull, former president of Commonwealth Edison and, at one time, leader of the largest power company in the country, was responsible for as much as 11% of all commercial electric production in the United States.

Insull's success derives from his ability to construct Commonwealth Edison rhetorically and materially as a solution to Chicago's political and energy ills. Replacing a system of small power companies producing limited, expensive power beholden to local politicians, Insull's company would appeal directly to the people and the state government of Illinois to argue the public's interest was best served by a single company beholden to the "public" and not specific individuals or communities. Much as AT&T would appeal to the people to avoid government ownership, or how Marshall Field's flagship store became a fixture of Chicago's modern cultural identity, Insull encapsulated Commonwealth Edison in rhetoric and orga-nizational structure—financial, material, and political—that was of the people rather than beholden to highly localized political elites [4].

Insull's ability to grow Commonwealth Edison and in doing so define the idea of the "pub-lic service company" in the American utility sector serves as a critical example, albeit flawed, of how power companies came to be publicly regulated and owned and what approaches may inform the current ongoing transitions. Transitioning to more sustainable forms of energy production, transmission, and ultimately use are no longer fringe conversations but understood by government, utilities, and much of the public as necessary to create vibrant, healthy, and just futures. What is lacking, and where the history of Insull and Commonwealth Edison provides guidance, is the core importance of creating a shared narrative and sense of ownership of alternative energy futures. This chapter concludes by exploring this point fur-ther and its implications for energy systems design for the public good.

2 A moral utility—Constructing the public service company

When Samuel Insull became chief executive of Chicago (now Commonwealth) Edison in 1892, the city had 18 central station power plants in operation and dozens of individual con-sumers, electric light companies, and various electric system configurations. Many of the "public service companies" covered small territories and were, in essence, a continuation of the previous system's political corruption and graft that had led to issues with Chicago's other utility services. Small utilities covering small territories also possessed significant financial and technical challenges—challenges that Insull sought to avoid through the process of seeking widespread public support. One of the most considerable challenges for early electric power companies was the company's "load factor," or the fraction of total power production used at any point in time. Small electric utility load factors rarely crossed 40%—causing higher power rates for all customers as the electric utility struggled to cover its operating, maintenance, and essential costs from loans on its power equipment.

Insull's solution to this problem was twofold: enrolling new and different customers in electricity use. First, Insull chose to purchase large power generation equipment—more costly upfront due to the loans required to make these purchases, but more efficient in power production. Second, to ameliorate the cost (principal and interest) of these large loans, Insull advocated for adopting a system of utility rate pricing where customers who used more power consistently paid less on per-kilowatt-hour basis [5]:

> For instance, take the two probably extreme classes of customers to whom the central-station company supplies electricity for lighting purposes ... . Your investment to take care of each of these customers is practically the same; therefore your total interest cost must be the same in both cases; but if you distribute this interest cost over the actual units consumed, you will find that the tenant of the office building costs you for interest per unit of energy sold many times more than does the occupant of the basement.

Building a territory of sufficient size to achieve lower rates and return on investments was an immediate challenge for Insull, given Chicago's system of local politicians controlling public service company contracts.

Rather than appeal to a specific alderperson or the city government directly, Insull chose to appeal to the public of Illinois writ large, advocating for electricity regulation by a body of state-level technical administrators (e.g., a public utility commission). To Insull, elected city officials lacked the commitment to both large-scale social transformation and effective regulation—conditions the long-term growth of his business required to be successful [6]:

> How are these industries [public service companies] regulated here? They are regulated in campaigns for the election of aldermen to the City Council when you come down to the finality of the thing. It is not a question of a man's ability to deal with the technical subjects that come before him; it is a question on the one side of a man being able to deliver the greatest number of speeches to get the greatest number of votes, and, on the other side, of proclaiming that he is the only honest man in the community.

The oversight of public service company management by state regulators suited Insull and his vision of public-utility relations. In a race to the bottom of rates, Insull's cost plus reasonable profit model of public regulation gave a critical advantage in growing the service area to the company that could enroll the largest and most diverse number of customer classes. Insull thus argued that it was rational and best served public interest for the electric utility to function as a natural monopoly with demarcated territories. Unlike smaller, earlier companies, this state-regulated utility would not be a political actor. Through this purification process of public regulation, the electrical utility would serve the public directly.

Insull's vision of an economical and state-regulated utility required the company to embrace a critical paradox. Freedom from local political manipulation and the ability to set stable rates and grow within a demarcated territory would necessitate coupling knowledge of the customer's behaviors with an organization-wide commitment to fostering goodwill between the power producer and power consumer. Though Insull would routinely argue for utility employees and executives to "keep out of politics all you possibly can" [7], his success was made through situating Commonwealth Edison as a political actor more readily committed to the visions of progress his customers shared than their local representatives.

Insull called directly on company managers and engineers to foster this goodwill by situating themselves as public servants [8]:

> We central-station managers ought to look upon ourselves as semi-public officials and so conduct our affairs with the community as to give us the advantage of a reputation for absolutely fair and impartial dealing. We should preach the same doctrine to our subordinates and insist upon the same policy being carried out in their dealings with the public. If such a course is pursued, we will not only be helping to improve the opinion of the community of corporations generally, but will be establishing our own business on so firm a basis as to add to the permanency of our investment and give promise of prosperity in the future.

Insull's call to public leadership highlights the importance he put on generating goodwill between the corporation and the public through separating its activities from the realm of traditional politics.

As utility employees were encouraged to actively seek out and understand the public's will as it intersected with the industry, Insull also advocated stock ownership in Commonwealth Edison. Insull sought to create political pressure through popular support in favor of policies friendly to his monopoly utility by making company success a matter of the public protecting its interests [9]:

> The second reason for putting into effect this plan [customer ownership] is the desire to influence public opinion. People have very great respect for the property they own … . In the state of Illinois, we have today 500,000 owners of utility securities. If you figure only four to a family, that means nearly one-third of the population of the state, to say nothing of all the banks and trust companies that own utility securities. One-third of the population of the state are interested in these utilities receiving fair treatment at the hands of the people's representatives in the Legislature.

In opposition to the self-interested and local-situated aldermen, municipal company, or tycoon, the utility was a form of property constructed as shared in both the political, financial, and material senses. Flowing from power lines and through cash infusions via the sales of securities, customers, and the utilities that served them, became actors engaged in building political alliances toward a culture of populist mass energy consumption.

3 Conclusions

Insull's vision of regulation in the public interest served to make durable and formalize a vision of mass-energy consumption, enabling the end to various urban and rural ills through standardized rate-making practices and cost models, industry-government communications, and public relations. These ills were products of what Insull saw as irrational self-interest in defiance of fiscal and technological management's best principles. A successful transformation of the energy public utility service system implied not only lower rates for all customer classes, but it also formalized Insull's belief that a class of individuals competent in the operation of utilities were best suited to provide the public multiple fundamental values: competence, judiciousness, social responsibility, fairness, transparency, and honesty.

Understanding the nature of public demand and enrolling the public in owning their energy source was an important rhetorical element of Insull's model that has, by and large, been pushed to the wayside by investor-owned utilities. The exception is municipal, cooperative, and other "public power" utilities that emphasize community ownership and responsibility [10]. Across the board, understanding customer preferences and energy use existed until very recently through customer service work and (typically) mandated energy efficiency programming. The emergence of demand-side management (DSM)—an approach where utilities can manage power demand at the individual home or business level—has shifted attention to understanding individual energy behaviors and technological preferences. In this way, electricity systems have come full circle, back to finding mechanisms for shaping how customers use and engage with energy at the individual level.

Yet, in this shift back toward recognizing the role of individualism in shaping overall electricity systems operations, in the United States, there is a lack of equal balance concerning the necessity of shaping systems of collective accountability present in Insull's rhetoric. As we continue to look at shaping individual consumption patterns, the parallel question should be, "how do we hold these new systems of relating people and power accountable to our democratic values?"

Insull's vision of a regulated public utility behaving in the public interest was by no means a perfect model, as no models are suited for every situation. Nevertheless, it is worth noting that Commonwealth Edison successfully brought the public into a vision of large-scale, state-regulated public service utilities that continues to inform how all major publicly traded utilities operate across the United States. In the face of our current and very uncertain energy future, Insull's successes orient us toward considering the importance of enrolling the broadest and most diverse swath of members of the public in a common vision. Through fostering a shared culture of responsibility to the public, such a model serves to shore up support against incumbent political powers reinforcing existing energy systems and their associated paradigms. Perhaps, by taking the best insights from Insull's successes and considering how existing alternative energy systems serve the public interest—and embody the values listed above—an equally as powerful and stable energy transition to a sustainable future may be possible.

References

[1] B. Bozeman, Public Values and Public Interest: Counterbalancing Economic Individualism, Georgetown University Press, Washington, DC, 2007.

[2] R.E. Flathman, The Public Interest, An Essay Concerning the Normative Discourse of Politics, John Wiley & Sons, New York, 1966.

[3] C.A. Miller, J. Richter, J. O'Leary, Socio-energy systems design: a policy framework for energy transitions, Energy Res. Soc. Sci. 6 (2015) 29–40.

[4] R. Marchand, Creating the Corporate Soul: The Rise of Public Relations and Corporate Imagery in American Big Business, University of California Press, Berkeley, CA, 1998.

[5] S. Insull, Standardization, cost system of rates, and public control, in: W.E. Kelly (Ed.), Central-Station Electric Service: Its Commercial Development and Economic Significance as Set Forth in the Public Addresses (1897–1914) of Samuel Insull, Private Publication, Chicago, IL, 1898, pp. 34–48.

[6]   S. Insull, A certain hostility to public service corporations, in: W.E. Kelly (Ed.), Central-Station Electric Service: Its Commercial Development and Economic Significance as Set Forth in the Public Addresses (1897–1914) of Samuel Insull, Private Publication, Chicago, IL, 1911, pp. 243–248.

[7]   S. Insull, The obligations of monopoly must be accepted, in: W.E. Kelly (Ed.), Central-Station Electric Service: Its Commercial Development and Economic Significance as Set Forth in the Public Addresses (1897–1914) of Samuel Insull, Private Publication, Chicago, IL, 1910, pp. 118–122.

[8]   S. Insull, Twenty-five years of central-station commercial development, in: W.E. Kelly (Ed.), Central-Station Electric Service: Its Commercial Development and Economic Significance as Set Forth in the Public Addresses (1897–1914) of Samuel Insull, Private Publication, Chicago, IL, 1910, pp. 144–157.

[9]   S. Insull, Production and Distribution of Electric Energy in the Central Portion of the Mississippi Valley, Private Publication, Chicago, IL, 1922.

[10] American Public Power Association, About APPA. https://www.publicpower.org/about. (Accessed 1 May 2021).

2

Governance and sustainability in distributed energy systems

Rudy Kahsar

Environmental Studies, University of Colorado Boulder, Boulder, CO, United States

1 Introduction

The energy sector is one of the largest and most important sectors of the world economy. Accounting for 8%–10% of world GDP, it is second only to health care in size and is equally pervasive in scope; there is no country or region on the Earth that does not rely on the fruits of energy systems to power transportation, heating, cooling, services, and appliances. Central to the form and expression of the energy sector is its vastly interconnected and globalized nature. Resources and raw materials are mined, shipped, processed, and distributed in supply chains that encircle the globe. This includes commodities such as oil and gas, niche minerals such as cobalt and neodymium, and, increasingly, electricity. While some degree of centralization is apparent in all parts of the energy sector, the electric power sector is particularly interesting for the ways in which governance of the sector may change as it becomes more distributed.

The modern electric power sector is by some accounts the largest machine ever built, consisting of generators, transformers, power lines, and distribution infrastructure that touches nearly all parts of the globe. The companies that oversee this infrastructure employ millions of workers who ensure that electricity is reliably delivered to match ever-changing demand in real-time over vast geographic regions. While the eastern interconnect of the United States is currently the world's largest, single frequency AC grid, increasing connections among European and North African states, as well as a potential Asian supergrid, suggest a future characterized by more interconnected electricity sectors.

Such interconnections provide many benefits, including greater system reliability, load balancing, renewable energy integration, and low cost. In fact, electric power infrastructure has been commercialized and institutionalized with such efficacy that, in the United States, retail electricity prices have stayed relatively constant in nominal terms from the 1980s through the 2010s.

Of course, such a windfall does not come without costs. Greenhouse gas emissions and local air quality are oft-cited and superficially obvious externalities [1], but other more hidden costs include a degree of centralization, standardization, and automation unparalleled in almost any other industry. As large supergrids expand to encompass more generators, regions, and consumers, control is increasingly concentrated in the few entities that coordinate this infrastructure.

The process rewards centralized, ordered, interlocking, and authoritarian systems of management and design, and in doing so, renders new ways of thinking, acting, and supplying electricity as nonstarters beneath the stranglehold of the status quo. Systems become a target for political influence and are frustrated by their inherent ossification. Technology and management become entrenched. Overturning the industry in one fell swoop would be, for reasons of cost, security, and human well-being an undesirable outcome; however, a gradual transition through new, distributed forms of technology and management could serve to turn the industry slowly away from its centralized system toward a decentralized model that reprioritizes local concerns and local decision making. This chapter will focus specifically on the electric power sector and will build on research from the fields of energy policy, social science, and philosophy to discuss how the transition from large centralized systems to small-scale distributed energy systems may serve to foster greater community engagement and democratic decision making.

2 Overcoming centralized systems

The electricity sector and its antecedents were not always so concentrated. In the agrarian past as well as during the first decades of the industrial revolution, energy was processed and used on site. Farms burned biomass, and until 1900, factories were mostly operated by on-site coal furnaces and steam turbines. The system was naturally distributed. Only as demand for electricity grew did investors and engineers realize that larger generators could produce electricity more efficiently and cost-effectively than a series of smaller dispersed generators. From 1900 to World War II, economies of scale drove the rise of electrification networks within cities. In the postwar boom, western countries invested heavily in new power lines, connecting cities and all the people in between to the grid system. Simultaneously, companies like General Electric and Siemens rolled out ever-larger generators. Today, coal and natural gas facilities commonly produce 1000 MW, enough to power 650,000 average homes, while the world's largest power plants produce many times that.

At first glance, the centralization of 20th-century energy technology systems makes sense. Not only would it be undesirable for each household to have a small coal power plant in their kitchen due to the emissions, it would also be far more expensive and thermodynamically inefficient (because of small sizes and significant ramping). Plus, it is much nicer to simply turn on a light with a switch than it is to have to stoke a fuel box or pump a generator. The outsourcing of electric power generation to large, centralized facilities solved all these problems. The only downside was that it required an extensively interconnected grid system to connect these generators to consumers. In the second of his six laws, the historian of technology Melvin Kranzberg wrote that, "Invention is the mother of necessity," and that every technical innovation seems to require additional technical advances in order to make it fully effective [2]. The grid system was exactly that—the necessary add-on that enabled the rise of large, centralized generators.

Today, after more than a hundred years of enabling the electric power sector as we know it, the grid itself has become a significant source of inertia. First, grids are still essential.

No developed countries operate their electricity systems without large-scale electric grids. Second, they are more static than ever. Because new transmission infrastructure is expensive to build and often fraught to approve [3], companies and investors avoid construction of new lines and often site large new generation facilities (including renewables) based on existing lines [4]. In this way, the inertia of the grid system often leads new generations to reinforce the grid system. The term "renewable energy" is often used synonymously with "distributed energy," and while the two terms are certainly related, they are, for this reason, not necessarily synonymous. Renewable siting decisions based on existing lines *passively* reinforce existing centralized systems.

Renewable energy can also *actively* drive further centralization. In early 2018, the State Grid Corporation of China began building a 2000 mile long 1.1 million volt D.C. transmission line to carry renewable energy from Xinjiang to load centers in the east, a clear move toward greater connectivity and centralization of control. New high voltage lines have been similarly advocated in the United States, for example, for taking wind energy from Wyoming to load centers in California [5]. Such interconnections can improve the reliability of renewable energy by aggregating resources over large geographic areas, thereby enabling greater uptake of renewable energy. But they also result in an ever-greater degree of centralization. Whereas Wyoming used to be, for all intents and purposes, independent from California, a more interconnected grid network will render politics, management, and infrastructure in both regions more interdependent and more subject to centralized decision making. In this case, Wyoming would be rendered a generation center for California, and a centralized authority would be given greater reign to determine outcomes in both regions.

But, neither active nor passive reinforcement of a centralized model is inevitable. Renewable energy is unique, in that it can flourish independently of the grid. Because of their modularity, solar generators can be built on rooftops and in neighborhoods, places that would have been previously unimaginable for traditional generators. Furthermore, as renewable energy has become more cost-competitive and widespread [6], wind and solar farms are more frequently being built in areas with the greatest renewable resource—places that are sufficiently windy or sunny—rather than along existing grid infrastructure and traditional generating corridors [7].

These shifts will put new pressures on the existing paradigm of centralized grid management. Because of the modularity and scalability of renewable energy, communities that come into conflict with the existing grid paradigm may begin to break free. The US state of Hawaii as well as a number of European areas have already been forced to remove incentives for renewable energy due to load balancing issues on their grids, and certain regions have implemented forced curtailment of renewable energy due to an overabundance of generation. Either actively or passively, small regions or individuals may, when faced with these constraints, eschew the centralized grid model in favor of the freedom to supply their own electricity through smaller, distributed electricity systems.

3 Governance of distributed energy systems

In contrast to the highly interconnected, centrally managed grid model, a more distributed electricity model will bring generation closer to the consumer. The most obvious way that it will do this is physical. Residents of a suburban neighborhood who may have been previously unaware that their electricity came from a coal plant on the other side of town will be brought back in touch with their electricity when it is locally sited, for example, in a solar garden at the end of the street. Physically, this much is obvious. But, perhaps more important than the physical closeness of a solar garden is the metaphysical way in which the electricity of the solar garden becomes "closer" to its end users.

The phenomenologist Martin Heidegger characterized the state of Being that humans occupy by its relationship with the world, noting that physical closeness is not always the most important factor in determining what we care about [8]. A loved one on the other side of the world, for example, may be more salient than all the people in between. Similarly, a community solar garden might foster a metaphysical closeness that was previously absent from centralized generators, even if those centralized generators were just as physically close by. Centralized generators are often justified, commissioned, and operated by executives hundreds or thousands of miles away. By contrast, a community solar garden is explicitly embedded in the community. From inception to operation, local citizens are forced to think about and interact with it. It might incite conversation about environmental values, serve as a point of geographic reference, or spark interest in the technology itself. Meanwhile, energy concepts become salient. Previously energy-illiterate neighbors who install new PV systems may begin using terms such as "kilowatt" or "solar azimuth angle." Whereas such a person may have previously said that their electricity came from "the wall," their new, personal relationship with their PV array has the power to generate a new relationship with their electricity.

One of the consequences of a phenomenology of "closeness" is that an individual is rendered more likely to "care" about something—i.e., they are more likely to have opinions, exercise brain energy, and act in accordance with or against the item in question [9]. If the generator at the end of the street emitted soot and mercury, a community would be exposed to those emissions firsthand and would likely have opinions over its use. Conversely, if the generators at the end of the street were something to be proud of, local citizens might rally around them. Regardless of whether their opinion was good or bad, the community would be far more inclined to *care* about the source of their electricity and as a result, be forced to come to communal decisions over how to balance its upsides and downsides. They might decide what type of electricity generators they want, where they should be built, and how much they are willing to pay for them. Each of the questions fosters local decision making, and it is in this way that the shift toward decentralized systems of electric generation returns power, control, and freedom to the consumers.

Of course, with an increase in freedom comes a tacit responsibility. A community tasked with ensuring the reliability of their power supply will not do so for long if they do not understand the technology they are working with. Indeed, it is probably this burden of responsibility and management that has, at least to some degree, made people so eager to give up their freedom to vertically integrated utilities in the past. Attention spans are short and generation technology is complicated. However, the constraints of technology are changing here as well. New firms now provide tailored technical resources and infrastructure through internet-connected devices and off-the-shelf adapters and control systems. These advances put the power back into the hands of ordinary citizens and empower communities to make changes beyond what has been possible in the vertically integrated utility paradigm.

The term "energy democratization" has been used to describe the increasing wealth of competition, technology, geographic dispersion, and community engagement in energy systems [10]. All of these qualities apply here. Whether the community at hand is located in a western democracy, a state-run dictatorship, or a developing country, the move to decentralized energy systems will necessarily boost community-based decision making and has the potential to free communities from the grip of both private utility companies and state-run enterprises. In America, rural communities are at last beginning to defect from long-term contracts of dirty energy at high prices. In Southeast Asia, communities are finally investing in local, distributed generation instead of capital-intensive centralized infrastructure. In Africa, communities are bypassing the centralized model altogether and making decisions about energy use at a community level. In this way, a movement toward a more distributed energy system is a movement toward a more democratic system of energy.

And because of this relationship between decentralization and local decision making, a country, region, or organization that advocates for democratic principles should necessarily advocate for distributed energy systems. These systems free communities from centralized subjugation to dirty energy, authoritarian interests, and centralized control. In some ways, this is not a new idea; those seeking to avoid highly centralized, authoritarian systems have long turned to renewable energy for its distributed qualities. In the 1970s and 1980s, long before the price of solar energy came down, anarchists and doomsday preppers put solar panels on their houses. Now, under the auspice of energy democratization, small communities can do the same.

4 Conclusions

After the 2008 financial crisis, governments around the world moved quickly to reinvest in their economies, often earmarking large sums to renewables. At the time, electricity from renewable energy was still quite a bit more expensive than conventional electricity and although the early 2010s were record years for renewable installations, they were also years of the considerable buildout of conventional energy. In the 2020s, a new era of renewable energy beckons. As energy generation and consumption are brought back to the community level, they have the power to foster local decision making and active democracy in ways that have been lost or muted in the era of globalization. The International Renewable Energy Agency (IRENA) has referred to this new era of clean, distributed energy as, among other things, "more boring" [11]. For an industry that has spent a hundred years under top-down control, it may indeed be more boring in board rooms and international centers of power, but in local communities, the most heated debates are surely just beginning.

References

[1] G. Energy, CO_2 Status Report—The Latest Trends in Energy and Emissions in 2018, International Energy Agency, 2019.

[2] M. Kranzberg, Technology and history: Kranzberg's laws, Technol. Cult. 27 (3) (1986) 544–560.

[3] P. Devine-Wright, Renewable Energy and the Public: From NIMBY to Participation, Routledge, 2014.

[4] D.A. King, Interregional coordination of electric transmission and its impact on Texas wind, Tex. J. Oil Gas Energy L. 8 (2012) 309.

[5] A.E. MacDonald, C.T. Clack, A. Alexander, A. Dunbar, J. Wilczak, Y. Xie, Future cost-competitive electricity systems and their impact on US CO_2 emissions, Nat. Clim. Change 6 (5) (2016) 526–531.

[6]   IRENA, Renewable Power Generation Costs in 2017, nternational Renewable Energy Agency, Abu Dhabi, 2018.

[7]   G. Kieffer, T.D. Couture, Renewable Energy Target Setting, International Renewable Energy Agency (IREA), UAE, 2015.

[8]   M. Heidegger, Being and Time, Suny Press, 2010.

[9]   M. Weiss, D. Guinard, Increasing energy awareness through web-enabled power outlets, in: Proceedings of the 9th International Conference on Mobile and Ubiquitous Multimedia, 2010, pp. 1–10.

[10] J.P. Tomain, Clean Power Politics: The Democratization of Energy, Cambridge University Press, 2017.

[11] H. Tricks, Clean power is shaking up the global geopolitics of energy, De Economist 15 (2018).

I. Imaginaries

3

Energy democracy's relationship to ecology

Andrea M. Feldpausch-Parker[a], Danielle Endres[b], and Tarla Rai Peterson[c]

[a]Environmental Studies, State University of New York College of Environmental Science and Forestry, Syracuse, NY, United States [b]Communication, University of Utah, Salt Lake City, UT, United States [c]Communication, University of Texas at El Paso, El Paso, TX, United States

1 Introduction

Politicians, practitioners, advocates, and scholars alike use terms such as energy security, independence, sovereignty, colonialism, justice, poverty, and even dominance to describe relations between energy democracy and the climate crisis. Although it began as a social movement [1,2], energy democracy is increasingly the focus of academic research, with scholarship both describing the tenets emerging in on-the-ground struggles and theorizing how to best perform the most democratic energy transitions possible. The movement was borne out of valuing democratic decision-making, local autonomy, and resisting regimes of environmental racism and energy colonialism [3,4]. Energy democracy research considers a wide range of potential actors, democratic values, and governance sites from energy production to everyday practices. In this chapter, we begin with Feldpausch-Parker et al.'s [5] definition of energy democracy as "an emergent social movement that re-imagines energy consumers as prosumers … who are involved in decisions at every stage, from energy production through consumption" (p. 2). We then share our energy democracy research framework and explain how ecology flows through its components of justice, participation, and power. Finally, we conclude with areas for future research.

2 An energy democracy framework

Our conceptual framework for analyzing energy democracy discourses includes three interconnected dimensions: justice, participation, and power [5]. The framework is a heuristic, "enabling examination of theoretical models, empirical examples of ongoing struggles over energy, and practical recommendations for communities engaged in promoting energy democracy" (p. 3).

Justice goes beyond long-standing social and environmental justice movements to include discourses of climate and energy justice, all of which address issues of marginalization and inequity in energy decision-making. Scholars and activists alike may promote justice by recognizing and attempting to ameliorate unequal distribution of environmental risks/harms and energy benefits through political processes in which all stakeholders have autonomy and the ability to participate in, and have a real impact on decision-making about energy in their own communities [6,7].

Participation, which is closely tied to justice, focuses on engaging in energy decision-making through a variety of democratic processes and practices. Although most nominally democratic nations mandate the existence of public participation processes, the degree and type of participation opportunities vary. While conventional benchmarks such as public hearings or meetings serve as a time-tested metric for public participation, energy democracy research and activism hold a broader view of participation [5]. For instance, sociotechnical transitions research may suggest (re)imagining ratepayers "as prosumer, or innovators, designers, and analysts who are involved in decisions at every stage" [5, p. 2]. Alternatively, one may foreground the formation and participation of social movements as key forces in shaping energy politics outside the formal avenues of public participation [5,8,9].

Power in this context refers to the ability to act to elicit change. Given that some individuals/beings wield more social power than others [10], power is thus intrinsically related to justice and participation. Burke and Stephens [11] contend that "central to an energy democracy agenda is a shift of power through democratic public and social ownership of the energy sector and a reversal of privatization and corporate control" (p. 38). An important aspect of power is that it may be simultaneously exerted as control over others and as a resistive force via "activism, grassroots democratic organizing, local governing structures, and public participation that seeks to change existing hierarchies and relationships" [5, p. 5]. This capacity is exemplified in Puerto Rico's budding energy democracy movement that resists the status quo of imported fossil fuel. Movement participants advocate transitioning toward distributed solar energy as valuable for its environmental impact and its contribution to community development and energy independence [3,12] (see also Sokol's chapter in Part II).

3 The challenge of integrating ecology/more-than-human into energy democracy

Ecology as a discipline is the study of organisms, populations, and communities. Begon, Harper, and Townsend [13] describe it as "peculiarly confronted with uniqueness: millions of different species, countless billions of genetically distinct individuals, all living and interacting in a varied and ever-changing world" (p. vii). Though most professional ecologists claim to focus on nature, they often forget that humans are organisms that interact with other biotic communities and abiotic systems. Scholars across the social sciences, biophysical sciences, and humanities have argued against this false dichotomy between humans and nature, and have advanced possibilities for thinking ecologically [14–17]. With regard to decisions about energy extraction, production, and consumption, an ecological approach would attune us to how these processes affect the ecosystems of humans, other animals, plants, lands, and more.

Humanity's existence as part of nature, however, does not mean human society is without its own governing systems. Therefore, distinguishing between humans and more-than-human [18] remains useful for sociotechnical decision-making. Abram describes the more-than-human realm as including humans as well as other sensing beings that inhabit Earth. He [18] urges humans to pay more attention to and engage in intersubjective relationships with the "active, animate, and, in some curious manner, alive" ecology of sensing beings from animals to trees to mountains (p. 55). If we do this, humans can "find ourselves in an expressive, gesturing landscape, in a world that *speaks*" (p. 81). Haraway [19], who theorizes that "beings constitute each other and themselves through their reaching into each other" (p. 6), provides further endorsement for consciously viewing the sociotechnical transition frequently named "energy democracy" as a transaction between dynamic and unremittingly changing elements that include the more-than-human world. This, however, poses a challenge in terms of how to represent more-than-human ecology within society's decision-making processes. From a systems perspective [20] human society cannot communicate directly with its environment. Instead, it uses observations, interpretations, and experiences to bring the more-than-human dimensions of ecology into decision-making. As such, it is a struggle to bring a full understanding of what ecology might say to the forefront of public discourse on energy system transitions.

Further, attempts to bring ecology into energy systems discourse can suffer from what we call "conversational drift" from ecological- to human-centered. This can happen as energy discussions shift from a broad ecological perspective to a more narrow focus, such as economics, simply because humans may not have the communicative capacity to communicate with and for the more-than-humans involved in an energy issue [21]. For example, the concept of ecosystem services was initially proposed to highlight the economic value of biodiversity to humanity. Unfortunately, in many cases, the service has replaced biodiversity as the focus (e.g., planting a monoculture of an exotic tree species for carbon sequestration to offset fossil fuel emissions). As Peterson et al. [22] point out, this conversational drift distorts the system to the point of erasing the ecosystem worker, or biodiversity. Explicitly building an ecological perspective into our model of energy democracy requires consciously *hearing* what the more-than-human world is *saying* and integrating that into human-constructed systems.

Despite these challenges, humans already communicate on behalf of more-than-human stakeholders, seeking to incorporate responses to their felt needs into society's decision-making processes [18,23–25]. Although some find it problematic for humans to take this role, Peters [24] observes that the practice of one social actor speaking on behalf of another is well established in human society already. He argues further that, although many people imagine communication as a sort of communion among humans that cannot exist between humans and more-than-humans (animals or trees), such a communion almost never occurs, making the distinction between communication with other humans and communication with more-than-humans fallacious. Rather, how close and how far we are from both humans and more-than-human others is a matter of degree.

4 Refining the framework

Because ecology is interconnected with human society and its technologies, it flows both heuristically and materially throughout our energy democracy framework. We now examine how ecology and more-than-human entities are intertwined with justice, participation, and power, and make explicit the importance of these relationships in discussions of democratic energy system transitions. We chose not to add ecology as a separate element of the framework as a means of resisting the artificial separation of nature from the culture critiqued above. In what follows, we envision ecology as spanning the entire energy democracy framework and explore how humans might speak for more-than-human entities in energy democracy.

In our model, justice serves to orient energy democracy theory, practices, and movements toward making choices to ameliorate the disproportionate burdens that underrepresented and marginalized populations face with the climate crisis and current fossil fuel dominated energy systems. Bringing ecology into the justice node means asking questions about (1) whether more-than-human beings can be considered underrepresented, marginalized, and disproportionately harmed populations and (2) how to ensure just models of participation that give voice to these communities (which will be discussed in the next section). Indigenous stances on environmental justice offer a precedent for considering the importance of justice for more-than-human ecology via Indigenous epistemologies that view the more-than-human world as animate and in relationship with humans [26–28]. Justice, while traditionally conceived as a human construct, can be applied to the more-than-human world. As Indigenous approaches indicate, this focus on justice for more-than-human beings in our ecologies does not have to trade-off with justice for human beings. Indeed, it would be problematic from an ecological systems perspective to simply shift our focus from human justice to more-than-human justice (for a different perspective, see chapters by Lennon and Dunphy, and by Selk and Kemmerzell, both in Part III). Rather, ecological justice would attend to the complex relations of justice affecting all aspects of the ecological system, human, and more-than-human alike. Of course, there will be situations in which (in)justices align across beings and when they are in conflict. Bringing an ecological approach to justice into the energy democracy framework then allows for consideration of justice and injustice across the spectrum of the more-than-human world. We might then ask questions about not only which people will benefit and which will be harmed from a particular energy process but also what more-than-human beings will benefit and be harmed. Asking these questions allows for a process in which the complex interactions within an ecology of human and more-than-human beings can be considered in an effort to maximize justice in efforts to make energy transitions and decision-making more democratic.

As the second component of our model, participation builds on our characterization of justice in that participation should include all stakeholders, both human and more-than-human. Senecah has often described the components of true participation in a decision-making process as access, standing, and influence, also known as the Trinity of Voice [29]. A process must go beyond technocratic discourses that privilege expert voices with technical knowledge for this to be fully realized for all relevant parties. As Peterson et al. [25] point out, these voices more often than not perpetuate the nature/culture divide. Instead, decision-makers have a responsibility, though one not often realized, to incorporate other forms of knowledge such as local and Indigenous knowledge that engage with or have the capacity to engage ecological systems in daily life. Peterson et al. [25], drawing from Aldo Leopold's land ethic, discuss this as a greater recognition of community that expands beyond humans to include "aquatic, atmospheric, and terrestrial subjects that work together to form a community" (p. 77). Continuing along this line of thought, more-than-human stakeholders do not have to be fully silent in a decision-making process. Callister's [30] land community participation (LCP) model proposes three intersecting continua—participation (indirect to direct), chronemics (ranging from object to narrative to spiritual), and power (ranging from hegemonic to emancipatory)—that enable consideration of more-than-human ecology in decision-making. In practice, this model encourages participants to go beyond dialog and deliberation indoors to include outside spaces as a means for more-than-human participation, which could be done in energy decision-making.

It should come as no surprise in regards to our final component that, as an element of energy democracy, the most crucial conceptualization of power is resistive, rather than power as maintenance of existing hegemonic configurations. Resistive power calls for a much more proactive and responsive attitude than maintenance power. It privileges the provisional over permanence, inclusion over exclusion, and simultaneous appreciation for individuality and symbiotic relationships over a preference for either [31]. To account for more-than-human ecology requires attunement to the ways that ecology can act as a form of resistance to the maintenance of power, particularly the systems of power that mark the separation between human and nature, and an ability for human actors to speak for more-than-human ecological resistance in human decision-making forums. As with any social movement, participants need to anticipate that, rather than producing an easy consensus, the policies they advocate will be opposed by influential entities that operate across a variety of hegemonic political configurations. Strategic communication is especially crucial for facilitating resistive power because the implementation of the changes proposed requires some level of public acceptance and engagement [32]. We follow Cox [33], in defining strategic communication as "an heuristic for identifying openings within networks of contingent relationships and the potential of certain communicative efforts to interrupt or leverage change within systems of power" (p. 122). Following from this definition, strategic communication includes developing a solid understanding of how existing power brokers have constituted ecological processes such as climate change, as well as the energy system itself, in ways that deny a voice to those very processes. That understanding can then provide an opening for reconstituting understandings of ecological processes, sociotechnical arrangements, and the societal consequences of those processes and sociotechnical systems through giving voice to what ecological systems and more-than-human beings are *saying*. Doing so contributes to perceptions of what is possible in energy communication. For example, meaningful discourse about solar energy need not be limited to technical jargon about mitigation of climate change or price points; it matters that sunshine is freely available to both human and more-than-human entities [34]. Ecological resistance, then, is a node of power that can and should be included in energy democracy, and strategic communication offers powerful guidance for navigating the challenges to more cooperative approaches to energy such as those identified by Wright (Part II), and may even enable resisting the "numbing" of public responses to the need for energy system transformation detailed by Ashworth and Witt (Part III).

5 Conclusions and future directions

This chapter sought to explicitly incorporate ecology into a framework for energy democracy, not as a component of the model, but by acknowledging its integrating function within the components of justice, participation, and power. The simple idea that humans are a part of nature, rather than apart from it, provides heuristic guidance for our approach to integrating energy democracy with more-than-human ecology. Human embeddedness in a more-than-human ecosystem is clearly articulated by Peters [24] when he explains that "communication is something we share with animals and computers, extraterrestrials and angels … The concept respects none of the metaphysical barriers that once protected human uniqueness, … traversing the bounds of species, machines, even divinity" (pp. 227–228). If serious consideration of how people might mesh with the more-than-human ecosystem includes giving up the notion of human exceptionalism, it is no wonder so many overseers of well-established sociotechnical systems such as the energy system are willing to expend great effort to maintain the status quo of technocratic decision-making that separates humans from nature.

While this ecologically inspired heuristic for composing energy democracy certainly relies on opportunities that have emerged from trends toward decarbonization and decentralization, it both extends and shapes these opportunities. For example, although the interaction of decarbonization and decentralization enabled by wind and solar energy offers possibilities for enhancing justice and opening participation to more citizens, the realization of those possibilities remains elusive. It requires targeted efforts to shift traditional economic patterns, such as efforts to decentralize market share that are detailed by Brisbois (Part I). It also directs us toward the importance of strategically communicating the need for change noted by Lennon and Dunphy (Part III), who point out problems with the term, "energy citizen," which trails the passive consumerist image into a weakened notion of energy democracy.

In order to move forward with the inclusion of more-than-human stakeholders in energy democracy, we need to ensure the creation of spaces for such engagement. Such inclusion does not have to be difficult or overthought. For example, Smith, Limburg, and Feldpausch-Parker (forthcoming) were influenced by Callister's [30] LCP model, having incorporated field trips to urban streams to re-engage participants with the living aquatic system to assist with decisions involving dam removals. Their participation structure of lectures engaging relevant knowledge, field trips engaging more-than-human stakeholders, and mediated modeling to incorporate systems thinking was an effective and inclusive way to pursue decision-making. This application of the LCP model can be easily analogized to energy systems as well as the incorporation of participant visits to proposed sites for energy production, be it a solar or wind farm or a mine and coal-fired power plant. Though this is just one example, it demonstrates possibilities for incorporating more-than-humans into the practice of energy democracy. And, perhaps more importantly, it demonstrates that, despite the legitimate concerns about greening the energy sector expressed by Selk and Kemmerzell (Part III), "gains in ecology [do not] come at the price of losses in democracy." Rather, gains in ecology are gains for human communities, as well as for the more-than-human world.

References

[1] S. Sweeney, Working toward energy democracy, in: Worldwatch Institute (Ed.), State of the World 2014: Governing for Sustainability, Island Press, Washington, DC, 2014, pp. 215–227.

[2] K. Szulecki, I. Overland, Energy democracy as a process, an outcome and a goal: a conceptual review, Energy Res. Soc. Sci. 69 (2020), 101768.

[3] C.M. de Onís, Energy colonialism powers the ongoing unnatural disaster in Puerto Rico, Front. Commun. (2018), https://doi.org/10.3389/fcomm.2018.00002.

[4] K. Szulecki, Conceptualizing energy democracy, Environ. Polit. 27 (1) (2018) 21–41.

[5] A.M. Feldpausch-Parker, D. Endres, T.R. Peterson, Editorial: A research agenda for energy democracy, Front. Commun. (2019), https://doi.org/10.3389/fcomm.2019.00053.

[6] G.A. Garcia-Lopez, The multiple layers of environmental injustice in contexts of (un) natural disasters: the case of Puerto Rico post-Hurricane Maria, Environ. Justice 11 (3) (2018) 101–108.

[7] A. Martin, S. McGuire, S. Sullivan, Global environmental justice and biodiversity conservation, Geogr. J. 179 (2) (2013) 122–131.

[8] B. Cozen, D. Endres, T.R. Peterson, C. Horton, J.T. Barnett, Energy communication: theory and praxis towards a sustainable energy future, Environ. Commun. 12 (3) (2018) 289–294.

[9] D. Endres, L.M. Sprain, T.R. Peterson, Social Movement to Address Climate Change: Local Steps for Global Action, Cambria Press, Amherst, New York, 2009.

[10] C. Mouffe, The Democratic Paradox, Verso, New York, NY, 2000.

[11] M.J. Burke, J.C. Stephens, Energy democracy: goals and policy instruments for sociotechnical transitions, Energy Res. Soc. Sci. 33 (2017) 35–48.

[12] C.M. de Onís, Fueling and delinking from energy coloniality in Puerto Rico, J. Appl. Commun. Res. 46 (5) (2018) 535–560.

[13] M. Begon, J.L. Harper, C.R. Townsend, Ecology, third ed., Blackwell Science, Malden, MA, 1996.

[14] W. Cronon, The trouble with wilderness; or getting back to the wrong nature, in: Uncommon Ground: Rethinking the Human Place in Nature, Norton & Company, New York, NY, 1996, pp. 69–90.

[15] B. McGreavy, J. Wells, G.F. McHendry Jr., S. Senda-Cook, Tracing Rhetoric and Material Life: Ecological Approaches, Springer, 2017.

[16] R.A. Rogers, Overcoming the objectification of nature in constitutive theories: toward a transhuman, materialist theory of communication, West. J. Commun. 62 (3) (1998) 244–272, https://doi.org/10.1080/10570319809374610.

[17] D. Haraway, Simians, Cyborgs, and Women: The Reinvention of Nature, Routledge, New York, NY, 1990.

[18] D. Abram, The Spell of the Sensuous: Perception and Language in a More-Than-Human World, Vintage, New York, NY, 1997.

[19] D. Haraway, The Companion Species Manifesto: Dogs, People, and Significant Otherness, Prickly Paradigm Press, Chicago, IL, 2003.

[20] N. Luhmann, Ecological Communication (J. Bednarz, Trans.), University of Chicago Press, Chicago, IL, 1989.

[21] L. Rickard, A.M. Feldpausch-Parker, Of sea lice and superfood: a comparison of regional and national news media coverage of aquaculture, Front. Commun. (2016), https://doi.org/10.3389/fcomm.2016.00014.

[22] M.J. Peterson, D.M. Hall, A.M. Feldpausch-Parker, T.R. Peterson, Obscuring ecosystem function with application of the ecosystem services concept, Conserv. Biol. 24 (1) (2010) 113–119, https://doi.org/10.1111/j.1523-1739.2009.01305.x.

[23] B. Latour, Politics of Nature: How to Bring the Sciences Into Democracy (C. Porter, trans.), Harvard University Press, Cambridge, MA, 2004.

[24] J.D. Peters, Speaking Into the Air: A History of the Idea of Communication, University of Chicago Press, Chicago, IL, 1999.

[25] M.N. Peterson, M.J. Peterson, T.R. Peterson, Environmental communication: why this crisis discipline should facilitate environmental democracy, Environ. Commun. 1 (1) (2007) 74–86.

[26] G. Cajete, A People's Ecology: Explorations in Sustainable Life, Clear Light, Santa Fe, New Mexico, 1999.

[27] R.W. Kimmerer, Braiding Sweetgrass: Indigenous Wisdom, Scientific Knowledge and the Teachings of Plants, Milkweed Editions, 2013.

[28] K. Whyte, Indigenous experience, environmental justice and settler colonialism, in: Nature and Experience: Phenomenology and the Environment, Rowman & Littlefield International, Lanham, 2016. https://papers.ssrn.com/abstract=2770058.

[29] S.L. Senecah, The trinity of voice: the role of practical theory in planning and evaluating the effectiveness of environmental participatory processes, in: S.P. Depoe, J.W. Delicath, M.-F.A. Elsenbeer (Eds.), Communication and Public Participation in Environmental Decision Making, State University of New York Press, Albany, NY, 2004, pp. 13–34.

[30] D.C. Callister, Land community participation: a new 'public' participation model, Environ. Commun. 7 (4) (2013) 435–455, https://doi.org/10.1080/17524032.2013.822408.

[31] T.R. Peterson, Paradox of trust in unsettled times: can scientists "Speak truth to power?", in: K.P. Hunt (Ed.), Understanding the Role of Trust and Credibility in Science Communication. Selected Proceedings of Iowa State University's 2018 Summer Symposium on Science Communication, 2018, https://doi.org/10.31274/sciencecommunication-181114-10. ISSN 2572-679X.

[32] T.R. Peterson, C.C. Horton, Communicating about solar energy and climate change, in: M. Nesbit (Ed.), The Oxford Research Encyclopedia of Climate Science, 2017. http://climatescience.oxfordre.com/.

[33] J.R. Cox, Beyond frames: recovering the strategic in climate communication, Environ. Commun. 4 (1) (2010) 122–133.

[34] D. Endres, B. Cozen, J.T. Barnett, M. O'Byrne, T.R. Peterson, Communicating energy in a climate (of) crisis, in: Communication Yearbook, 40, Routledge, New York, NY, 2016, pp. 419–447.

4

Utopias and dystopias of renewable energy imaginaries

Karen Hudlet-Vazquez[a], Melissa Bollman[a], Jessica Craigg[b], and James McCarthy[a]

[a]Graduate School of Geography, Clark University, Worcester, MA, United States [b]Energy & Resources Group, University of California—Berkeley, Berkeley, CA, United States

1 Introduction

While renewable energy technologies open new possibilities for more distributed, decentralized, and self-sufficient energy systems, it remains an open question whether their mass adoption will reduce or replicate the asymmetries, divisions, and violence engendered by and through existing energy infrastructures [1–3]. The concept of sociotechnical imaginaries, defined as "collectively held, institutionally stabilized and publicly performed visions of desirable futures (or of resistance against the undesirable), animated by shared understandings of forms of social life and social order, and supportive of advances in science and technology" [4, p. 4], has gained traction in energy studies as scholars increasingly attend to the political possibilities related to renewable energy system transitions. Sociotechnical imaginaries are visions over science, technology, the state, and society and can operate to preserve current energy regimes or present a rupture. For example, Longhurst and Chilvers [5] describe sociotechnical imaginaries as visions on renewable energy that can result in alternative practices and, in turn, allow for sociotechnical transformations.

Much of the energy imaginaries literature implicitly or explicitly draws upon notions of energy democracy to understand how energy systems could be otherwise. Energy democracy provides a heuristic for reflecting on what a more just energy system might look like in terms of collective control, participatory decision-making, and fairer distributions [6], as well as on how energy systems facilitate the expansion of democracy [7]. Energy democracy movements have also interrogated the roles of different actors (the state, corporations, civil society, consumers, and users) in energy system change and have illuminated the possibility (and even necessity) for cultivating alternative values. As an umbrella concept, energy democracy can draw inspiration from anti-politics—such as resisting fossil fuels, the market-driven green agenda, solely government and expert decision-making on energy politics [8] that propose new ways of producing and consuming energy, such as some collective and politically motivated community energy projects [9], degrowth, citizen-led energy projects, and diverse economies. Power is at the heart of all imaginaries, shaping how the political visions associated with renewable energy systems develop and take hold [10]. A core assumption of this chapter is that energy systems are always political and as such are produced by and influence societal, cultural, and economic forms of organization [7,11–13]. Discussions of sociotechnical imaginaries tend to theorize power as hierarchical and institutionally located rather than relational and distributed [14]. This chapter attempts to bring in power as a central rather than marginal component of analyzing sociotechnical imaginaries obstructing and promoting change.

To understand the relationship between imaginaries and power we rely on conceptions of power as productive and circulating. For Foucault, power is "neither possessed nor located" but a tenuous practice which supposes that subjugated social groups are continuously challenging the hegemony of the dominant ones [15, p. 3]. Domination and resistance are always intertwined in expressions of power such that there is resistance in domination and domination in resistance [15,16]. As energy systems—and the decision-making processes that inform them—are based on relationships between actors, we can conceptualize social power as distributed through, manifest within, and co-produced with the discursive and material formations of energy landscapes [12]. Power can inhibit alternative energy imaginaries through specific discourses that contribute to rigidifying energy visions such as eco-modernization. Building on Foucault's theorizations of biopower, Boyer's [11,17] the concept of "energopower" illustrates how electricity and petroleum have been integral to state techniques designed to manage life and populations. Hardt and Negri [18] revise biopower to imagine potential new relations, subjectivities, and practices by which, through collective desires and capacities, new ways of living may emerge against the attempts of capitalism to commodify life. This chapter reviews the existing literature on renewable energy imaginaries to better understand how power, in its many forms, underlies visions of renewable energy systems and energy democracy. In addition, it identifies the processes through which renewable energy imaginaries become integrated into new material infrastructures, discourses, practices, and subject positions. In doing so it aims to address the nonlinear process of envisioning, performing (or bringing into being), and consolidating sociotechnical imaginaries described by Marquardt and Delina [19]. Finally, we take up Hajer and Pelzer's [20] call to go beyond explaining expected futures and into an approach for constructing desirable ones through new narratives and practices.

2 Imagining renewable energy

For Foucault, the possibility of resistance is ubiquitous because power is diffuse and the targets of resistance are not entities, but mentalities and practices associated with discourses and norms. These new discourses and regimes of practice can influence the construction of reality [21] and energy landscapes. By connecting normative framings and epistemic orders with technologies and practices, sociotechnical imaginaries provide the vocabulary for thinking about possibilities for reforming the current energy system [22], as well as alternatives to it. The materialities of renewable energy technology have the potential to trigger alternative imaginaries of control, decision-making, and governance. This is exemplified through moves toward community energy that aim to relocate centers of control over the energy system and distance themselves from corporate-owned energy systems [23]. More diffuse energy systems could challenge dominant energy systems by reducing the distance between sites of production and consumption.

There are multiple sociotechnical imaginaries within what we refer to as "dominant" and "alternative" visions. Alternative imaginaries encompass those that challenge hegemonic (eco-modern) visions of energy transition that center on the use of new technology to perpetuate the current capitalist energy system in what has been framed as "win-win" situations [23] and "all hands on deck" approaches in which technological fixes can sustain our current energy consumption [24]. Thus, this imaginary belief is that the energy system can be reformed through technology to avoid the worst consequences of the fossil economy. However, technological advances also invite us to view the energy transition from outside capitalism. This position between reforming the systems from inside and creating anew can be understood through Escobar's [25] distinction between alternative development and alternatives to development. Additionally, alternative imaginaries resemble Hage's [26] distinction between radical political and anthropological imaginaries: The former as the act of stepping outside for a critical view of the current political, social, and economic institutions, and the latter as the ability to envision alternatives outside the system. The reflective practice of imagining alternatives not only invites us to see our social world with a critical eye but also haunts us with other possibilities. Simmet [27] describes how the dominant and intertwined visions on energy and development, based on expert knowledge and the need for industrialization, marginalize other local social imaginations on the possibilities of energy, spelling a need to depart from them in imagining alternatives.

Alternative sociotechnical imaginaries play a role in creating something different. Imaginaries are linked to desires and fantasies [28]; likewise, speculation on energy infrastructures is linked to the politics of the subject who is imagining [29,30]. For instance, Lennon [30] presents a proposal for democratizing and decolonizing energy through the Black Lives Matter movement. In destabilizing dominant and hegemonic visions of energy, he views energy as a force, a vitality, through which people have the power to transform energy beyond the scope of top-down energy policies. Williams [31] describes how practices of speculation can shape political desires contributing to the formation of a mobilizing effect at the community level. Overall, alternative imaginaries do not see a solution in the existing economic, social, and political models that dominate modern energy systems and thus propose a new way of being. In fact, Wilson [32] and Norgaard [33] urge us toward new ways of being using other knowledge systems, such as Indigenous and feminist, given the limits of Western ontologies.

Nevertheless, because of the tensions between how alternative and dominant imaginaries interact and relate to each other, the former may fit snugly into the latter [34,35]. Lennon [30] describes how energy transitions framed under colorblind climate change visions can obscure racial grief, result in continued labor exploitation in poor communities of color, and perpetuate carbon-intensive consumption. Community economies, as well as community energy, can reproduce neoliberal values in relation to the roles of the state, the individual and participation [36]. Renewable energy imaginaries, even alternative ones, can reproduce settler colonialism [37]. Utopian visions of desired futures can therefore have the unexpected consequence of strengthening existing power and privilege. These case studies point to the complexities in the relationship between alternative and dominant imaginaries.

In addition to their focus on futuring, imaginaries are also influenced by the past. New energy systems are deployed in already uneven landscapes [38] co-produced by modern bio-political "technologies" such as racism, imperialism, and colonialism. International and national imaginaries often obscure nuances and collective memories of a place, ignoring that all sociotechnical visions are situated [39]. In this sense, territorial trajectories and the history of a place—as well as local development trajectories, experiences with previous development projects, and the context of struggles for autonomy—shape and influence energy imaginaries [40]. Thus, there is a need to broaden the temporal frame in which imaginaries are analyzed [41]. Moreover, "the resulting politics of science and technology may shape not only the narrow issues surrounding those specific enterprises but also wider social and political understandings about a nation's past, present, and future" [42, p. 124]. In this way, sociotechnical imaginaries are entwined with notions of nationhood, collective identity, and a national "good life," with the result that challenging dominant imaginaries appears as unpatriotic or backward [4,7,12].

3 Performing and practicing renewable energy imaginaries

"Techniques of futuring" [20] focus on the importance of performativity and practices. This allows for bringing together actors around an existing imaginary and motivating them to action, thus transforming the status quo and creating different imaginaries [16]. Performativity brings attention to how imaginaries come into being by enacting regimes of practices and contra-practices [17,43]. This section analyzes the relationship between hegemonic and alternative imaginaries and practices of governing, resisting, and becoming within diverse entanglements of stakeholders (authorities, companies, civil society) at different scales. Energy infrastructures (both real and imagined) reflect broader discourses of what energy systems "should be" more than simply the structural position of decision-makers or "responsible" technological choices. It has been well established that our modern energy systems embody political choices regarding what forms of governance, types of subjects, and configurations of infrastructures best support a carbon-based economy and society [6,9], and the same could be true of the renewable energy systems being imagined presently.

In their review of energy futures, Delina and Janetos [44] argue that the process of negotiating imaginaries can obscure minority futures. There is thus a need for understanding the evolution and plurality of energy futures—including local ones—to fully grasp the diversity of visions based on cultural, economic, and political influence. In this vein, there has been a research on how local, nonexpert visions that challenge the national imaginary become bounded imaginaries tied to a place due to a relative lack of local political power [45]. Mega-narratives such as national development [4], universalizing Western narratives of economic growth through technology and industrialization [46], and the Westernization of lifestyles [27] can overtake alternative local views. Universal visions of energy futures can also ignore the lifeworlds and particular meanings and practices that people associate with energy [47], as well as different interpretations of justice [48,49]. Nevertheless, others have described how contradictory imaginaries of energy futures can be complementary and converge toward a common project [50], while different national imaginaries and local expectations can be managed through ambiguity and negotiation for common goals [51].

The scale of imaginaries is relevant when theorizing energy transitions and identifying who imagines what. Likewise, the setting is relevant in shaping the emergence of particular visions that become entangled in the co-production of visions [5]. Boundaries between local, national, and international imaginaries are not necessarily clear-cut; they interact and co-construct each other as new imaginaries emerge to challenge the dominant ones [52]. The heuristic of global assemblages[a] might provide an alternative way to conceptualize imaginaries at different scales and to explore how dominant views of energy and development have contributed to shaping national imaginaries [45].

Rethinking the way in which energy imaginaries are performed and promoted at specific sites opens the possibility for taking seriously local desires and aspirations through more direct forms of participation, such as practical recognition [40], focus groups for co-constructing knowledge [54], citizen-led energy projects [43], and direct forms of democracy [55]. Activism, specifically prefigurative activism in which the visions and values promoted are enacted through social experimentations, can foster egalitarian renewable energy systems and more sustainable futures by performing contra-narratives against large energy infrastructure and oil [15]. Another strategy for resisting or reshaping dominant imaginaries is through the creation of spaces of hesitation [56]. For example, Gabrys [57] proposes making energy visible in order to expose how technologies perform distinct materialities and political goals. The resulting space of hesitation allows for interrogation of the business-as-usual procedures of energy systems. Such a "cosmopolitics of energy" [57] presents how practices can reveal entanglements of different imaginaries or visions without necessarily arriving at a final solution; however, change can be generated through "hesitating practices" [57] and creative sites of material and political struggle. Gabrys argues that diverging materialities of energy articulate the possibility of alternative connections and practices among many actors, including the more-than-human (e.g., devices, infrastructure). Similarly, what Brennan [58] calls "visionary infrastructure," such as solar street lighting, can challenge previous exclusions and lead to imagining the world differently.

4 Renewable energy and the cultivation of new subjects

Novel energy infrastructures and practices open the possibility for alternative subjectivities and sociopolitical arrangements. This section explores some of these potentialities in the context of ongoing low-carbon energy transitions, under the assumptions that (1) sociotechnical imaginaries constitute a "technology of power" [59,60] implicated in the production of low-carbon subject positions, such as the "smart" energy user or renewable energy prosumer; and (2) alternative energy system imaginaries can arise from these new subjectivities and their associated practices like reflexive energy conservation and peer-to-peer energy trading.

---

[a] The concept of assemblage broadly describes heterogeneous actors provisionally coming together to produce emerging properties by their interactions. Assemblages are contingent and often dynamic [53].

A Foucauldian governmentality approach to low-carbon transitions helps foreground how novel devices (distributed solar energy systems; smart meters), infrastructures (microgrids; high-voltage power lines), discourses ("soft energy paths" [61]; "carbon neutrality"), and imaginaries ("green growth"; "solar communism" [62]) act as political "technologies" for producing a variety of subjectivities. Foucauldian perspectives understand subjectification as self-governance expressed through individual behaviors and the assumption of social roles in accordance with perceived truths [15]. As a result, the subjectification process can result in different outcomes depending on whether an individual acts in congruence with the prevailing "governmentality" (dominant political rationality) or resists it by engaging in "counterconducts" or non-conforming self-governing behaviors [63].

A large subset of the energy governmentality research to date focuses on how "smart" energy technologies, i.e., data-driven, Internet-connected, and semiautonomous grids, meters, and appliances act as conduits for productive power. Empirical studies examining "smart" device deployment in the Global North asserts these technologies train subjects to self-regulate their energy production and consumption in service of modern state imperatives, such as grid stability and market regulation [64], or to cultivate neoliberal subject positions by stripping away the human agency typically involved in energy decision-making [65]. Research also shows that imaginaries shape the political rationalities that influence low-carbon device design. For instance, ethnographic investigations into the mechanics of smart technology design find smart devices are configured to compensate for an imagined public's presumed inability to effectively manage their own energy use [66,67]. The documented practice of smart technology creators designing for, and not with, energy users suggests that emerging low-carbon technologies may be maintaining, if not widening, the white-collar energy professional-layperson divide that originated in the fossil fuel era [30,68].

On the other hand, other research finds subjects resisting the "carbon governmentality" embedded in smart energy meters and renewable generation infrastructure by adopting "mongrel identities" that blend neoliberal and eco-modern frameworks with traditional social norms and values [69,70]. More specifically, "Hargreaves" [71] research found many UK householders did not fully adopt the carbon conscious mentality associated with smart energy monitors and continued to make consumption choices within a more familiar "pounds and pence" cognitive framework. Meanwhile, residents of the US state of Texas reportedly welcomed the economic benefits of wind energy development without adopting the guiding "ecological rationality" [72]. While additional research is needed to clarify precisely how and why subjects enact or resist "sociotechnical scripts" embedded in low-carbon energy devices [73], these preliminary findings indicate energy technologies can problematize or create a wide variety of low-carbon subject positions, including some that contradict the technocratic ecomodernism that undergirds many energy transitions pathways.

Whereas Foucauldian governmentality perspectives have been useful for thinking through how dominant imaginaries can influence technology production and subject formation during low-carbon transitions, Negri's concept of "biopotenza," or capacity of the human multitude to create social life anew [74], shifts focus to how alternative subjectivities and practices could perform alternative energy fantasies. According to Hardt and Negri [75], the actualization of any technological potentiality, such as an alternative energy system, requires the previous existence of an enabling subjectivity born from antagonistic social conflict, for example, the struggle between capital and labor. As such, the production of alternative subjectivity—which is a joint subjectivity of a multitude of bodies—represents a liberatory act of reclaiming the world and generating alternative ways of being [76].

In the energy context, such an alternative subjectivity might manifest in energy practices like prosumership or peer-to-peer trading that open possibilities for contesting dominant political rationalities and providing subjects with more purposeful, egalitarian, and stable social identities than incumbent systems [77]. Visions of "open-source" renewable energy systems based on technologically mediated networks of cooperation (i.e., peer-to-peer energy trading enabled by small-scale distributed systems and ICTs) share a great deal in common with the production system Hardt and Negri imagine will support a new global democracy in the conclusion of *Multitude* [75]. Both systems are expected to arise from outside the state as an ungovernable, creative, immaterial excess that will ultimately liberate subjects from undesirable physical toil and immiseration [77]. Though the creation of a global peer-to-peer energy system remains, for now, an unrealized dream, an upsurge of citizen-led renewables projects, such as those described by Matsuzawa in this volume [43], demonstrates some localities are striving to shift the locus of energy decision-making away from the nation-state by affording more opportunities for everyday people to actively participate in energy governance, thus potentially helping them transcend the tightly circumscribed "citizen-as-consumer" subject position characteristic of neoliberal energy regimes [78].

5 Conclusions

This chapter has reflected on renewable energy sociotechnical imaginaries as a site of possibility for sociopolitical change. The potentialities of renewable energy technologies have sparked the imagination of a variety of energy transition actors and produced the opportunity for alternative political systems, social relations, and subject positions. The energy imaginaries literature describes the nonlinear, ambiguous, and uncertain processes of envisioning, promoting, and performing renewable energy futures in various contexts. Though not a comprehensive review, the diversity of imaginaries explored in the chapter illustrates the complexities associated with challenging dominant energy regimes and instituting more participatory, collectively controlled and just systems in their place. However, there is a need for further research and theorizing on alternative energy imaginaries based on distributed and productive understandings of power.

References

[1] C. Kunze, S. Becker, Collective ownership in renewable energy and opportunities for sustainable degrowth, Sustain. Sci. 10 (3) (2015) 425–437.
[2] Y. Rumpala, Alternative forms of energy production and political reconfigurations: exploring alternative energies as potentialities of collective reorganization, Bull. Sci. Technol. Soc. 37 (2) (2017) 85–96.
[3] G. Bridge, B. Özkaynak, E. Turhan, Energy infrastructure and the fate of the nation: introduction to special issue, Energy Res. Soc. Sci. 41 (5) (2018) 1–11.
[4] S. Jasanoff, S.-H. Kim, Dreamscapes of Modernity: Sociotechnical Imaginaries and the Fabrication of Power, University of Chicago Press, Chicago and London, 2015.
[5] N. Longhurst, J. Chilvers, Mapping diverse visions of energy transitions: co-producing sociotechnical imaginaries, Sustain. Sci. 14 (4) (2019) 973–990.

[6] S. Becker, M. Naumann, Energy democracy: mapping the debate on energy alternatives, Geogr. Compass 11 (8) (2017) 1–13.

[7] T. Mitchell, Carbon Democracy: Political Power in the Age of Oil, Verso, London and New York, 2011.

[8] K. Szulecki, I. Overland, Energy democracy as a process, an outcome and a goal: a conceptual review, Energy Res. Soc. Sci. 69 (101768) (2020).

[9] S. Becker, C. Kunze, Transcending community energy: collective and politically motivated projects in renewable energy (CPE) across Europe, People, Place Policy Online 8 (3) (2014) 180–191.

[10] J.H. Tidwell, A.S.D. Tidwell, Energy ideals, visions, narratives, and rhetoric: examining sociotechnical imaginaries theory and methodology in energy research, Energy Res. Soc. Sci. 39 (11) (2018) 103–107.

[11] D. Boyer, Energopolitics and the anthropology of energy, Anthropology News 52 (5) (2011) 5–7.

[12] M.T. Huber, Lifeblood: Oil, Freedom, and the Forces of Capital, University of Minnesota Press, Minneapolis and London, 2013.

[13] A. Malm, Fossil Capital: The Rise of Steam Power and the Roots of Global Warming, Verso, London and New York, 2016.

[14] J. Chilvers, H. Pallett, Energy democracies and publics in the making: a relational agenda for research and practice, Front. Commun. 3 (14) (2018) 1–16.

[15] L. Gailing, Transforming energy systems by transforming power relations. Insights from dispositive thinking and governmentality studies, Innovat. Eur. J. Soc. Sci. Res. 29 (3) (2016) 243–261.

[16] J. Sharp, Geographies of domination/resistance, in: R. Paddison, C. Philo, P. Routledge, J. Sharp (Eds.), Entanglements of Power: Geographies of Domination/Resistance, Routledge, Oxford and New York, 2002, pp. 1–42.

[17] D. Boyer, Energopower: an introduction, Anthropol. Q. 87 (2) (2014) 309–333.

[18] M. Hardt, A. Negri, Commonwealth, Harvard University Press, Cambridge, 2009.

[19] J. Marquardt, L.L. Delina, Reimagining energy futures: contributions from community sustainable energy transitions in Thailand and the Philippines, Energy Res. Soc. Sci. 49 (5) (2019) 91–102.

[20] M. Hajer, P. Pelzer, 2050—An energetic odyssey: understanding "techniques of futuring" in the transition towards renewable energy, Energy Res. Soc. Sci. 44 (1) (2018) 222–231.

[21] N. Ettlinger, Governmentality as epistemology, Ann. Assoc. Am. Geogr. 101 (3) (2011) 537–560.

[22] A. Cofiño, De la resistencia en las montañas a la autogestión y la defensa de los bienes comunes: Construcción de la hidroeléctrica comunitaria Luz de los Héroes y Mártires de la Resistencia, en la Zona Reina, Quiché, Guatemala, Revista Pueblos Y Fronteras Digital 9 (17) (2014) 21–33.

[23] M. Burke, J. Stephens, Political power and renewable energy futures: a critical review, Energy Res. Soc. Sci. 35 (11) (2018) 78–93.

[24] T. Hale, "All hands on deck": the Paris agreement and nonstate climate action, Glob. Environ. Polit. 16 (3) (2016) 12–22.

[25] A. Escobar, Territories of Difference: Place, Movements, Life, Redes, Duke University Press, Durham, 2008.

[26] G. Hage, Critical anthropological thought and the radical political imaginary today, Crit. Anthropol. 32 (3) (2012) 285–308.

[27] H.R. Simmet, "Lighting a dark continent": imaginaries of energy transition in Senegal, Energy Res. Soc. Sci. 40 (2018) 71–81.

[28] C. Strauss, The imaginary, Anthropol. Theory 6 (3) (2006) 322–344.

[29] A.H. Sorman, E. Turhan, M. Rosas-Casals, Democratizing energy, energizing democracy: central dimensions surfacing in the debate, Front. Energy Res. 8 (2020) 1–7.

[30] M. Lennon, Decolonizing energy: black lives matter and technoscientific expertise amid solar transitions, Energy Res. Soc. Sci. 30 (8) (2017) 18–27.

[31] R. Williams, "This shining confluence of magic and technology": Solarpunk, energy imaginaries, and infrastructures of solarity, Open Libr. Humanit. 5 (1) (2019) 1–35.

[32] S. Wilson, Energy imaginaries: feminist and decolonial futures, in: B.R. Bellany, J. Diamanti (Eds.), Materialism and the Critique of Energy, MCMPrime Press, Chicago and Edmonton, 2018, pp. 377–412.

[33] K.M. Norgaard, Whose energy future? Whose imagination? Revitalizing sociological theory in the service of human survival, Soc. Nat. Resour. 32 (1) (2020) 1–8.

[34] J. McCarthy, Rural geography: alternative rural economies-the search for alterity in forests, fisheries, food, and fair trade, Prog. Hum. Geogr. 30 (6) (2016) 803–811.

[35] V. Selk, J. Kemmerzell, Worse than its reputation? Shortcomings of "energy democracy,", in: J. Keahey, M. Pasqualetti, M. Nadesan (Eds.), Democratizing Energy: Imaginaries, Transitions, and Risks, Elsevier, Amsterdam, 2021.

[36] L. Arguelles, I. Anguelovski, E. Dinnie, Power and privilege in alternative civic practices: examining imaginaries of change and embedded rationalities in community economies, Geoforum 86 (11) (2018) 30–41.

[37] R.D. Stefanelli, C. Walker, D. Kornelsen, D. Lewis, D.H. Martin, J. Masuda, C.A.M. Richmond, E. Root, H. Tait Neufeld, H. Castleden, Renewable energy and energy autonomy: how indigenous peoples in Canada are shaping an energy future, Environ. Rev. 27 (2) (2018) 95–105.

[38] C. Harrison, Race, space, and electric power: Jim crow and the 1934 North Carolina rural electrification survey, Ann. Am. Assoc. Geogr. 106 (4) (2016) 909–931.

[39] W.M. Eaton, S.P. Gasteyer, L. Busch, Bioenergy futures: framing sociotechnical imaginaries in local places, Rural. Sociol. 79 (2) (2014) 227–256.

[40] G. Blanco-Wells, The social life of energy: notes for the territorialized study of energy transitions, Sociologias 21 (51) (2019) 160–185.

[41] H.-L. Lai, Situating community energy in development history: place-making and identity politics in the Taromak 100% green energy tribe initiative, Geoforum 100 (3) (2019) 176–187.

[42] S. Jasanoff, S.-H. Kim, Containing the atom: sociotechnical imaginaries and nuclear power in the United States and South Korea, Minerva 47 (2) (2009) 119–146.

[43] S. Matsuzawa, Energy democracy movements in Japan, in: J. Keahey, M. Pasqualetti, M. Nadesan (Eds.), Democratizing Energy: Imaginaries, Transitions, and Risks, Elsevier, Amsterdam, 2021.

[44] L. Delina, A. Janetos, Cosmopolitan, dynamic, and contested energy futures: navigating the pluralities and polarities in the energy systems of tomorrow, Energy Res. Soc. Sci. 35 (1) (2019) 1–10.

[45] J.M. Smith, A.S.D. Tidwell, The everyday lives of energy transitions: contested sociotechnical imaginaries in the American West, Soc. Stud. Sci. 46 (3) (2016) 327–350.

[46] S. Movik, J. Allouche, States of power: energy imaginaries and transnational assemblages in Norway, Nepal and Tanzania, Energy Res. Soc. Sci. 67 (9) (2020) 1–11.

[47] F. Boamah, E. Rothfuß, From technical innovations towards social practices and socio-technical transition? Rethinking the transition to decentralised solar PV electrification in Africa, Energy Res. Soc. Sci. 42 (2) (2018) 1–10.

[48] H.P. Bedi, "Our energy, our rights": national extraction legacies and contested energy justice futures in Bangladesh, Energy Res. Soc. Sci. 41 (6) (2018) 168–175.

[49] G. Bombaerts, K. Jenkins, Y.A. Sanusi, W. Guoyu, Expanding ethics justice across borders: the role of global philosophy, in: G. Bombaerts, K. Jenkins, Y.A. Sanusi, W. Guoyu (Eds.), Energy Justice Across Borders, Springer, 2019, pp. 3–21.

[50] G. Trencher, J. van der Heijden, Contradictory but also complementary: national and local imaginaries in Japan and Fukushima around transitions to hydrogen and renewables, Energy Res. Soc. Sci. 49 (10) (2019) 209–218.

[51] M. Korsnes, Ambition and ambiguity: expectations and imaginaries developing offshore wind in China, Technol. Forecast. Soc. Chang. 107 (6) (2016) 50–58.

[52] A.M. Levenda, J. Richter, T. Miller, E. Fisher, Regional sociotechnical imaginaries and the governance of energy innovations, Futures 109 (1) (2019) 181–190.

[53] B. Anderson, C. McFarlane, Assemblage and geography, Area 43 (2) (2011) 124–127.

[54] M. Hurlbert, M. Osazuwa-Peters, J. Rayner, D. Reiner, P. Baranovskiy, Diverse community energy futures in Saskatchewan, Canada, Clean Techn. Environ. Policy 22 (5) (2020) 1157–1172.

[55] J. Angel, Towards an energy politics in-against-and-beyond the state: Berlin's struggle for energy democracy, Antipode 49 (3) (2017) 557–576.

[56] I. Stengers, The cosmopolitical proposal, in: B. Latour, P. Weibel (Eds.), Making Things Public: Atmospheres of Democracy, MIT Press, Cambridge, 2005, pp. 994–1003.

[57] J. Gabrys, A cosmopolitics of energy: diverging materialities and hesitating practices, Environ. Plan. A 46 (9) (2014) 2095–2109.

[58] S. Brennan, Visionary infrastructure: community solar streetlights in Highland Park, J. Vis. Cult. 16 (2) (2017) 167–189.

[59] M. Foucault, The History of Sexuality, vol. I, Allen Lane, London, 1978.

[60] M. Foucault, Discipline and Punish, Penguin Books, London, 1991.

[61] A.B. Lovins, Soft Energy Paths: Toward a Durable Peace, Friends of the Earth International, 1977.

[62] D. Schwartzman, Solar communism, Sci. Soc. 60 (3) (1996) 307–331.

[63] M. Foucault, The subject and power, Crit. Inq. 8 (4) (1982) 777–795.

[64] F. Klauser, T. Paasche, O. Söderström, Michel Foucault and the smart city: power dynamics inherent in contemporary governing through code, Environment and Planning D: Society and Space 32 (5) (2014) 869–885.

I. Imaginaries

[65] J. Sadowski, A.M. Levenda, The anti-politics of smart energy regimes, Polit. Geogr. 81 (102202) (2020).

[66] T.M. Skjølsvold, C. Lindkvist, Ambivalence, designing users and user imaginaries in the European smart grid: insights from an interdisciplinary demonstration project, Energy Res. Soc. Sci. 9 (9) (2015) 43–50.

[67] C. Cherry, C. Hopfe, B. MacGillivray, N. Pigeon, Homes as machines: exploring expert and public imaginaries of low carbon housing futures in the United Kingdom, Energy Res. Soc. Sci. 23 (1) (2017) 36–45.

[68] L. Nader, Barriers to thinking new about energy, Phys. Today 34 (2) (1981) 9.

[69] Y. Strengers, L. Nichols, Convenience and energy consumption in the smart home of the future: industry visions from Australia and beyond, Energy Res. Soc. Sci. 32 (10) (2017) 86–93.

[70] R. Defila, A. Di Giulio, C.R. Schweizer, Two souls are dwelling in my breast: uncovering how individuals in their dual role as consumer-citizen perceive future energy policies, Energy Res. Soc. Sci. 35 (1) (2018) 152–162.

[71] T. Hargreaves, Smart meters and the governance of energy use in the household, in: J. Stripple, H. Bulkeley (Eds.), Governing the Climate: New Approaches to Rationality, Power and Politics, Cambridge University Press, Cambridge, 2013, pp. 127–143.

[72] W. Jepson, C. Brannstrom, N. Persons, "We don't take the pledge": environmentality and environmental skepticism at the epicenter of US wind energy development, Geoforum 43 (4) (2012) 851–863.

[73] H. Ahlborg, Changing energy geographies: the political effects of a small-scale electrification project, Geoforum 97 (2018) 268–280.

[74] M. Coleman, K. Grove, Biopolitics, biopower, and the return of sovereignty, Environ. Plan. D: Soc. Space 27 (3) (2009) 489–507.

[75] M. Hardt, A. Negri, Multitude: War and Democracy in the Age of Empire, Penguin Press, New York, 2004.

[76] C.T. Wolfe, Materialism and temporality: on Antonio Negri's "constitutive" ontology, in: T. Murphy, A.-K. Mustapha (Eds.), The Philosophy of Antonio Negri: Revolution in Theory, Pluto Press, Sterling, 2007, pp. 198–220.

[77] J. Ruotsalainen, J. Karjalainen, M. Child, S. Heinonen, Culture, values, lifestyles, and power in energy futures: a critical peer-to-peer vision for renewable energy, Energy Res. Soc. Sci. 34 (12) (2017) 231–239.

[78] B. Lennon, N.P. Dunphy, Mind the gap: citizens, consumers, and unequal participation in global energy transitions, in: J. Keahey, M. Pasqualetti, M. Nadesan (Eds.), Democratizing Energy: Imaginaries, Transitions, and Risks, Elsevier, Amsterdam, 2021.

5

Technoregions of insurrection: Decentralizing energy infrastructures and manifesting change at scale

Ry Brennan

Department of Sociology, University of California, Santa Barbara, CA, United States

1 Introduction

Santa Barbara County sits at the end of two fraying extension cords. Located on California's Central Coast, where the Pacific Gas & Electric (PG&E) transmission line from the north stretches toward the Southern California Edison (SCE) line from the south, Santa Barbara County is surrounded by extreme fire risk zones.

From mid-2014 to 2017, investor-owned utilities (IOUs) caused over 2000 wildfires in California [1]. In Northern California, PG&E's negligence was made infamous by the 2018 destruction of Paradise, California [2]. Further to the south, in Santa Barbara and Ventura Counties, SCE paid $360 million in 2019 to quietly settle lawsuits for starting the Thomas and Woolsey Fires, which burned 440 square miles and precipitated a deadly landslide [3]. San Diego Gas & Electric (SDG&E) ignited three fires in 2007 which destroyed 1300 homes, and the company expects they will be responsible for another major fire within 20 years [4].

My fieldwork studying the nexus of energy and wildfires in Santa Barbara County reveals the infrastructural and organizational dysfunctions of the centralized energy system. In this chapter, I use these dysfunctions as a starting point for a broader critique of centralization inspired by anarchist and regionalist theory and the call for resilient systems. In so doing, I propose a reconceptualization of the *technoregion* that attends to the liberatory potential of technology and realizes the utility of the region as a spatial organizing principle. This concept—and its accordant strategic epistemology *technoregionalism*—equips readers with a sense of how insurrectionary change toward decentralist systems can advance at scale through technoregional reorganization.

I begin with the observation that wildfires expose the infrastructural fragilities of centralized energy infrastructures. These systems comprise centralized generation facilities and vast networks of IOU transmission lines running through wilderness. These lines are often poorly maintained, and when they slap together in high winds, sag in heat, or suffer falling branches, they can send sparks into desiccated surroundings and ignite fires. In response to this danger, IOUs often suspend services through public safety power shut-offs (PSPSs). Because most communities lack local generation capacity, those living downstream from fire risk zones are then left without power. When IOUs generate power far away from end-users, the result is a fragile system vulnerable to even small disturbances along supply chains. Wildfires and PSPSs make clear that communities can only rely on energy generated close to home.

Organizational centralization is homologous to infrastructural centralization and produces similar dysfunctions. Management and regulation of energy systems are centralized at the state level and insulated from democratic procedures. Energy management is dominated by IOUs that own grid infrastructure and have historically been tasked with procuring energy. IOUs operate as private companies, but they secure so-called "natural monopolies" by submitting to regulation by state-level Public Utilities Commissions [5]. A revolving door between regulatory bodies and IOUs, together with a general belief that failing utilities would spell infrastructural disaster, ensure that regulation remains protective [6]. IOUs experience no performance incentives for maintaining their transmission lines, and they are bailed out or extended insurance funds when their systems fail [7].

By grounding energy democracy in anarchist and regionalist thought, I argue that systems that are resilient to wildfires, PSPSs, and other socio-ecological shocks must be decentralized and managed locally precisely because resilience is a fundamentally local capacity. In the context of energy systems, resilience entails each community's ability to generate enough power to meet their energy needs locally and operate their grids independently of external shocks [8]. Energy decentralization is thus fundamental to resilience in the face of wildfires and PSPSs.

2 From decentralism to technoregionalism

Theoretically, I ground my argument for energy decentralization in anarchism. First, anarchism as a critical theory emerges in response to the inherent unfairness of all centralized systems: centralized systems redistribute power and resources away from locales and toward the increasingly authoritarian center while dismantling local decision-making structures [9]. Such anarchist critiques have prefigured those of environmental justice scholars and critical race theorists who have similarly argued that redistribution schemes do not effectively manifest equality, but instead conjure up social problems, such as by creating expendable populations and concentrating them into sacrifice zones [10–12]. In the context of energy, inequalities created by centralized structures leave some communities without power while also encumbering other communities with the burden of generating power for more affluent zones.

Second, agency and autonomy are definitionally local capacities, and anarchist scholars center these goods for practical and normative reasons. Practically, decentralization allows local publics to shape systems, including energy systems, to their own needs and affordances [13]. Further, decentralization makes it possible for local people to stage social experiments [14]. Among such social experiments must be new energy and grid technologies. Normatively, insofar as democracy involves robust public participation [15–17], people are best able to practice democracy when decision-making systems are decentralized [18]. Moreover, local decision-making is essential to community self-determination and the cultivation of *amitas* (communal friendship) and *mētis* (practical knowledge embedded in local experience and mutualistic social praxis) [9,19,20]. Energy questions, as matters of public concern, properly belong amid the suite of public matters worthy of democratic deliberation.

Finally, whereas centralization engenders mass ignorance and alienation from the most important structures and infrastructures that condition life chances [21], decentralization encourages people to develop their capacities and proficiencies and share local knowledge [13,22]. This connection among decentralization, public involvement, and local knowledge resonates with both decolonial scholars, who have asserted the justice implications of centering local priorities and knowledge regarding energy questions [23–25], and with science and technology scholars, who illustrate the substantive, instrumental, and normative benefits of bringing publics into upstream conversations about technological development [26–28].

Anarchist theory demands decentralism in practice. To best realize the virtues of decentralization, a sharp distinction must be made between decentralism and isolationism. Critical environmental justice scholars have recognized the multi-scalar impacts of ecological crises, which transgress boundaries and thus defy isolationism [12]. Whether localities claim sovereignty to excuse their own extractionism or erect borders to create external sacrifice zones, I argue it is these artificial boundaries, not decentralism itself, which enables the Hobbesian war of all against all that supposedly necessitates centralized governance.

It is clear that decentralist politics must retain the agency of local actors in decision-making processes, be responsive to multi-scalar effects of ecological crisis, and reject borders' relevance regarding ecological effects. I argue that these three concerns may be addressed by pursuing energy transformations at the level of the *technoregion*. As I reintroduce the term here, a technoregion is distinct from the definition offered by Timothy Luke, who defines technoregions as "artificially generated" zones that "ignore almost all concrete cultural ties to local land, water, plants, animals, climate, and people in order to respecify social space technoeconomically, according to the demands of global capitalist exchange" [29, p. 72]. Since Luke's technoregions operate with "little regard for place, tradition, or ecosystem" [29, p. 72], the two concepts contained within the term—"techno" and "region"—are contradictory. If regions, as they are canonically understood in geography, are places united by substantive similarities [30], Luke's technoregions subsume places to global capitalist flows. While Luke's concept may be useful to describe some phenomena of globalization, associating these artificial zones with technology itself veers precariously toward technophobia while leaving little room to explore technology's liberatory potential outside of capitalism.

Applying a decentralist, multi-scalar politics to energy questions requires recuperating the concept of the technoregion. In my use, a technoregion refers simply to the geographic area defined by the common use of a given technology (or assemblage of technologies); prominent examples of technoregions are areas served by the same wastewater treatment facility, cell tower, transit authority, or power plant. Clarifying a liberatory conceptualization of technoregions and demonstrating how they can be used for insurrectionary purposes will be the focus of the remainder of this chapter. I argue that the liberatory potential of defining technoregions in this way begins in affirming technology's humanistic potential and its responsiveness—rather than its antagonism—to local place. The concept of the technoregion in turn encourages *technoregionalism*, a strategic epistemology that involves looking at technology as regionally patterned. I advocate technoregionalism as a framework for addressing energy questions through a decentralist politics that also acknowledges how systems unfold across multiple scales.

3 Technology, humanity, and ecology

Seizing a liberatory conceptualization of technoregions must involve reframing technology itself. The Aristotelian concept of *techné*, reintroduced here through social ecologist Murray Bookchin [31], elucidates the social and ethical significance of technology. The concept of *techné* encompasses not just how but why technology is created. Technology thus comes laden with ethical responsibility, which implies rational limits on the use of technology according to its social purpose. Further, because *techné* concerns not just the product of technological works but also the creativity and rationality of the producer themself, *techné* cultivates "potency" through causing change in the "thing or in the artist himself considered as other" [32, p. 305]. Understood in this way, *techné* is fundamentally humanistic, not in its assertion of a Promethean Man, but because it is aimed toward the flourishing of human potential. Whether a given technology meets the standard of *techné* is an incessant question, but this ethical framing of technology extricates us from the trap of technophobia that springs when we come to conflate technology *simpliciter* with particular technologies enlisted by authoritarian structures.

Humanistic technology requires that common people be able to interact with technologies and direct them to social purposes. Resonant with the aforementioned anarchist theory, such interaction requires decentralized and democratic management. Writing in the 1960s, Bookchin saw tremendous potential in contemporary technologies. Versatile and efficient, they could be scaled down so that sites of production could be brought closer to sites of end-use [13]. Communities could enjoy participatory engagement with technologies and experience local autonomy. Liberation from centralized control structures was made possible, though never guaranteed, by local technological mastery. Contemporary readers may regard the development of microgrid technologies, explored later in this chapter, as a profound demonstration of how technological advancement manifests the possibility for liberatory decentralism. Reclaiming this liberatory potential of technology articulates the importance of challenging centralization and insisting on the strategic epistemology of technoregionalism.

While technologies are often directed against place, all technologies must account for place, and many technologies arise precisely in response to and in interaction with their surroundings [33,34]. Such interaction may ultimately be anti-ecological, but technologies do not emerge from blindness, ignorance, or disregard of ecology. An ocean barge is not ignorant of the distance between Hong Kong and San Pedro, just as the Hoover Dam is not ignorant of the Colorado River.

For Bookchin, the antagonistic relationship of technology to ecology is not the result of technology itself, but of the centralized form it takes when deployed by authoritarian social structures. Centralization encourages the extraction of resources and the production of technologies that outstrip natural limitations. As economic life is decentralized, with more resources produced closer to home, humanity's dependence upon locally unique ecologies will become a more visible and lively part of human culture. As such, human communities will become at once more respectful of natural limitations and more nourished by local ecological relations [13]. Rather than sharing an injurious relationship with ecology, here technology emerges at the intersection of humanity and their broader ecologies and constitutes their responsible interaction.

Bringing Bookchin's interpretation of technology through *techné* into the concept of the technoregion enriches our understanding of the interaction among humans, ecology, and technology. Anticipating environmental justice scholarship, his work transcends the false choice between human betterment and ecological resilience; for Bookchin, technology plays a central role in their mutual enrichment. He argues, provocatively today, that these technologies promise to integrate us with the more-than-human world: not only with the androids but with the dandelions as well.

4 Committing to regionalism

Decentralization enables both liberatory technology and resilience. Energy resilience is a local capacity that must be met through the local management of energy technologies. Put another way, energy technologies are best managed regionally. I argue that regionalism lends decentralist politics a framework for retaining the primacy of decentralized places while scaling out to be responsive to multi-scalar phenomena, thus greatly contributing to technoregional theory and practice.

Regions are fitting spatial units for decentralist politics because they are defined not by artificial boundaries, but by their interiors. While geographical theories about regions have shifted tremendously over the past century, the dominant conception of "region" today is rooted in the localities school [30]. The localities school rehabilitated the concept of regions so that they were no longer definitionally understood as existing somewhere between the nation-state and the urban center, but as communities and places [35]. Here, regions have to do with recognizing context: in short, where things happen affects how things happen. Anarchist thinkers have been preoccupied with this anthropological notion of place from Pyotr Kropotkin [36] to David Graeber [37], but this thinking is also a result of long traditions in geography dating back to Paul Vidal de La Blache, whose work focusing on the French *pays* emphasized the dialectical relationship between culture and nature that produces *genres de vie* [38].

Regions are distinct from political units, especially states, because they do not fetishize borders. State activity often revolves around the drawing, redrawing, and maintenance of borders, which limit bureaucratic domination and distinguish taxable, draftable citizens from foreigners [19]. Border fetishism thus assists in the accumulation of centralized bureaucratic power. Borders are also important to states insofar as sovereignty protects internal dysfunctions, including those that result in multi-scalar ecological harms. In contrast, because regions are multi-scalar, borders themselves cannot settle disputes among neighboring regions: affairs impacting two geographic regions must be managed at a scalar level that encompasses both parties.

In his work on community resilience, Lorenzo Kristov offers a model of multi-scalar energy planning that I argue resonates with these regionalist insights and further assists in clarifying the concept of the technoregion. Kristov bases his model on the familiar ecological hierarchy—which places individuals at the center of concentric circles fanning out to communities, populations, and ecosystems—adapted to meet a range of community needs from local food access to electricity management [8]. Concentric circles of influence and exchange center the resilience needs of localized actors while encouraging multi-scalar collaboration depending on the objective being pursued. Crucially, the resilience needs of each ecological center are never subsumed to benefit centralized structures; further, collaboration at each scale is never presumed, but arises out of necessity and utility. Kristov's model resonates with a multi-scalar technoregional approach because, whereas political units can conjure borders to externalize ills and protect dysfunction, the materiality of shared technology—such as a power plant—prompts organized collaboration at that technoregional scale. I offer technoregionalism as just such a strategic epistemology that involves choosing to look at technology as regionally patterned.

In this sense, technoregionalism is close kin to bioregionalism. The perspective of bioregionalism encourages rehabilitating the interaction between bio-geographic terrain and the human "terrain of consciousness" [39, p. 218]. There are many important criticisms to be made of bioregionalists: some interpret the relationship between environment and consciousness too deterministically (e.g., Ref. [40]), while in general, the tendency to assert an unmediated relationship between humanity and the rest of the biotic community (e.g., Ref. [41]) deserves greater interrogation. A more sober reading of bioregionalism might indicate that, while the relationship between human consciousness and geographic terrain is underdetermined, the usefulness and even virtues of that relationship are clear. Moreover, the relationship between terrain and consciousness is mediated by culture [42], as well as other types of overlapping regions.

In recognizing, with Bookchin, that technologies always exist at the intersection of humanity and the more-than-human world, I argue it is most useful to understand bioregions and technoregions as overlapping and interactive rather than necessarily competitive. It may even be useful to evaluate technoregions by their fidelity to surrounding bioregions, or by whether mediation of ecologies is harmonious, alienated, or even playful or perverse. Beyond simply expressing a technophilic analogue to bioregionalism, technoregionalism offers a new way to clarify how human consciousness relates to geographic terrain.

The normative thrust of bioregionalism, as it is clarified by Marc McGinnis [41], is to encourage us to reinhabit place. Technoregionalism has an analogous normative thrust: to give decentralist politics a framework for claiming—sometimes for the first time—popular ownership of technologies. Grasping our energy futures requires the same attention to technoregionalism as past generations have paid to bioregionalism. The challenge and promise of technoregionalism that distinguishes it from bioregionalism is that technoregions can be created, their maps redrawn. They are an act of human creation. It is to that creation that I turn now to conclude this chapter.

5 Technoregions of insurrection

Technoregions, like bioregions, already exist. Technoregionalism points to a novel epistemic strategy that impels actors to dismantle centralized management and seize control of energy futures through decentralist, multi-scalar politics. To realize their liberatory potential, technoregions must be transformed into technoregions of insurrection. Akin to a dual power strategy, technoregional changes are made alongside the existing system to erode its necessity and legitimacy. For local actors, this work involves redrawing technoregions so that they may be decentralized, democratized, and made locally resilient.

Here, I offer a brief overview of an insurrectionary technoregional strategy, taking the implementation of a microgrid as an example, and use my experience as a scholar-activist in Santa Barbara County to ground my recommendations. Microgrids are comprised of generation, storage, and demand response systems that can be localized to the level of the building, then built out modularly through two-way interconnections [43]. By sharing loads horizontally, microgrids can both weave together multi-scalar units and island themselves to operate autonomously [8]. As such, they are particularly suited to the cause of technoregionalism.

First, in insurrectionary fashion, actors should begin with an appreciation for existing technological affordances. To plan a microgrid, actors should map the locations of power plants, transmission lines, and substations that constitute local energy infrastructures, as well as the commercial and industrial facilities that draw the most power. Actors should also assess whether their region is a net producer or net consumer of energy, and thus whether to prioritize bolstering local resilience or evaluating their possible exploitation by neighboring net consumers.

Second, actors should learn how their technoregional microgrid might overlap with and bootstrap onto other kinds of regions. Bioregional features will indicate possible supplies of wind, wave, geothermal, and solar power, as well as natural limits such as mountain ranges or large stretches of wilderness. Mapping organizational regions will show what kinds of local advocacy coalitions may be forged; here, attention should be paid not only to organizational skillsets, but also to the geographic range and relative autonomy of organizations, including trade union locals, social justice and environmental chapters, and Indigenous groups. Politico-regions may lend support to technoregional microgrids, especially where public agencies such as community choice aggregation entities exist; alternatively, politico-regions may undermine microgrid implementation by forcing false choices between economical and just construction, impeding local experimentation, or obstructing prompt proceedings through bureaucratic inefficiencies.

Third, actors should assist other locales in establishing their own resilient technoregional microgrids. Insurrectionary technoregions can share best practices with newcomers by creating easily adaptable policy templates or holding regional skillshare summits. They can coordinate by uniting into larger interest groups on the state level, enabling other insurrectionary technoregions to assert themselves and effectively bring down the drawbridge for upstarts. Microgrids can also scale out by collaborating with their neighbors to share resources horizontally. While each technoregion remains responsible to the many centers of its ecological hierarchy, microgrids are also capable of scaling out to optimize their collective performance. Unlike IOUs operating centralized grids, local managers of horizontally integrated microgrids have a clear incentive to ensure safe and mutually beneficial energy sharing.

In this chapter, I have drawn from a broader empirical study on centralized energy infrastructures to illustrate the social and ecological problems that they create, and to begin the process of imagining a decentralized praxis of energy democracy. In so doing, I have used anarchist and regionalist theory to develop a liberatory understanding of technoregions that supports energy resilience. I assert that a technoregional perspective is essential to creating resilient energy futures suited to human and ecological flourishing. The current energy system characterized by centralized management and government-protected IOUs is incapable of repairing itself precisely because it was never built to make decentralized communities resilient. Reform is untenable so long as the levers of reform rest with those committed to centralization and the maintenance of IOUs. Grasping energy resilience requires reclaiming and rebuilding our technologies, moving technoregion by technoregion until the energy system owned by IOUs and guarded by state regulation is fully supplanted by a patchwork of decentralized, democratic, and resilient technoregions. We will have much to learn through this practice of communal management: our electrical system, unlike water and food systems, has never been managed through properly decentralized, democratic politics. We have never had the chance to experiment with these technologies where we live, allowing cowardice to dominate energy planning while centralized energy systems reproduce social and ecological calamity. It is time we cultivate the conviction that we should have a hand in the ethical use of technologies and build an infrastructure that makes this possible.

References

[1] T. Luna, California utility equipment sparked more than 2,000 fires in over three years, Los Angeles Times (2019). 28 January.

[2] I. Penn, P. Eavis, PG&E pleads guilty to 84 counts of manslaughter in camp fire case, New York Times (2020). 18 June.

[3] J. Serna, Southern California Edison strikes $360-million settlement over wildfires and mudslide, Los Angeles Times (2019). 13 November.

[4] R. Rivard, SDG&E says there's a 100% chance it'll start or contribute to a major wildfire, Voice of San Diego (2019). 10 June.

[5] R. Hirsh, Power Loss, MIT Press, Cambridge, MA, 1999.

[6] H. Wasserman, The Last Energy War, Seven Stories Press, New York, 1999.

[7] L. Kristov, Transitioning to a safe and clean electrical grid, Solutions News (2020). 4 September.

[8] L. Kristov, Resilient community, Draft for Discussion, Academia.edu, 2018.

[9] P. Kropotkin, The State: Its Historic Role, Freedom Pamphlets, London, 1898.

[10] B. Sovacool, M. Dworkin, Energy justice: conceptual insights and practical applications, Appl. Energy 142 (2015) 435–444.

[11] T.B. Voyles, Wastelanding, University of Minnesota Press, Minneapolis, 2017.

[12] D. Pellow, What Is Critical Environmental Justice?, Polity Press, Medford, MA, 2018.

[13] M. Bookchin, Post-Scarcity Anarchism, Black Rose Books, Buffalo, NY, 1986.

[14] D. Abad de Santillán, The libertarian revolution (1937), in: R. Graham (Ed.), Anarchism: A Documentary History of Libertarian Ideas, vol. 1, Black Rose Books, New York, 2005, pp. 475–477.

[15] C.B. Macpherson, The Life and Times of Liberal Democracy, Oxford University Press, New York, 1977.

[16] C. Pateman, Participation and Democratic Theory, Cambridge University Press, New York, 1970.

[17] G. Alperovitz, America Beyond Capitalism, John Wiley and Sons, Hoboken, NJ, 2005.

[18] C. Milstein, Deciding for Ourselves, AK Press, Chico, CA, 2020.

[19] J. Scott, Seeing Like a State, Yale University Press, New Haven, 1998.

[20] D. Graeber, Debt, Melville House, Brooklyn, 2011.

[21] P. Goodman, People or Personnel?, Random House, New York, 1965.

[22] L. Galleani, The end of anarchism (1907), in: R. Graham (Ed.), Anarchism: A Documentary History of Libertarian Ideas, vol. 1, Black Rose Books, New York, 2005, pp. 119–124.

[23] C.M. de Onís, Fueling and delinking from energy colonially in Puerto Rico, J. Appl. Commun. Res. 46 (5) (2018) 535–560.

[24] S. Avila-Colero, Contesting energy transitions: wind power and conflicts in the Isthmus of Tehuantepec, J. Political Ecol. 24 (2017) 993–1012.

[25] L. Álvarez, B. Coolsaet, Decolonizing environmental justice studies: a Latin American perspective, Capital. Nat. Social. 31 (2) (2018) 50–69.

[26] D. Fiorino, Citizen participation and environmental risk: a survey of institutional mechanisms, Sci. Technol. Hum. Values 15 (2) (1990) 226–243.

[27] S. Jasanoff, Technologies of humility: citizen participation in governing science, Minerva 41 (2003) 223–244.

[28] J. Wilsdon, R. Willis, See-Through Science: Why Public Engagement Needs to Move Upstream, Demos, London, 2004.

[29] T. Luke, Art and the environmental crisis, Art J. 51 (2) (1992) 72–76.

[30] B. Warf, Regional geography, in: B. Warf (Ed.), Encyclopedia of Human Geography, SAGE Publications, Thousand Oaks, CA, 2006, pp. 405–407.

[31] M. Bookchin, The Ecology of Freedom, AK Press, Oakland, CA, 2005.

[32] M. Bookchin, The Ecology of Freedom, AK Press, Oakland, CA, 2005 (Quote from Aristotle *Metaphysics* 1046b).

[33] L. Parks, "Stuff you can kick": toward a theory of media infrastructures, in: P. Svensson, D.T. Goldberg (Eds.), Between Humanities and the Digital, MIT Press, Cambridge, MA, 2015, pp. 355–373.

[34] N. Starosielski, Pipeline ecologies: rural entanglements of fiber-optic cables, in: N. Starosielski, J. Walker (Eds.), Sustainable Media, Routledge, New York, 2016, pp. 38–55.

[35] A. Jonas, A new regional geography of localities? R. Geogr. Soc. 20 (2) (1988) 101–110.

[36] P. Kropotkin, The Conquest of Bread and Other Writings, Cambridge University Press, New York, 1995 (1892).

[37] D. Graeber, The Utopia of Rules, Melville House, Brooklyn, 2015.

[38] P. Vidal de La Blache, Principles of Human Geography, Henry Holt and Company, New York, 1926.

[39] P. Berg, R. Dasmann, Reinhabiting California, in: P. Berg (Ed.), Reinhabiting a Separate Country, Planet Drum, San Francisco, 1978, pp. 217–220.

[40] K. Sale, Dwellers in the Land, Sierra Club, San Francisco, 1985.

[41] M.V. McGinnis, Bioregionalism, Routledge, New York, 1999.

[42] L. Mumford, The Culture of Cities, Harcourt Brace Jovanovich, New York, 1970 (1938).

[43] A. Weinrub, Democratizing municipal-scale power, in: D. Fairchild, A. Weinrub (Eds.), Energy Democracy, Island Press, Washington, DC, 2017, pp. 139–171.

Assemblages of energy and equity: Rearticulating Illich

J. Macgregor Wise

School of Social and Behavioral Sciences, Arizona State University, Glendale, AZ, United States

1 Introduction

In a 1974 research report, *Energy and Equity*, Ivan Illich argued that when the average energy use of a population exceeds a certain limit, that society becomes inequitable. Excess energy has a corrupting influence on the society as a whole and on the freedom of each individual: "[b]eyond a certain median per capita energy level, the political system and cultural context of any society must decay" [1, pp. 18–19]. Not only this, that society tends toward technocracy (government by technological experts) and rapidly expands means of social control. In addition, he argues that excessive energy use becomes addicting and sets a society on a path of always wanting more.

The case study that he pursues to exemplify this argument is about modern traffic. Speed is taken as a measure of energy use in society. The analysis of Illich and his colleagues concludes that "[o]nce some public utility went faster than ± 15 m.p.h., equity declined and the scarcity of both time and space increased" [1, p. 23].

> Beyond a critical speed, no one can save time without forcing another to lose it. The man [sic] who claims a seat in a faster vehicle insists that his time is worth more than that of the passenger in a slower one. Beyond a certain velocity, passengers become consumers of other people's time, and accelerating vehicles become the means for effecting a net transfer of lifetime [1, p. 42].

Basically, once you go faster than 15 m.p.h., you begin to consume more than your fair share of resources; additional infrastructure needs to be built to support your speed; you are afforded higher social status, and others must be inconvenienced if not shunted aside to let you do so.

Illich's argument regarding energy and equity, and its detailed case study, is grounded in his theory of the place of technology in society and a method, which he calls *counterfoil research*, for evaluating that balance. While his argument regarding energy explicitly is not

---

about any particular technology, institution, individual, or economy (capitalist or socialist), it is important to understand how his foundational concept of *conviviality* undergirds his approach to participatory democracy and counterfoil research.

2 Conviviality and counterfoil research

Conviviality is Illich's term to describe tools that work to enhance the agency and creativity of individuals (tools that people use, not tools that use people). He writes: "I consider conviviality to be individual freedom realized in personal interdependence and, as such, an intrinsic ethical value" [2, p. 11]. The term also describes a society where people and tools live in balance and equity. As Illich puts it "A convivial society would be the result of social arrangements that guarantee for each member the most ample and free access to the tools of the community and limit this freedom only in favor of another member's equal freedom" [2, p. 12]. Here, the emphasis is on the just distribution of resources (including energy) and personal realization in balance with others in the community. I should note that his notion of a convivial tool is not a romantic return to "small" or "primitive" technologies. Many such technologies were brutal and inequitable, and many convivial technologies could only be achieved recently with advances in efficient electronics and manufacturing.

Illich's goal is not to dictate what a convivial society must look like, but to set out a methodology whereby a community could recognize the balance and imbalance of its relationships, including relationships with tools and resources; to recognize when means have turned into ends [2, p. 14]. Counterfoil research is both a "dimensional analysis" and a "political process" [2, p. 18]. The former is "rational research on the dimensions within which technology can be used by concrete communities to implement their aspirations without frustrating equivalent aspirations by others" [2, p. 84]. In other words, the purpose is to find the tipping point when the good that a tool or resource (including forms of energy) provides becomes something socially harmful. The second part is a political process for a community to decide what social conditions it is willing to tolerate and what forms of inequity are allowed for particular purposes. This is the call for participatory democracy [3]; for the community to decide.

It is also important that a community examine every artifact, resource, and relationship. Illich argues that one of the products of industrial society is a limited imagination, the inability to see beyond the processes and products that we have been offered. He refers to this as *radical monopoly*: a monopoly not of a particular brand but of a type of product and a particular way of doing things "[W]e must focus our attention on the industrially determined shape of our expectations" [2, p. 21]. This includes the assumption that more energy for everyone is always a good idea.

3 Assemblages

There are a number of limitations to Illich's argument, found in the anthropocentric assumptions from which he works. I will highlight two here: the idea of natural balance and the emphasis on the rights of a liberal individual. Several times in *Tools for Conviviality*, Illich makes mention of a "natural balance" beyond which a technology, institution, or process inevitably transforms without considering the ways that balance could be mitigated and instantiated by cultural and social conditions. There are also assumptions about the natural abilities of humans (though he does recognize briefly the need to address disability). These become problematic universalizing statements and reveal an unquestioned normative view of the human. Secondly, the core actor in his society is the individual citizen. He recognizes that such an individual is necessarily constrained by the needs of others, but still, the notion of the autonomous, liberal individual is one that has been widely critiqued. Likewise, the primary concern is human agency, freedom, and life.

What to do with the provocations of Illich? The productive prodding, impassioned critique, radical questioning, and overall emphasis on community self-determination are important, but just as Illich asks us to see beyond "the industrially determined shape of our expectations" [2, p. 21], we need to see beyond his anthropocentric perspective.

For Illich, humans and tools are distinct entities, and so the first step is to problematize the taken for granted categories of both tool and human and consider them from the perspective of assemblage, as arrangements and becomings [4–6]. As Andy Clark has pointed out, it is our relations with tools, objects, and environments that make us human in the first place [7]. A properly posthuman [8] perspective questions distinctions of human/nonhuman on the one hand and critiques the perspective of anthropocentrism on the other [9,10]. Key to this perspective for Rosi Braidotti [9] are the concepts of relationality and affect (which Deleuze and Guattari draw from Spinoza's use of the term as the "ability to affect and be affected" [9,11]). Such a perspective is not indifferent to humans, or to questions of politics, justice, and equity, it just focuses on collective relations with "the multi-layered and multi-scalar ecologies to which they belong" [9, p. 54].

For example, consider the difference between a human-book assemblage and a human-e-reader (e.g., Kindle) assemblage in terms of energy. Apart from the energy used to produce and distribute the book, energy in the human-book assemblage is corporeal (muscle and cognition), whereas energy in the human-e-reader assemblage is supplemented by batteries and the assemblage of the power grid (traced back to its sources in solar, coal, etc.). The perceived affordances of the latter assemblage (including online communicative capabilities) and the relatively inexpensive, almost invisible, energy cost encourages a proliferation of similar products and assemblages (smart devices and the Internet of Things), a result of what Illich would refer to as an addiction to energy (and he might ask who is not receiving this energy so that this smart device can work).

To grasp an assemblage is to see the elements as a functioning whole (it is not human + book, but human becoming book and book becoming human). This has ethical implications for the constituents of the assemblage (e.g., the human is implicated in the environmental costs of the energy grid of which they are a part, see Ref. [12]), on the one hand, and affording a different relation to space and environment on the other (at multiple scales: home, neighborhood, region). And an assemblage is not just about an individual human with an individual device, but a vast assemblage of devices, humans, grids, and environments (at multiple scales) where the generation, storage, and circulation of energy are part of its affective capacity (what the assemblage can do).

4 Conclusions

Posthuman counterfoil research means that determinations of conviviality and equity cannot rely on the criterion of individual agency but on the relations of multiple dimensions of ecology (humans, machines, animals, plants, and other forms of life) and multiple scales. The "multiple balance" that Illich calls for in *Tools for Conviviality* needs to go beyond his brief attention to the environmental crisis to address the health and agency of multiple human and nonhuman agents. Posthuman counterfoil research can become a form of what Rosi Braidotti has called posthuman knowledge, a scholarship that is "engaged with nonhuman, *zoe*/geo/ technocentered and post-natural objects, themes, and topics" [9, p. 79]. Such thinking is always radically collaborative and interdisciplinary. As she writes:

> [T]hinking in posthuman times is about increasing the capacity to take in the intensity of the world and to take on its objectionable aspects. Thinking is about increasing our relational capacity, so as to enhance our power (*potentia*) for freedom and resistance [9, p. 79].

A posthuman counterfoil energy audit cannot be limited to a survey of the impact of particular fuels and their infrastructures but needs to draw the line (as Deleuze might say [13, p. 137]) between these forms of energy and the general capacity to affect and be affected expressed in our assemblages (the line between power and power) at different scales. A detailed exercise is mapping not just the generation of energy—wind turbines, landscapes, wildlife, storage, and transmission infrastructure—but also the uses of energy by each device and household and how these patterns encourage or inhibit equitable social relations, with the social encompassing nonhumans as well as humans.

Empowered humans have an increased ability to affect the environment. In Phoenix, Arizona, inequities in income, resources, and built environment (e.g., shade) mean that neighborhoods that cannot afford power bills, HVAC equipment, or repairs, face lethal levels of heat each summer in ways that more affluent neighborhoods do not [14]. Illich's analysis in *Energy and Equity* was about the increases in power in a society and their inequitable distribution in terms of the agential capacities for different individuals and groups (some are empowered and some disempowered). We need to radically expand the notion of community to include the posthuman assemblages of which we are all a part. Convivial assemblages of energy are ones where the agency of humans, nonhumans, and ecologies are all elements of the multiple balance.

References

[1] I. Illich, Energy & Equity, Marion Boyars, London, 1974.
[2] I. Illich, Tools for Conviviality, Harper & Row, New York, 1973.
[3] R. Sclove, Democracy and Technology, Guilford, New York, 1995.
[4] G. Deleuze, F. Guattari, A Thousand Plateaus: Capitalism and Schizophrenia, University of Minnesota Press, Minneapolis, 1987 (B. Massumi, Trans.; Original work published 1980).
[5] J.M. Wise, Assemblage, in: C. Stivale (Ed.), Gilles Deleuze: Key Concepts, Routledge, New York, 2014, pp. 91–102.
[6] J.D. Slack, J.M. Wise, Culture and Technology. A Primer, second ed., Peter Lang, New York, 2015.
[7] A. Clark, Natural Born Cyborgs: Minds, Technologies, and the Future of Human Intelligence, Oxford University Press, New York, 2004.

[8]   There are distinct variations on the idea of posthumanism. This essay follows the perspective advocated by Rosi Braidotti which follows the philosophy of immanence of Gilles Deleuze. Braidotti distinguishes this approach from other perspectives, such as Actor-Network Theory, Object-Oriented Ontology, and Transhumanism [9].

[9]   R. Braidotti, Posthuman Knowledge, Polity, Medford, 2019.

[10] D. Haraway, A Cyborg Manifesto: science, technology, and socialist-feminism in the late twentieth century, in: Simians, Cyborgs, and Women: The Reinvention of Nature, Routledge, New York, 1990, pp. 149–182.

[11] B. Massumi, Notes on the translation and acknowledgements, in: G. Deleuze, F. Guattari (Eds.), A Thousand Plateaus: Capitalism and Schizophrenia, University of Minnesota Press, Minneapolis, 1987, pp. xvi–xix (B. Massumi, Trans.).

[12] J.M. Wise, A hole in the hand: assemblages of attention and mobile screens, in: J. Hadlaw, A. Herman, T. Swiss (Eds.), Theories of the Mobile Internet: Materialities and Imaginaries, Routledge, New York, 2015, pp. 212–231.

[13] G. Deleuze, C. Parnet, Dialogues, Columbia University Press, New York, 1987 (H. Tomlinson and B. Habberjam, Trans.; Original work published 1977).

[14] W. Stone, Heat Is Killing More People Than Ever—What Phoenix Is Trying To Do About It, 2018. https://kjzz.org/content/667969/heat-killing-more-people-ever-%E2%80%94-what-phoenix-trying-do-about-it. Accessed 12 April 2021.

I. Imaginaries

Futures

Re-imagining energy-society relations: An interactive framework for social movement-based energy-society transformation

Joohee Lee[a,b], John Byrne[a,b], and Jeongseok Seo[c]

[a]Center for Energy & Environmental Policy, University of Delaware, Newark, DE, United States [b]Foundation for Renewable Energy & Environment, New York, NY, United States [c]National Assembly, Seoul, Republic of Korea

1 Introduction

Achieving energy transitions requires more than just a one-dimensional change in our energy path. For example, scientific breakthroughs and inventions are needed to drive change in energy systems. Historical incidents can also strongly motivate people to reevaluate existing energy systems and consider alternatives. Most importantly, a paradigm shift in energy systems and governance becomes possible when members of society see the need for change and participate in energy transition movements at multiple scales [1,2].

How can individuals and social collectives meaningfully contribute to the process of energy transitions? While there are many distinctive socio-political layers to the issue of energy transitions, valuable lessons can be learned from crises in other sectors and at other times. For example, two crucial lessons can be learned from the ongoing global crises—the COVID-19 pandemic and climate change. One lesson is that there is deep-seated social injustice relating to how those most at risk have little influence on the governance of the crisis. The second lesson is that sudden, unprecedented social threats usually overwhelm the planning and technical infrastructures that are intended to avert immediate panic. Whether it is hospital systems in the case of the pandemic or disaster management systems in the case of climate change, our ability to cope in the near term has evaporated.

Many studies have found similarities between the climate-energy crisis and the coronavirus pandemic where systems in both fields grapple with stranded individuals and communities [3–5]. With the COVID-19 outbreak, countries are realizing the value and limits of *setting science-based goals* (e.g., flattening the curve of positive cases and disseminating vaccines) and *mobilizing collective action* (e.g., wearing a mask and social distancing) for fighting a global emergency. The former—the what-to-achieve part of the question—can be established in a relatively straightforward way by science research. However, the latter—the how-to-achieve part of the question—can work successfully only when there is public awareness and engagement. Likewise, climate-energy crises require significant citizen involvement if the transformative change is to take place.

In this chapter, we suggest a conceptual framework for social movement-based energy-society transformation, building upon the growing literature on democratic energy transitions. Three interactive sets of principles—*motivating*, *operational*, and *organizing* principles—are defined as conceptual lenses for challenging the industrialist understanding of energy as a commodity, which leaves energy decisions in the hands of technocrats, political elites, and the market mechanisms [6–8]. We argue that the transformation of existing society-energy relations cannot be fully realized by only replacing fossil fuels with carbon-free, renewable energy sources. Instead, we conceive social movements as key drivers of energy transitions that reframe energy as commons and redistribute governing power and ownership of energy to communities and citizens. For illustration, two re-imagination experiments—Sustainable Energy Utility and One Less Nuclear Power Plant—are reviewed as examples of localized efforts that have successfully integrated and practiced these principles.

2 Beyond technical goal setting: What makes just and democratic energy transitions possible?

In 2018, the Intergovernmental Panel on Climate Change (IPCC)'s Special Report [9] delivered a clear message to the global community: we need an urgent and deep transformation of the carbon-intensive economy in order to keep the global temperature rising below 1.5°C. The report warned that letting the temperature increase to 2°C or higher (which is more likely to happen at the current speed of temperature rise) would engender socio-ecological consequences on an unprecedented scale. The problem is that achieving this least-worst scenario (2°C target) seems already highly challenging, with a number of major economies still hesitant about making timely and meaningful changes in their pursuit of "growth without end" [7].

Shortly before the release of the alarming 2018 IPCC report, Jakob and Hilaire [10] estimated the volume of fossil fuels that must remain unused in order to curb the global temperature rise under 2°C. The so-called "unburnable fossil-fuel reserves" include 80% of coal reserves, 50% of gas reserves, and 30% of oil reserves worldwide—these would add 11,000 gigatons of CO_2 to the already carbon-saturated skies if used. Even if we succeed in completely ceasing greenhouse gas emissions, it is predicted that climate change effects will continue to persist for at least 1000 years [11]. This irreversibility of the climate crisis indicates that we cannot afford any further inaction. Nevertheless, climate action has not occurred at the speed and scale that we need.

The energy sector is one of the areas that is most targeted for climate action. Many national energy plans emphasize how a country's energy sector will be transformed into a climate-friendly one while maintaining a healthy economy. However, a deep transformation of the "hard energy path" [12] requires more than just defining a target for carbon reduction or renewable energy deployment (e.g., 100% renewable-sourced electricity power generation by 2050). As some energy social scientists point out, energy transition schemes too often rely on large-scale technical solutions that promise decarbonization while the root causes of the problem are neglected or glossed over [6].

For example, the pressing need to rapidly embrace and diffuse nuclear power technology as a climate solution worries some social scientists, as the power of institutional commitments to unlimited economic growth is left under-checked in pro-nuclear, low-carbon energy paths [7,12]. Well-intended decarbonization, achieved by rapid diffusion of nuclear power technology, may fulfill certain chemistry and technology readiness criteria but may not meet the standards of energy justice. Unfortunately, energy justice concerns are often treated as externalities that deserve compensatory relief rather than as key social criteria that should define the political and economic architecture of energy transitions [1,7,13]. This tendency also applies to the case of renewable energy development. Capturing this concern with the term "Green Titans," Byrne et al. [7] rightly pointed out that industry-led, mega-scale renewable energy projects often avoid questioning the phenomenon of energy obesity. Without challenging the cornucopian view of the climate-energy crisis, we may only repeat key failings of the energy system we are trying to replace.

Given the projections and estimates provided by the 2018 IPCC report [9] and Jakob and Hilaire [10], we recognize two critical criteria that should be met in future energy transition strategies—urgency (time) and scale (space) of action. As history records, the momentum for rapid and transformative change in societal norms and systems often comes from social movements and collective action (e.g., women's rights, anti-racism, and environmental justice movements). Relying solely on bureaucratic-minded strategies (e.g., incremental policy interventions, market regulations, and incentives) can hardly anchor sociotechnical transformations at the speed and magnitude we need. Social movements led by polycentric and pluralistic actors will almost certainly play the role as triggers and accelerators of innovations that redefine energy-society dynamics [14–18].

As discussed in the energy democracy literature [19,20], social movement-driven energy transitions emphasize the *resisting, reclaiming,* and *restructuring* processes of change. Burke and Stephens [19] highlighted three goals of energy democratization: (1) "resist the dominant fossil-fuel energy agenda," (2) "reclaim social and public control over the energy sector," and (3) "restructure the energy sector to better support democratic processes, social justice and inclusion, and environmental sustainability" (p. 37). Benefiting from previous studies [19–21], we propose three sets of guiding principles that support and enable each process of movement-based energy transitions. First, motivating principles offer normative guidance for outlining a *democratic, justice-driven,* and *commons-based* energy future that resists existing energy-society relations (see Chapter 13 for an in-depth discussion of "resistance" in energy democratization). Second, the operational principles of interest for this chapter's re-imagining of energy-society relations are the pursuit of a *commonwealth* principle in the design of the new economy and a *community trust* principle as guidance for the rollout of the new political system. These new political economy tools can reclaim the democratization process for

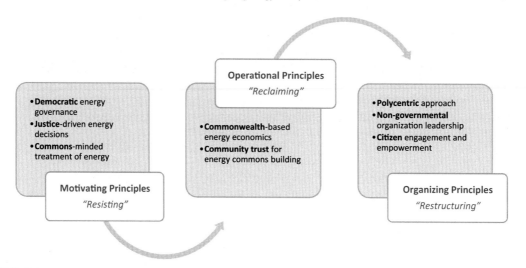

FIG. 7.1   An interactive framework to depict accelerating, movement-based energy-society transformation.

communities and citizens. Finally, organizing principles guide practical restructuring strategies, and it is our observation that the most progressive strategies are often pursued by *polycentric, non-governmental,* and *citizen-empowering* tools. The overarching goal of this energy commons movement is to reframe energy decisions as a province of localities and citizens [18,22]. Taken together, Fig. 7.1 visualizes our interactive framework of principles that can accelerate the anticipated transformation in modern energy to address the climate emergency.[a]

3 Toward a political economy of transformative change in modern energy

Modern energy systems can be characterized as large-scale and centralized; they tend to take high technological risks to achieve the goal of producing more energy services at minimal costs [23]. The energy sector's prioritization of efficiency over socio-environmental values connects back to the typical industrialist vision of modern society growing without end [7]. Unlike other ecological services we need for survival (e.g., air and water), energy services are produced and distributed almost entirely through industrial-scale energy infrastructures, such as power plants and electric grids, which are traditionally governed by private or government-owned utilities. For this reason, individuals have no choice but to passively consume the energy services provided by decision-makers and operators of existing energy systems even if the energy services they use may contradict their environmental stewardship [24]. Consumers, therefore, are given no significant role in determining or contributing to their own energy future.

[a] Our framework and its principles are by no means comprehensive for explaining the complete process of energy democratization and decarbonization. We invite practitioners in the energy sector to make suggestions that would contribute to establishing a workable roadmap for overarching energy sustainability and justice goals.

Despite the energy systems and structures that are fabricated by private and government-owned utilities, there have been innovative experiments conducted by smaller-scale actors—civil society, local governments, and communities—that set their own goals and action plans in response to pressing climate-energy crises. These bottom-up approaches have grown quickly over the past decade, generating success stories and revealing real-world challenges [16,25–28] (also see Chapters 10 and 14). Meanwhile, discussions about energy justice and energy democracy have added an important theoretical basis to this polycentric and localized practice of climate-energy action. These two strands of scholarship frame citizens and communities as essential contributors to the democratization and decarbonization of energy systems, not as barriers to change [1,29].

In what follows, we discuss how the multidimensional principles of our interactive framework can galvanize the process of redefining energy-society relations and empowering non-traditional actors (especially citizens, communities, and localities that have the agency to resist, reclaim, and restructure the relationship). In this way, we can imagine *and* realize a new political economy built on the operating principles of *energy commons* and *community energy governance*.

3.1 Shifting to an energy-as-commons approach

Access to natural resources, including primary energy sources, is essential for humans to survive and thrive. What distinguishes energy services (e.g., electricity, heating fuel, and transportation fuel) from other ecosystem services (e.g., water, air, and biodiversity) is the heavy involvement of human-made infrastructures in the entire process of production (e.g., mining and power plants), distribution (e.g., electric grids), consumption (e.g., appliances), and waste treatment (e.g., radioactive waste management). In a fossil fuel-based industrial economy, every stage in the process of harnessing and utilizing energy for human use requires infrastructure-scale investment and construction, which can hardly be done at the individual level. Modern energy system designs and technologies have reflected industrial society's aspiration for limitless growth, prioritizing the absolute efficiency of supply and demand chains above the impacts of these normative priorities on the democracy and justice dimensions of social development. As a result, the modern energy landscape [30] (in both its geographic and social senses) was predominantly shaped by large energy companies and national governments to meet their interests, such as profitability and national security, without leaving much space for other types of energy systems to compete with them.

With the recent emergence of cost-competitive, decentralizable renewable technologies, traditional energy industries are now facing fundamental challenges to their dominance in the energy sector. Indeed, decentralized energy projects are growing fast, which calls into question the old belief that energy systems work best in large-scale and centralized forms that are controlled by industries and political elites. One of the core elements that make community-based energy systems distinctive from conventional industry-led ones is the sharing of the costs and benefits of energy use—in other words, a commonwealth economy. As Byrne et al. [7] observed, pursuing commonwealth in energy systems becomes possible when members of communities share the principle of using less energy when possible and harnessing locally available renewable energy sources when needed. The notion of using less is the direct opposite of industrialist energy system models. However, community-level energy projects benefit from using less because the ultimate goal of their energy systems is not to maximize profits from sales or cost reductions but to enhance overall human well-being (including both the economic and non-economic dimensions of well-being) through realizing people's long-term energy capabilities [31]. And using renewable energy contests the industrial norm of efficiency as sustainability exposes the climate and other defects of optimality [1,6,32,33].

By treating energy (especially renewables) as a social as well as an ecological commons, rather than as a commodity that could only be accessed through private markets, citizens can choose to stop continuing on the energy path controlled by investor-owned utilities and to build momentum to replace cornucopian energy economics with a commonwealth form [1,32,34,35]. This bottom-up approach to re-imagining energy systems matches the concept of "commoning," which Albareda and Sison [36] defined as a process of "experimenting with new organizational designs to produce and distribute new commons, generating new labor divisions and allocations to support governance with collective action principles defined by the community members" (p. 731). An important goal of commoning energy is to invite people to treat energy concerns beyond the traditional common-pool resource framing. That is, energy commons creation is not only about localizing resource management but is also about empowering members of the community, regardless of their socio-economic status, to rethink the relationship between society and energy and identify their roles in reorganizing it [37]. By doing so, participants across layers of society can contribute to shaping the meaning of sustainability and justice so the voices of the voiceless, in addition to those who have influence, are at least to some degree reflected in the energy system.

3.2 Shifting to a community energy governance approach

The creation of a local energy commons requires not only a commonwealth-minded framing of energy but also a new political perspective on how to govern energy. Polycentrism is one of the innovative approaches to rethinking climate-energy governance. While climate-energy issues were once thought to be solved primarily by national and international institutions and industries, the goal of polycentrism is to redistribute governing power to non-traditional actors in dealing with such issues [38]. Modern energy systems are sociotechnical constructs that have long been controlled by small groups of people with power and knowledge. In contrast, local energy governance systems like community solar [39] offer platforms and resources (e.g., information, knowledge, space, tools, collective agency, etc.) that individuals and households can use to actively contribute to delinking the strong relationship between centralized energy systems and regulatory institutions.

The practice of polycentric, commons-minded energy governance is, however, inherently a challenging process [7]. For instance, important energy decisions are to be made through participatory and deliberative processes that involve a collection of heterogeneous opinions from stakeholders and debates among them. Empirical studies have found that the development of community trust among members of a community (or society) makes it possible to move forward even when there are conflicting voices and visions [40–42]. In other words, trust is a precondition for empowering people to behave as active energy decision-makers even with the existence of uncertainty and obstacles.

To elaborate on the core value of trust-based self-governance of resources, we want to draw an analogy between public libraries and local energy systems.[b] One of the primary functions of public libraries is to provide users with access to information resources in various forms (e.g., books, videos, and computers) and spaces for community activities [43]. Rather than purchasing or renting these intellectual resources on the private market, citizens can utilize library services to access a broad spectrum of knowledge and art and equip themselves for deeper engagement in social issues that affect their lives. The recent movement of transforming libraries from "information commons" to "learning commons" reflects the expanding role of public libraries in sustaining *everyday democracy* by enhancing citizens' opportunities to learn [44,45]. Local-level energy systems share the same goal. The primary functions of local energy projects are to serve and empower people interested in energy transitions that, literally, energize creativity and innovation [46]. They are not about practicing best (or most efficient) management of the resources but about identifying community energy needs and realizing their shared climate-energy visions.

In this sense, both public libraries and local energy projects function as platforms that provide citizens with opportunities to self-govern intellectual and natural resources and harness them on behalf of commoning experiments. Citizen engagement and empowerment are indispensable for sustaining a democratic and just society. These innovative commons systems are, therefore, more than properties [47]; their functions cannot be replaced entirely with conventional privatized systems because commons-based systems do not just provide material services (e.g., books and electric power) but also humanize the production and consumption of those resources. This happens through encouraging activities and interactions among people that lead them to rethink the sustainability issues affecting their and others' lives in relation to those resources. For this reason, the mission of localized energy systems is not merely to meet predetermined goals but also to continuously promote self-governing energy cultures among participants based on community trust and responsibility (also see the "closeness" effects of small-scale distributed energy systems discussed in Chapter 2).

Successful creation of energy commons and self-governance approaches requires us to meet the underlying standards of democracy (e.g., transparency and public participation) and justice (e.g., fairness and accountability). However, satisfying these motivating principles is not enough for rapid movement toward an energy-society transformation. The operating and organizing drivers discussed in this section should work together to support social movement efforts to reclaim and restructure modern energy. In the next section, we examine two pioneering experiments that have successfully materialized and integrated these interactive principles of energy transitions.

4 Re-imagination experiments: Sustainable Energy Utility and One Less Nuclear Power Plant initiative

This section briefly reviews two cases of social movement-led energy transformation experiments driven by the motivating, operational, and organizing principles discussed above.

---

[b] This analogy is based on a conversation between Dr. John Byrne and Delaware State Senator Harris McDowell.

The Sustainable Energy Utility model in use in the United States illustrates a commons-based re-imagination of public utilities that is designed to serve community values and visions instead of shareholder profits. Seoul's One Less Nuclear Power Plant initiative challenges central government biases toward large-scale, technocratic solutions by energizing citizen movements and business allies to design and govern a new political economy of local energy and local creativity. Both cases exemplify the growing leadership of the city and sub-national governments in citizen-engaging energy transitions. Both contain elements of resistance to traditional central government-led, top-down strategies. And both offer portraits of reclaiming and restructuring on behalf of a sustainable and just energy democracy.

Empirical studies have positively evaluated the technical potential of urban energy transitions in both the case countries—United States and Korea [48–50]. Using the case of New York City, Byrne and Taminiau [51,52], for instance, found that the city-wide deployment of rooftop solar photovoltaic (PV) systems (~ 10 GWp potential) could generate enough electricity to cover up to 53% of the city's daylight hour consumption. It is also expected that low-income areas of New York City could potentially earn revenue from their solar PV systems by exporting excess electric power to areas of the city that consume more power, such as Manhattan. The findings show that a "solar city" strategy could achieve significant peak shaving and create economic and environmental benefits for low-income communities.

Advancing the methodology they used in the New York City case, Taminiau et al. [53] conducted a more comprehensive analysis of sustainable urban energy metabolism using the case of Daejeon, South Korea. In addition to the solar city strategy studied in Refs. [51, 52], the paper introduced the concept of a "savings city," which promotes city-wide energy efficiency measures to use less energy. When combined with energy efficiency interventions, solar city Daejeon could self-supply 56% of its electricity demand. Like the New York City case, many areas of Daejeon can produce more solar power than needed and, therefore, could potentially export the surplus to parts of the city that consume more energy. The two components of energy commonwealth highlighted in these two city case studies—shaving the city's energy demand through efficiency improvements and utilizing local renewable energy—are possible in many more cities around the world. Of course, the question is how to encourage such changes to happen rapidly.

4.1 Experiment 1: Sustainable Energy Utility

The Sustainable Energy Utility (SEU) is a non-profit community utility model created to help energy users consume less energy and access local, sustainable energy sources. The idea of an SEU was first introduced by researchers at the Center for Energy & Environmental Policy at the University of Delaware [32]. In 2007, the State of Delaware became home to the first SEU after electricity rates increased more than 56%. The cause of this change in rates was utility deregulation policy, which removed price caps on residential and small business charges that had been frozen for 10 years.

Born out of a resistance movement to energy policy convention, the passage of eight bills in the Delaware Legislature led to a new utility form driven by energy justice and decarbonization principles. Since then, similar efforts have been made in other parts of the United States and beyond to form SEUs [54–56]. As the 2009 special issue of the *Bulletin of Science, Technology & Society* [57] highlighted, the structure of the SEU model can be applied internationally to challenge conventional energy systems built on commodity principles. Deep energy efficiency and renewablization strategies can provide the means to reduce the cost of energy services and empower communities to redirect energy capital to sustainable and equitable purposes. The SEU's commons-minded programs prioritize sustainability and equity over optimality in local energy production and consumption [7,32]. Commonwealth and community trust have been the essential drivers of operationalizing this SEU approach.

One Delaware SEU program highlights sustainable energy financing for energy efficiency and on-site solar generation projects. In 2011, the Delaware SEU issued a $72.5 million state-wide tax-exempt bond for public and university energy efficiency and solar energy projects, guaranteeing $148 million in energy savings from efficiency and solar measures [32]. This public bond financing strategy for energy efficiency and renewable energy demonstrates the practice of commonwealth energy economics in the sense that the investment was made not to generate profits but to achieve shared climate-energy goals [58].

The design features of the Delaware SEU were adopted in other locations, especially with the creation of the DC SEU, the Pennsylvania Sustainable Energy Financing (PennSEF) program, and the Ann Arbor SEU. Each jurisdiction works with the team at the University of Delaware that designed the original application. In the case of the PennSEF program, the Foundation for Renewable Energy & Environment [59] collaborates with the Pennsylvania Treasury Department on energy efficiency and on-site renewable energy projects for municipalities [35]. PennSEF pooled 35 municipalities in the state to replace street lighting, thereby cutting municipal electricity use and budgets and resulting in a 50% reduction in carbon emissions from local government operations [60]. PennSEF illustrates how transparent operation and trust-based project management are essential factors if the will and action of municipalities are to be aggregated into a common sustainable energy development effort.

Functioning as a trusted advisor, the SEU model in both forms provides communities and localities with tools and networks that allow a wide range of non-traditional actors (e.g., citizens, communities, small businesses, and environmental and faith groups) to be part of the climate-energy transformation.

4.2 Experiment 2: Seoul's "One Less Nuclear Power Plant" initiative

One Less Nuclear Power Plant (OLNPP) is a city-scale, justice-focused climate-energy initiative introduced by the Seoul Metropolitan Government (SMG) in 2012. As found in its name, OLNPP explicitly challenges the uneven distribution of "privatized risks" [61] created by South Korea's nuclear power network, especially in the wake of the Fukushima accident. This distinctive transition strategy illustrates how commons-minded energy practices can help alleviate the systemic injustices inherent in large-scale, centralized, and high-risk energy development [23].

The initiative is established on democratic and justice principles. Since its launch, OLNPP has aimed to increase the city's local energy self-sufficiency and address the inequitable distribution of risks from the national nuclear power network, specifically by promoting public engagement in decision-making and policy implementation [62]. Seoul's re-imagination of energy encompasses not only the development of new political and economic principles but also their operationalization in practical forms, including finance, ownership, and the forging of a prosumer relationship between energy production and consumption. As of 2019, OLNPP saved an amount of energy equivalent to 40% of Seoul's annual energy consumption through energy conservation, energy efficiency, and renewablization measures, reducing CO_2 emissions by 13.6 million tons [63].

OLNPP is internationally applauded for pioneering non-traditional approaches to energy governance that align well with the concept of an innovation commons. Indeed, the power of the operational and organizing principles is reflected in the scale of citizen engagement in this movement. For instance, citizen-engaging programs like the Eco-Mileage Program [64,65], Residential Mini-PV Program [66–68], and Energy Self-Sufficient Village Program [69] demonstrate the on-the-ground success of the initiative in convincing citizens and businesses to support and meet deep energy savings goals. SMG reported 5 million of its more than 10 million citizens had participated in various efforts made under the OLNPP initiative between 2012 and 2019 [63]. SMG has recognized citizens as decision-makers by involving them in the city's energy policymaking process in the form of a Citizen Council; actively engaging civil society and activists in its Working Committee also reflects OLNPP's inclusive character [70]. Most impressive, in just about 2 years, the initiative's energy conservation and local solar programs reduced the city's energy use by an amount equal to the power generated by one nuclear power plant [63].

The OLNPP initiative's re-imagining of the uneven energy landscape has shown the potential for the city government's polycentric leadership to engage and empower citizens through movement-based programs and activities. We argue that the impacts of climate-energy transition movements at the OLNPP scale will be crucial for the future direction of energy democratization.

5 Conclusions

The case studies highlighted here provide a lesson. In order to resist energy policy convention, reclaim governance of energy in society, and restructure energy-society relations around democratic, sustainable, and just development, we will need to mobilize individuals and social collectives. Acceleration of change is only possible with the mobilization of social movements. In short, the scale of change is the province of social movements. Civil rights, women's rights, and voting rights movements precede and make possible laws that are later passed by governments, not the other way around.

We have learned essential lessons from science in mapping our complex challenges. We are now at the stage where framing the climate-energy challenge requires rethinking industrialist energy-society relations, incumbent actors and institutions, and broadly, its political economy. It is wishful to think that we can affect change without knowing the science *and* also the social dimensions of our problem.

That said, our next effort is to find paths of transformative change. An interactive framework of principles propelling social change on a scale now required by the climate crisis is suggested. It is intended to assist us in operationalizing the changes we must motivate. This step necessarily entails re-imagination. But we wish to stress that this re-imagination process is already underway. Rather than a call for new hypotheses or new technologies, we propose that our next step is to learn how to create a political economy of energy commons and community-based governance. We argue that practical lessons on this score are in our possession: an expansion of re-imagination experiments like SEU and OLNPP (and many, many more that we know about), which can disarm the crisis. We cannot afford to waste another day on defeatist thinking and studied inaction.

References

[1] J. Byrne, N. Toly, L. Glover, Energy as a social project: recovering a discourse, in: J. Byrne, N. Toly, L. Glover (Eds.), Transforming Power: Energy, Environment, and Society, Routledge, 2006, pp. 1–32.

[2] B.K. Sovacool, et al., Sociotechnical agendas: reviewing future directions for energy and climate research, Energy Res. Soc. Sci. 70 (2020). Article 101617.

[3] K.M. Perkins, et al., COVID-19 pandemic lessons to facilitate future engagement in the global climate crisis, J. Clean. Prod. 290 (2021). Article 125178.

[4] R.D. Manzanedo, P. Manning, COVID-19: lessons for the climate change emergency, Sci. Total Environ. 742 (2020). Article 140563.

[5] D. Klenert, et al., Five lessons from COVID-19 for advancing climate change mitigation, Environ. Resour. Econ. 76 (4) (2020) 751–778.

[6] J. Byrne, L. Glover, C. Martinez, The production of unequal nature, in: J. Byrne, L. Glover, C. Martinez (Eds.), Environmental Justice: Discourses in International Political Economy, Routledge, New York, NY, 2002, pp. 261–291.

[7] J. Byrne, C. Martinez, C. Ruggero, Relocating energy in the social commons: ideas for a sustainable energy utility, Bull. Sci. Technol. Soc. 29 (2) (2009) 81–94.

[8] C. Martinez, From commodification to the commons: charting the pathway for energy democracy, in: D. Fairchild, A. Weinrub (Eds.), Energy Democracy: Advancing Equity in Clean Energy Solutions, Island Press, Washington, DC, 2017, pp. 21–36.

[9] IPCC, Summary for policymakers, in: V. Masson-Delmotte, et al. (Eds.), Global Warming of 1.5 °C. An IPCC Special Report on the Impacts of Global Warming of 1.5°C Above Pre-Industrial Levels and Related Global Greenhouse Gas Emission Pathways, in the Context of Strengthening the Global Response to the Threat of Climate Change, Sustainable Development, and Efforts to Eradicate Poverty, World Meteorological Organization, Geneva, Switzerland, 2018, pp. 1–24.

[10] M. Jakob, J. Hilaire, Unburnable fossil-fuel reserves, Nature 517 (7533) (2015) 150–151.

[11] S. Solomon, et al., Irreversible climate change due to carbon dioxide emissions, Proc. Natl. Acad. Sci. 106 (6) (2009) 1704–1709.

[12] A.B. Lovins, Soft Energy Paths: Toward a Durable Peace, Friends of the Earth International, Cambridge, MA, 1977.

[13] K. Jenkins, D. McCauley, C.R. Warren, Attributing responsibility for energy justice: a case study of the Hinkley Point Nuclear Complex, Energy Policy 108 (2017) 836–843.

[14] J.S. Gregg, et al., Collective action and social innovation in the energy sector: a mobilization model perspective, Energies 13 (3) (2020).

[15] D.J. Hess, Energy democracy and social movements: a multi-coalition perspective on the politics of sustainability transitions, Energy Res. Soc. Sci. 40 (2018) 177–189.

[16] G. Seyfang, A. Haxeltine, Growing grassroots innovations: exploring the role of community-based initiatives in governing sustainable energy transitions, Eviron. Plann. C. Gov. Policy 30 (3) (2012) 381–400.

[17] I.D. Smith, Energy transition and social movements: the rise of a community choice movement in California, in: D. Kurochkin, E.V. Shabliy, E. Shittu (Eds.), Renewable Energy: International Perspectives on Sustainability, Springer International Publishing, Cham, 2019, pp. 91–129.

[18] E. Greenberg, C. McKendry, Contested power: energy democracy and the repoliticization of electricity in the western U.S, Energy Res. Soc. Sci. 73 (2021). Article 101942.

[19] M.J. Burke, J.C. Stephens, Energy democracy: goals and policy instruments for sociotechnical transitions, Energy Res. Soc. Sci. 33 (2017) 35–48.

[20] K. Szulecki, I. Overland, Energy democracy as a process, an outcome and a goal: a conceptual review, Energy Res. Soc. Sci. 69 (2020). Article 101768.

[21] A.H. Sorman, E. Turhan, M. Rosas-Casals, Democratizing energy, energizing democracy: central dimensions surfacing in the debate, Front. Energy Res. 8 (2020). Article 499888.

[22] J.C. Stephens, Energy democracy: redistributing power to the people through renewable transformation, Environ. Sci. Policy Sustain. Dev. 61 (2) (2019) 4–13.

[23] J. Lee, J. Byrne, Expanding the conceptual and analytical basis of energy justice: beyond the three-tenet framework, Front. Energy Res. 7 (2019). Article 99.

[24] M.V. Mathai, Towards a sustainable synergy: end-use energy planning, development as freedom, inclusive institutions and democratic technics, in: I. Oosterlaken, J. van den Hoven (Eds.), The Capability Approach, Technology and Design, Springer Netherlands, Dordrecht, 2012, pp. 87–112.

[25] N. Forrest, A. Wiek, Learning from success—toward evidence-informed sustainability transitions in communi-
      ties, Environ. Innov. Soc. Transit. 12 (2014) 66–88.
[26] G. Walker, What are the barriers and incentives for community-owned means of energy production and use?
      Energy Policy 36 (12) (2008) 4401–4405.
[27] D.N.-Y. Mah, Community solar energy initiatives in urban energy transitions: a comparative study of Foshan,
      China and Seoul, South Korea, Energy Res. Soc. Sci. 50 (2019) 129–142.
[28] D.J. Hess, D. Lee, Energy decentralization in California and New York: conflicts in the politics of shared solar
      and community choice, Renew. Sustain. Energy Rev. 121 (2020). Article 109716.
[29] C.A. Miller, A. Iles, C.F. Jones, The social dimensions of energy transitions, Sci. Cult. 22 (2) (2013) 135–148.
[30] M. Pasqualetti, S. Stremke, Energy landscapes in a crowded world: a first typology of origins and expressions,
      Energy Res. Soc. Sci. 36 (2018) 94–105.
[31] J. Lee, H. Kim, J. Byrne, Operationalising capability thinking in the assessment of energy poverty relief poli-
      cies: moving from compensation-based to empowerment-focused policy strategies, J. Hum. Dev. Capab. 22 (2)
      (2021) 292–315.
[32] J. Byrne, J. Taminiau, A review of sustainable energy utility and energy service utility concepts and applica-
      tions: realizing ecological and social sustainability with a community utility, WIREs Energy Environ. 5 (2)
      (2015) 136–154.
[33] J.C. Pezzey, M.A. Toman, Progress and problems in the economics of sustainability, in: T. Tietenberg, H. Folmer
      (Eds.), The International Yearbook of Environmental and Resource Economics 2002/2003, Elgar, Cheltenham,
      2002, pp. 165–232.
[34] M. Wolsink, Distributed energy systems as common goods: socio-political acceptance of renewables in intelli-
      gent microgrids, Renew. Sustain. Energy Rev. 127 (2020). Article 109841.
[35] J. Taminiau, et al., Advancing transformative sustainability: a comparative analysis of electricity service and
      supply innovators in the United States, WIREs Energy Environ. 8 (4) (2019) e337.
[36] L. Albareda, A.J.G. Sison, Commons organizing: embedding common good and institutions for collective
      action. Insights from ethics and economics, J. Bus. Ethics 166 (4) (2020) 727–743.
[37] J. Euler, Conceptualizing the commons: moving beyond the goods-based definition by introducing the social
      practices of commoning as vital determinant, Ecol. Econ. 143 (2018) 10–16.
[38] A. Goldthau, Rethinking the governance of energy infrastructure: scale, decentralization and polycentrism,
      Energy Res. Soc. Sci. 1 (2014) 134–140.
[39] J. Byrne, et al., Navigating the Changing Landscape of Community Solar in Delaware: Policy Designs
      and Governance Frameworks to Support Community-Owned Sustainable Energy, Center for Energy &
      Environmental Policy, University of Delaware, Newark, DE, 2020.
[40] C. Büscher, P. Sumpf, "Trust" and "confidence" as socio-technical problems in the transformation of energy
      systems, Energy Sustain. Soc. 5 (1) (2015). Articel 34.
[41] G. Walker, et al., Trust and community: exploring the meanings, contexts and dynamics of community renew-
      able energy, Energy Policy 38 (6) (2010) 2655–2663.
[42] B.P. Koirala, et al., Trust, awareness, and independence: insights from a socio-psychological factor analysis of
      citizen knowledge and participation in community energy systems, Energy Res. Soc. Sci. 38 (2018) 33–40.
[43] J.M. Budd, Public libraries, political speech, and the possibility of a commons, Public Libr. Q. 38 (2) (2019)
      147–159.
[44] J. Blewitt, Public libraries, citizens and the democracy, Power Educ. 6 (1) (2014) 84–98.
[45] J. Buschman, Everyday life, everyday democracy in libraries: toward articulating the relationship, Pol. Libr. 4
      (1) (2018). Article 10.
[46] C.A. Miller, J. Eschrich (Eds.), Cities of Light: A Collection of Solar Futures, ASU Center for Science and the
      Imagination, 2021.
[47] M.J. Williams, Urban commons are more-than-property, Geogr. Res. 56 (1) (2018) 16–25.
[48] J. Byrne, et al., A solar city strategy applied to six municipalities: integrating market, finance, and policy fac-
      tors for infrastructure-scale photovoltaic development in Amsterdam, London, Munich, New York, Seoul, and
      Tokyo, WIREs Energy Environ. 5 (1) (2016) 68–88.
[49] J. Byrne, et al., Multivariate analysis of solar city economics: impact of energy prices, policy, finance, and cost
      on urban photovoltaic power plant implementation, WIREs Energy Environ. 6 (4) (2017) e241.
[50] J. Byrne, et al., Are solar cities feasible? A review of current research, Int. J. Urban Sci. 21 (3) (2017) 239–256.

[51] J. Byrne, J. Taminiau, Utilizing the urban fabric as the solar power plant of the future, in: P. Droege (Ed.), Urban Energy Transition, second ed., Elsevier, 2018, pp. 31–49.

[52] J. Taminiau, J. Byrne, City-scale urban sustainability: spatiotemporal mapping of distributed solar power for New York City, WIREs Energy Environ. 9 (5) (2020) e374.

[53] J. Taminiau, et al., Infrastructure-scale sustainable energy planning in the cityscape: transforming urban energy metabolism in East Asia, WIREs Energy Environ. (2021) e397.

[54] L. Teron, S.S. Ekoh, Energy democracy and the city: evaluating the practice and potential of municipal sustainability planning, Front. Commun. 3 (2018). Article 8.

[55] M.V. Mathai, Elements of an alternative to nuclear power as a response to the energy-environment crisis in India: development as freedom and a sustainable energy utility, Bull. Sci. Technol. Soc. 29 (2) (2009) 139–150.

[56] J.-M. Yu, The restoration of a local energy regime amid trends of power liberalization in East Asia: the Seoul sustainable energy utility, Bull. Sci. Technol. Soc. 29 (2) (2009) 124–138.

[57] H.B. McDowell, S. Finnigan, Introduction to the special issue, Bull. Sci. Technol. Soc. 29 (2) (2009) 79–80.

[58] J. Houck, W. Rickerson, The sustainable energy utility (SEU) model for energy service delivery, Bull. Sci. Technol. Soc. 29 (2) (2009) 95–107.

[59] Foundation for Renewable Energy & Environment, Available from: https://freefutures.org/.

[60] Foundation for Renewable Energy & Environment, PennSEF Pioneers Clean Energy Financing for 35 Municipalities, 2017, (cited 2021 March 14) Available from: https://freefutures.org/announcement/pennsef-pioneers-clean-energy-financing-for-35-municipalities/.

[61] M. Nadesan, Fukushima and the Privatization of Risk, Springer, 2013.

[62] T. Lee, T. Lee, Y. Lee, An experiment for urban energy autonomy in Seoul: the one 'less' nuclear power plant policy, Energy Policy 74 (2014) 311–318.

[63] Seoul Metropolitan Government, One Less Nuclear Power Plant, 2021, [cited 2021 April 5]. Available from: https://news.seoul.go.kr/env/environment/climate-energy/one-less-nuclear-power-plant.

[64] J. Kim, T.Y. Jung, Y.G. Kim, Multilevel analysis of civic engagement and effectiveness of energy transition policy in Seoul: the Seoul Eco-Mileage Program, Sustainability 12 (23) (2020). Article 9905.

[65] Y.D. Ko, B.D. Song, Sustainable service design and revenue management for electric tour bus systems: Seoul city tour bus service and the Eco-Mileage Program, J. Sustain. Tour. 27 (3) (2019) 308–326.

[66] S. Yang, W. Chen, H. Kim, Building energy commons: three mini-PV installation cases in apartment complexes in Seoul, Energies 14 (1) (2021). Article 249.

[67] M.-H. Kim, T.-H.T. Gim, Spatial characteristics of the diffusion of residential solar photovoltaics in urban areas: a case of Seoul, South Korea, Int. J. Environ. Res. Public Health 18 (2) (2021). Article 644.

[68] J.-S. Lee, J.W. Kim, The factors of local energy transition in the Seoul Metropolitan Government: the case of mini-PV plants, Sustainability 9 (3) (2017). Article 386.

[69] H. Kim, A community energy transition model for urban areas: the energy Self-Reliant Village Program in Seoul, South Korea, Sustainability 9 (7) (2017). Article 1260.

[70] J. Seo, An Integrated Assessment Framework for Energy Governance: Toward a Sustainable, Equitable, and Democratic Energy Transition, Doctoral Dissertation, University of Delaware, 2018.

Democratic governance of fossil fuel decline

Johanna Bozuwa[a], *Matthew Burke*[b], *Stan Cox*[c], *and Carla Santos Skandier*[d]

[a]Climate+Community Project, San Francisco, CA, United States [b]Leadership for the Ecozoic, University of Vermont, Burlington, VT, United States [c]Ecosphere Studies, The Land Institute, Salina, KS, United States [d]Climate and Energy Program, The Democracy Collaborative, Washington, DC, United States

1 Introduction

Every passing year reveals new realities of a warmer world, with record-breaking hurricanes and wildfires. These climatic changes have propelled demands to end the production and use of fossil fuels, but proposals still largely rely on inadequate market-based mechanisms that offer no coherent means for providing for essential needs and no clear plan for effectively and fairly distributing the remaining fossil fuel budget. Large fossil fuel companies have meanwhile issued a series of reports stating their "climate ambitions," none of which states an end date for oil and gas extraction nor commits to supporting their labor force confronted with job losses. Their plans are a far cry from what must be done to transition in time, let alone in a just way for workers and communities [1].

In this chapter, we advocate for the public ownership of the fossil fuel industry and explore ways it can be democratized through the implementation of non-market decision-making processes that allow for a managed and just decline of the fossil fuel era to spur the needed societal transformation. Through a review of the theoretical and empirical literature, we examine eight different participatory and deliberative governance structures to understand their democratic capacities and potentials for transforming existing energy assemblages. Through this review, we argue that alternative modes of governance are available, and their democratic potential should be elevated. Although these examples do not provide the entirety of a comprehensive plan for a managed decline, they do demonstrate the many possibilities at local, regional, national, and international levels that are already in practice and available to ensure an energy transition that is just, equitable, reparative, and sustainable.

The world's industrial societies evolved during an era of abundant, versatile energy in the form of oil, gas, and coal. Similarly, the development of alternative energy capacity and other technological advances alone will be far from sufficient to either accomplish or adapt to the deliberate cessation of fossil-fuel extraction. Achieving, and then adapting to, the voluntary tightening and eventual elimination of fossil fuel supplies will require a thoroughgoing physical, economic, and cultural transformation of society. For this we need democratization and coordinated planning throughout the energy sector, features often external to markets and the industry business model. Claiming a livable future means decisively ending extraction and shepherding in a new system based on principles of energy democracy by bringing key decisions on fossil fuel supply within public and community control.

Nationalizing the fossil fuel industry is one clear way to organize non-market systems for making key decisions around the remaining fossil fuel supply, transitioning local economies, and cleaning up the mess of decades of extraction. Nationalization allows for combining moratoria on fossil fuel development with the establishment of declining production caps each year. It also ensures a wind-down that is centered on clear and specific climate, labor, and social mandates. This contrasts directly with a more marketized, discriminatory system in which investors and financial elites decide where, when, and by whom the last reserves are extracted, leaving individuals and communities to compete based on purchasing power for these diminishing supplies [2]. In a regime of nationalized, capped, declining fossil-fuel supplies, fuels should be treated no longer as private goods but as scarce common-pool resources. For this to occur, public ownership and control of the fossil fuel industry must be purposefully tailored to avoid replicating past experiences of extraction and patterns of private industry, particularly through the adoption of democratic decision-making mechanisms.

2 Living examples of democratic decision-making mechanisms

While scholars and the public at large struggle to envision decision-making processes that can be truly participatory, transparent, and accountable to the public, there are plenty of living examples that can contribute to a vision and practice for a democratic and socially just transition from fossil fuels. Because democratic models are needed across scales, this chapter offers examples functioning from the local to global levels. Specifically, this section introduces and briefly summarizes historical and contemporary approaches available for non-market distribution and allocation of fossil fuels, organized around needs.

The section first addresses those decision-making mechanisms that are already widely implemented or readily available, especially at the local level, including participatory budgeting, worker-owned cooperatives, water districts, and community fuel banks and energy trusts. It then goes on to address mechanisms that require more attention and coordination at regional, national, and multinational levels to support their uptake and implementation, including public regional planning, rationing boards, Indigenous energy sovereignty, and regulation of international trade.

2.1 Local mechanisms

Many democratic decision-making mechanisms operate primarily at the local level, providing diverse working examples to inform a broader uptake of non-market, needs-based fossil fuel allocation. Local mechanisms offer potential advantages in terms of local administration and accountability, ease of access and participation among non-experts, and flexibility and responsiveness to local conditions, yet local processes may also suffer from inadequate capacity and expertise as well as local power imbalances, demonstrating the need to share resources, experiences, and accountability across communities. Local decision-making mechanisms may be more effective and fairer if they have functional connections with local efforts elsewhere or are nested within regional, national, or international systems while maintaining local autonomy.

2.1.1 Participatory budgeting

The growing experience with participatory budgeting offers an important alternative for governing fossil fuel decline. Participatory budgeting is a democratic, deliberative process that allows the active involvement of citizens in decisions regarding the prioritization of public resources [3]. Its modern origins follow from experiences in the city of Porto Alegre, Brazil in the late 1980s as a response to steep inequalities and rising poverty. Participatory budgeting has since spread to thousands of locations worldwide, allowing communities to have input on budget decisions at all levels [4].

Several municipal examples of participatory budgeting organized around green investments are now in practice in cities such as New York and Lisbon [5]. Social movements such as the Movement for Black Lives (M4BL) and People's Budget Los Angeles aim to broaden participatory budgeting to ensure the engagement of impacted communities with decisions around energy use for community needs, reversing decades of disinvestment and pushing beyond limited engagement on discretionary expenditures [6]. Although specific models vary, participatory budgeting offers a mechanism for collective consideration of the many ways to connect energy services to human needs, opening space for assessing alternative modes of provisioning, a necessary aspect given the multiple uncertainties and challenges of living within limits [7].

Participatory budgeting of declining fossil fuel supplies would build from these experiences to enable participants to collaboratively negotiate and design processes for deciding the purpose and destination of existing fossil fuel energy and energy funds [8]. This mode of energy planning would also enable community education and engagement while building needed capacities and empowerment for long-term public participation in energy system decision-making processes [9,10] (see also Chapter 9).

2.1.2 Worker-owned cooperatives

Workers are a key constituency of energy democracy when considering a fossil fuel winddown. Instead of distant, profit-driven shareholders deciding their fate, the energy transition away from fossil fuels must care for and empower workers at the frontlines of change. Energy transitions are particularly difficult given the potential to create massive economic shifts, as extractive communities tend to lack a diversified economy. Fossil fuel jobs are also more often unionized and better compensated than many alternatives [11]. Despite these challenges, transitions can be successful when workers have a seat at the decision-making table.

Worker cooperative structures, businesses owned and controlled by their employees through shares and equal representation, illustrate the potential of labor self-determination and empowerment [12]. The Mondragon worker cooperative network provides a particular example of how labor-driven governance can be key during challenging times. For more than 60 years, the network has created a diversified, regional economy in Spain's Basque region, where it employs more than 70,000 workers, a third of whom are also owners [13]. During the mortgage crisis, Spain's unemployment reached over 25%, with several of Mondragon's member businesses facing struggles [14]. To maximize employment, its Cooperative Congress, with direct worker representation, adopted measures such as transfer of employees across member businesses, salary cuts when strictly necessary with managers taking the biggest cuts, and access to additional capital through network solidarity funds. When Fagor Electrodomésticos, one of Mondragon's largest cooperatives, filed for bankruptcy, these measures ensured that most of its workforce remained employed [13].

Although worker cooperatives face obstacles including lack of financing and pressure to mimic profit-focused businesses, the Mondragon example demonstrates the benefits of positioning workers at the center of decision-making, especially during difficult times. Converting soon-to-be-obsolete fossil fuel companies into worker coops would not be a desirable path. However, worker coops' democratic governance and cooperation efforts can lay out the types of safety nets that transitioning workers should have and illustrate the forms of organization that should be encouraged as society works to build a diversified, resilient, and renewable 21st-century economy.

2.1.3 Water districts

Water districts also provide relevant examples of non-market solutions to allocation based on needs. For example, the West Basin Water Association, a non-governmental group, was formed in 1948 to prevent the depletion of scarce groundwater and resultant saltwater intrusion south of Los Angeles. Its board was composed of water users within the basin, public and private water providers, and officials from municipal agencies. The association declared a target cap on the number of acre-feet of water to be extracted per year in the basin and developed a plan for rationing the supply. Board members rarely took a vote on an issue until they had first discussed it thoroughly and reached a consensus.

Although the association had no enforcement power, relying instead upon discussion, education, and lobbying to convince official agencies to take action, its efforts were largely successful. By 1963, groundwater extraction had been reduced to its rate of 1943, before the development boom that had led to over-pumping in the first place. At the group's urging, an official water replenishment district was created in 1959 to take over regulation and rationing of water in the basin. In the words of political economist Elinor Ostrom, the association found a way of "organizing new enterprises to secure appropriate forms of community action in providing common goods and services," emblematic of "public entrepreneurship" [15], providing a powerful model for allocating scarce local energy resources [16].

2.1.4 Community fuel banks and energy trusts

Community fuel banks and energy trusts build on the experiences of institutions designed to coordinate the distribution of essential goods and services based on need. Food banks operate with the goal of distributing food to those people and families for whom market-based food systems fail to provide sufficient, quality food. Food banks may either provide food directly to hungry people or supply food to local intermediaries, including soup kitchens and frontline community organizations. Community fuel banks operate in a similar fashion, typically providing emergency fuel assistance in the form of seasonal funds for low-income community members to ensure fuel delivery or prevent electricity shut-offs. Some organizations combine efforts; for example, Feeding Britain partnered with the Fuel Bank Foundation to address the dual needs of hunger and warmth [17]. Energy trusts such as those used in New Zealand provide a way for multiple communities to participate in and control energy distribution networks, independently or in combination with local governments [18].

These systems of distribution have important limitations, including reliance on charity rather than the responsibility of governments, as demonstrated by their proliferation following cuts to social welfare programs. However, these examples suggest a framework for moving beyond commodification of energy services while deploying resources to provide for needs, particularly when the supply of fossil fuels is declining over time. Further, green public service banks and related community financial institutions could enable decision-making and local accountability regarding the financing of community-based energy projects [9]. In the context of fossil fuel decline, these financing decisions could be legally mandated to prioritize grants and funds needed to support community fuel banks and energy trusts.

2.2 Regional mechanisms and beyond

While remaining accountable to local communities, a just decline of fossil fuels must extend beyond the local level. Though fewer examples exist in practice, democratic mechanisms are needed for governing fossil fuel decline at scale to coordinate decision-making across diverse geographies, extended supply chains, and increasing technical complexity. These mechanisms can draw on larger pools of expertise and resources and help bind together varied communities and jurisdictions in the context of our collective planetary crisis.

2.2.1 Public regional planning and the Tennessee Valley Authority

We will need both decentralized decision-making and large-scale intervention and planning, as well as mechanisms that can stitch the different levels of governance together. Regional planning institutions can meet this need, as they reflect the wisdom of local communities while integrating federal climate policy.

One clear historical example of regional planning is the Tennessee Valley Authority (TVA), a massive feat of federal investment in regional development. By 1942, with less than 10 years of existence, TVA was responsible for creating almost 28,000 jobs and electrifying an entire rural region that had long been left in the dark. Up to this day, TVA is in charge of supplying electricity to roughly 10 million people across seven states [19]. Although these benefits transformed the region in a number of positive ways, its planning legacy is marred by the racially discriminatory Jim Crow politics that shaped policy and legislation throughout the south of the country at the time [20]. Not only were Black farmers geographically separated from white communities, but TVA also disproportionately targeted Black farmland for the siting of massive hydroelectric dams and reservoirs, devaluing the land as less desirable [21]. Even today TVA continues to operate a governance structure that is neither democratic nor transparent.

Designing new forms of publicly accountable regional development requires explicit design principles for a just transition. Planning processes should be directly informed by workers, frontline, and fence-line communities through the promotion of leadership that includes representation from Black, Indigenous, and People of Color (BIPOC) since so much of fossil fuel infrastructure's burden has fallen on them. Clear standards for success are needed that do not perpetuate forms of economic extraction but instead build community wealth for the long term. Regional planning provides possibilities for integrating more grassroots-based decision-making, like participatory budgeting, to make collective decisions and empower community members in the process of coordinating across like regions.

2.2.2 *Rationing boards*

When markets fail in the allocation of scarce essential goods, governments often resort to rationing. It was during World War II that the United States adopted its strictest and most comprehensive rationing plan, imposing national limits on household consumption of resources, including gasoline and heating fuels. The administration was carried out by approximately 5600 local rationing boards. Every household was entitled to a standard ration of certain essential goods; in the case of fuels, consumers could apply to the local board for supplemental rations on the basis of need. The boards had a high degree of discretion in responding to applications as long as these did not exceed their federally assigned monthly quota for each product. The boards were not democratically constituted; members were appointed, not elected. By law, each board was to be representative of its community, and with respect to gender, class, and occupation, they generally were. However, racial equity was a different matter: glaringly, only 0.7% of board members nationally were Black [22], illustrating the racial caste system that governed all aspects of social life at that point in US history.

Like rationing boards of the World War II years, future governance of scarce fossil fuel resources could be done at the community level through the creation of local boards, this time making sure that membership is democratically determined by, representative of, and responsible to the community in all its dimensions. In that vein, the distribution would be based on needs along with reparations for decades of unequal distribution, particularly among communities that have disproportionately experienced the burdens of fossil fuel production. These experiences include both the loss of holdings to the development of energy infrastructure as well as excessive exposure to toxic pollutants resulting from the placement of fossil fuel refineries and petrochemical plants in BIPOC communities [23]. In addition to equitable per-household rations, the plan should include a community pool of fuel allowances to be allocated collectively through deliberative democratic processes analogous to the participatory budgeting of local public funds described previously.

2.2.3 *Indigenous energy sovereignty*

Indigenous peoples continue to experience violent alienation from their traditional lands due to energy extractivism. Centering justice in fossil fuel decline means confronting unjust colonial legacies of extraction and development among diverse Indigenous peoples and enabling energy governance organized around Indigenous sovereignty. Indigenous energy sovereignty involves reclamation of Indigenous jurisdiction on traditional territories and regrounds energy decision-making based on rights, autonomy, self-determination, and respect for diverse cultures and worldviews [24,25].

Building upon the principle of Free Prior Informed Consent, Indigenous peoples can lead consent-based practices for determining energy supply to meet self-defined needs according to their own legal orders, priorities, and values. Indigenous energy sovereignty is especially important given the significant energy reserves within existing and traditional native lands, as well as the opportunities for beneficial impacts for the lives of some of the world's most vulnerable, economically constrained, and historically oppressed communities [26]. Indigenous energy sovereignty offers no guarantee of a smooth transition from fossil fuels, as shown by the actual experiences of Indigenous communities, including the Southern Utes, the Navajo Nation, and the Māori people. Yet these and other unique experiences demonstrate the diversity of practices of self-determination and the tangible improvements achievable for the lives and livelihoods of Indigenous peoples worldwide [27,28] (see also Chapter 23).

In view of the need for real alternatives, Indigenous energy sovereignty enables diverse decision-making processes regarding energy supply with and beyond state-centric approaches, balancing the need for economic development, the duty to protect traditional lands and waters, and the obligation to support future generations and planetary life [29]. The role of Indigenous peoples must not be confined to that of junior partners in the state- and corporate-led resource extraction, or worse, serving as the recipient of the fossil fuel industry's unwanted liabilities. Advancing Indigenous energy sovereignty must then involve reversing legacies of colonialism by recognizing land tenure and resource rights, implementing nation-to-nation agreements for reparations and land reclamation, and ensuring procedural and financial capacities for Indigenous nations to engage in comprehensive, strategic energy planning [30,31].

2.2.4 *International trade and distribution*

As fossil fuel supply chains extend across borders, multinational and global agreements are needed for democratically governing the decline of fossil fuels [32]. Moratoria and bans on fossil fuels are potentially the most effective supply-side constraints, but their increasing uptake remains uneven, is often implemented temporarily, can lead to shifting of production elsewhere, and requires better coordination with demand-side policies [33]. Such mechanisms would ideally apply upstream in the supply chain, reshaping international trade conventions of fossil fuels to conform with the Paris agreement and ensuring priority use for those locations most in need. A proposed Fossil Fuel Non-proliferation Treaty offers an active example of this approach, based on collective commitments to non-proliferation, disarmament, and peaceful use [34].

Such a treaty would begin with an assessment and registry of existing reserves and an agreement on principles of equity and burden-sharing for phasing declines across countries and fuel types, accounting for historical emissions, providing financial support for renewable development in poor nations, and creating systems of monitoring and compliance [35]. Phasing out fossil fuel stocks and production requires limiting extraction, regulating supplies, and eliminating fossil fuel subsidies. This approach also affirms the Lofoten Declaration, which obligates wealthy nations to lead the effort to cease further extraction, dismantle fossil fuel infrastructures, and enact associated industrial, economic, and employment policies [36]. Implementation could be led by the International Energy Agency, a United Nations agency, or other international institutions in combination with national committees responsible for ensuring emissions commitments [37].

The challenges of intergovernmental coordination and enforcement notwithstanding, experience with multilateral environmental agreements that use trade restrictions as key tools for health and environmental protection, including the Basel and Stockholm Conventions

[38,39], can help guide the design and implementation of transparent and enforceable international trade restrictions, ideally involving a global trade ban on fossil fuels [40,41].

3 Conclusions

Failures of conventional methods of the energy transition, combined with rising levels of inequality and harms of climate change, underscore the need for a socially and ecologically just transition from fossil fuels. As energy supplies become increasingly precarious over time, continued reliance on market mechanisms cannot ensure an equitable and sustainable energy transition. Effectively managing the decline of fossil fuels requires shifting ownership and control of fossil fuel sources as well as restructuring processes of their distribution and allocation around democratic planning and decision-making. The examples described in this chapter provide needed points of reference for democratizing energy systems, demonstrating actual methods and institutions for the non-market, participatory distribution of remaining fossil fuels organized around worker and community needs rather than profit and accumulation. By recognizing the injustices of relying upon the market-based allocation of remaining fossil fuel stocks and demonstrating the availability of democratic alternatives, this chapter aims to improve the capacity of communities, particularly those historically oppressed, to re-imagine, experiment, and implement a just transition. Going forward, increasing societal and community capacity will require experimentation and development of these models in specific locations and contexts, with support from workers, policymakers, researchers, and social movements.

References

[1] D. Tong, Big Oil Reality Check: Assessing Oil and Gas Company Climate Plans, Oil Change International, 2020. http://priceofoil.org/2020/09/23/big-oil-companies-still-failing-on-climate/. (Accessed 7 October 2020).
[2] L. Edwards, S. Cox, Cap and adapt: a failsafe policy for the climate emergency, Solutions 11 (3) (2020) 22–31.
[3] Y. Cabannes, The contribution of participatory budgeting to the achievement of the sustainable development goals: lessons for policy in commonwealth countries, Commonw. J. Local Gov. 21 (2019). ID 6707.
[4] The Participatory Budgeting Project. https://www.participatorybudgeting.org. (Accessed 9 May 2020).
[5] N. Kumar, With Green Participatory Budgeting, Racial Justice and Climate Justice Go Hand in Hand, Gotham Gazette, June 27, 2020. https://www.gothamgazette.com/opinion/130-opinion/9541-green-participatory-budgeting-racial-justice-climate-justice-new-york-city.
[6] People's Budget LA. https://peoplesbudgetla.com/. (Accessed 9 May 2020).
[7] L.I. Brand-Correa, J. Martin-Ortega, J.K. Steinberger, Human scale energy services: untangling a golden thread, Energy Res. Soc. Sci. 38 (2018) 178–187.
[8] A. Capaccioli, G. Poderi, M. Bettega, V. D'Andrea, Exploring participatory energy budgeting as a policy instrument to foster energy justice, Energy Policy 107 (2017) 621–630.
[9] M.J. Burke, J.C. Stephens, Energy democracy: goals and policy instruments for sociotechnical transitions, Energy Res. Soc. Sci. 33 (2017) 35–48.
[10] E.A. Moallemi, S. Malekpour, A participatory exploratory modelling approach for long-term planning in energy transitions, Energy Res. Soc. Sci. 35 (2018) 205–216.
[11] B. Callaci, M. Paul, A Brief History of Displacement: How Workers Are Fighting Back Amid the Looming Climate Crisis, Data for Progress, 2019. https://filesforprogress.org/memos/a-history-of-worker-displacement.pdf. (Accessed 7 October 2020).
[12] Democracy at Work Institute, What Is a Worker Cooperative? https://institute.coop/what-worker-cooperative. (Accessed 10 July 2020).

[13] S. Kasmir, The Mondragon cooperatives: successes and challenges, Global Dialogue 6 (1) (2016). https://globaldialogue.isa-sociology.org/the-mondragon-cooperatives-successes-and-challenges/. (Accessed 7 October 2020).

[14] R. Matthews, The Mondragon Model: How a Basque Cooperative Defied Spain's Economic Crisis, The Conversation, 2012. https://theconversation.com/the-mondragon-model-how-a-basque-cooperative-de-fied-spains-economic-crisis-10193. (Accessed 7 October 2020).

[15] E. Ostrom, Public Entrepreneurship: A Case Study in Ground Water Basin Management (Ph.D. dissertation), University of California, Los Angeles, 1965.

[16] E. Ostrom, Beyond markets and states: polycentric governance of complex economic systems, Am. Econ. Rev. 100 (2010) 641–672.

[17] Feeding Britain, Fuel Banks. https://feedingbritain.org/what-we-do/flagship-projects/fuel-banks/. (Accessed 5 September 2020).

[18] C.E. Hoicka, J.L. MacArthur, From tip to toes: mapping community energy models in Canada and New Zealand, Energy Policy 121 (2018) 162–174.

[19] D.E. Lilienthal, TVA: Democracy on the March, Harper & Brothers, New York City, 1944.

[20] A People's History of the Tennessee Valley Authority, Appalachian Voices, 2019. https://www.tiki-toki.com/timeline/entry/1330290/A-Peoples-History-of-the-Tennessee-Valley-Authority/. (Accessed 7 October 2020).

[21] M. Walker, African Americans and TVA reservoir property removal: race in a New Deal program, Agric. Hist. 72 (2) (1998) 417–428.

[22] E. Redford, Field Administration of Wartime Rationing, U.S. Office of Price Administration, Washington, DC, 1947.

[23] I.G. Castellón, Cancer Alley and the fight against environmental racism, Vill. Envtl. L. J. 32 (1) (2021) 15–43.

[24] C. Schelly, D. Bessette, K. Brosemer, V. Gagnon, K.L. Arola, A. Fiss, et al., Energy policy for energy sovereignty: can policy tools enhance energy sovereignty? Sol. Energy 205 (2020) 109–112.

[25] Land Back, Yellowhead Institute, Ryerson University, Toronto, Ontario, 2019. redpaper.yellowheadinstitute.org.

[26] S.L. Smith, B. Frehner (Eds.), Indians and Energy: Exploitation and Opportunity in the American Southwest, School for Advanced Research Press, Santa Fe, New Mexico, 2010.

[27] J. MacArthur, S. Matthewman, Populist resistance and alternative transitions: indigenous ownership of energy infrastructure in Aotearoa New Zealand, Energy Res. Soc. Sci. 43 (2018) 16–24.

[28] L. Necefer, G. Wong-Parodi, P. Jaramillo, M.J. Small, Energy development and Native Americans: values and beliefs about energy from the Navajo Nation, Energy Res. Soc. Sci. 7 (2015) 1–11.

[29] A.A. Smith, D.N. Scott, 'Energy Without Injustice'? Indigenous Ownership of Renewable Energy Generation, Social Science Research Network, 2018. https://papers.ssrn.com/abstract=3251922.

[30] D. Brookshire, N. Kaza, Planning for seven generations: energy planning of American Indian tribes, Energy Policy 62 (2013) 1506–1514.

[31] R. Rakshit, C. Shahi, M.A. (Peggy) Smith, A. Cornwell, Community capacity building for energy sovereignty: a First Nation case study, Sustain. Environ. Res. 3 (2) (2018) 177.

[32] G. Piggot, C. Verkuijl, H. van Asselt, M. Lazarus, Curbing fossil fuel supply to achieve climate goals, Clim. Pol. 20 (8) (2020) 881–887.

[33] N. Gaulin, P.L. Billon, Climate change and fossil fuel production cuts: assessing global supply-side constraints and policy implications, Clim. Pol. 20 (8) (2020) 888–901, https://doi.org/10.1080/14693062.2020.1725409.

[34] P. Newell, A. Simms, Towards a fossil fuel non-proliferation treaty, Clim. Pol. (2019) 1–12.

[35] G. Muttitt, S. Kartha, Equity, climate justice and fossil fuel extraction: principles for a managed phase out, Clim. Pol. (2020) 1–19.

[36] The Fossil Fuel Non-Proliferation Treaty. https://www.fossilfueltreaty.org/. (Accessed 15 September 2020).

[37] H. van Asselt, Governing the transition away from fossil fuels: The role of international institutions, SEI Working 07, 2014. https://www.sei.org/publications/governing-the-transition-away-from-fossil-fuels-the-role-of-international-institutions/.

[38] Secretariat of the Basel Convention, Basel Convention on the Control of Transboundary Movements of Hazardous Wastes and Their Disposal, United Nations Environment Programme (UNEP), Geneva, Switzerland, 2020, www.basel.int/.

I. Imaginaries

[39] Secretariat of the Stockholm Convention, Stockholm Convention on Persistent Organic Pollutants (POPS), United Nations Environment Programme (UNEP), Geneva, Switzerland, 2020. www.pops.int/.

[40] S. Ahmad Khan, Clearly hazardous, obscurely regulated: lessons from the Basel convention on waste trade, AJIL Unbound 114 (2020) 200–205.

[41] H. Fiedler, R. Kallenborn, J. de Boer, L.K. Sydnes, The Stockholm convention: a tool for the global regulation of persistent organic pollutants, Chem. Int. 41 (2) (2019) 4–11.

9

Decentralizing energy systems: Political power and shifting power relations in energy ownership

Marie Claire Brisbois

Science Policy Research Unit, University of Sussex, Brighton, United Kingdom

1 Introduction

Energy systems are in the midst of a rapid, yet uncertain, transformation. Driven by the overarching need to decarbonize, rapid technological development is creating a web of previously impossible energy system pathways. Many of these pathways depend on the evolution of decentralized generation, storage, and use of energy by new actors: cities, community organizations, cooperatives, private individuals, and others. Most modern Western energy systems, and their associated technical, institutional, social, and political arrangements, are built around, and often locked into, centralized energy systems [1]. This means that emergent changes in energy systems will likely have profound implications for social, political, and economic life as we know it.

Political power, the ability to have your interests realized in political settings, arises from diverse sources. One of the most enduring sources, related to economic prioritization of profit and economic growth, is market share [2]. In sectors such as energy that provide essential services, control over market share is of interest to political decision-makers not just because profits contribute to growth, but also because elections can be won or lost on the provision of these services. If a sitting government fails to ensure that enough of the electorate is able to turn on the lights, or reliably travel to work, there are likely to be consequences at the ballot box. This means that energy companies in centralized systems tend to exercise considerable political power.

As global energy systems transform, emerging actors, many with motivations beyond corporate profit maximization and economic growth, are beginning to control some aspects of the energy system. This is particularly apparent in the Western world where these changes are making it increasingly possible to reshape who is sitting at political decision-making tables [3]. These changes also mean that actors with very different motivations than status quo profit maximization can increasingly make their voices heard. This chapter focuses on dynamics in industrialized democracies (i.e., the global North) and examines how these new actors are entering the energy sector, the ways in which energy interests are heard, and what changes in system structure and ownership could mean for energy and environmental policy.

2 Decarbonization and decentralization of electricity

Electricity is a central pillar of energy systems (e.g., for lights, appliances, computers). However, as heating, cooling, and transportation systems electrify, this is becoming ever more important [4]. This means that those who control the production and distribution of electricity are poised to become even more important as we act to address climate change and decarbonize energy systems.

Most Western economies have evolved to depend on centralized generation for electricity [5]. This usually comes from a handful of large, centralized sources: coal or gas generation plants, nuclear reactors, hydroelectric dams, and, increasingly, wind and solar farms. These installations generate large amounts of electricity that are fed out through transmission and distribution lines (i.e., the electricity grid). Most countries in the West (and some in the Global South) have undergone processes of economic liberalization. This means that these installations are usually, although not always, operated for profit and owned and controlled by a relatively small number of corporate interests. The grid also lends itself to monopoly ownership and operation and is again often, although not always, operated for profit [6].

The development of renewables has allowed a shift to higher proportions of clean energy generation in electricity systems in countries around the world [5]. While many new companies have formed around large-scale wind and solar installations, these new interests usually operate according to the same market and centralized system logics that characterize the old system: large scale, centralized generation with distribution through a centrally operated grid [7]. While this is feasible and fits with old ways of knowing and doing it is becoming increasingly clear that there are other possible system architectures.

Advancements in renewables have opened up possibilities for a major shift in both where electricity generation is done, and by whom. As prices for renewable technologies have dropped, and efficiencies improved [4,8], it has become both practical and feasible for a raft of new actors from cities, to community organizations, to cooperatives, to private individuals, to become generators and storers of electricity, rather than just consumers. Physically, this will look very different from the status quo. In addition, dividing market shares amongst what could feasibly become millions of actors complicates what used to be a relatively locked-in centralized political and economic situation. This has significant political implications.

3 Political power in electricity systems

There are a few typical ways that political power in electricity systems manifests. By examining these, it is possible to see how the introduction of new actors into electricity systems can impact energy and related decisions. In some Western nations (e.g., the Netherlands, Germany, Australia) these outcomes are already emerging [9]. However, in most places, the transition to a significant proportion of decentralized generation is still nascent [8].

Research on political power directs attention to some of the key activities and conditions that allow different actor groups to pursue their interests. These can be grouped into overlapping categories including actions (or non-actions) and structures that coerce those who would otherwise act differently, allow the effective exercise of influence, or shape the ideas and ideologies defining what is viewed as necessary or desirable [10]. Table 9.1 provides an overview of how these dynamics are manifesting in current energy-focused political systems, with further discussion of each dynamic below. The first column describes the dynamic. The

TABLE 9.1 Key political power dynamics in the electricity sector.

Dynamic		Incumbent political power	Emergent political power
Political lobbying and elite access	Source(s)	Oil and gas companies currently provide significant amounts of electricity generation	Government support (on occasion) for lobbying, emergent capacity development, social legitimacy
	Example	Oil and gas lobbying against electrification and decentralization of electricity systems through institutionalized channels and personal connections (Ref. [11] on Ontario, see Ref. [12] on United Kingdom, Ref. [13] on Germany)	Community energy groups lobbying to win market access and ownership rights in national climate and energy policy (see Ref. [14] on the Netherlands)
Control over relevant technical information and knowledge	Source(s)	Incumbent knowledge of the technical capacity and operations of the grid	Increased experience, ability to engage in collaborative network building enhanced by high social legitimacy
	Example	Grid companies using their position, knowledge and expertise to shape regulation in their favor (Ref. [15] on the United Kingdom, see Ref. [16] on Japan)	Increased technical capacity of new energy actors, or ability to bring in external expertise, to interpret and engage with technical information (see Ref. [17] on Ontario, Ref. [18] on Denmark)
Control over resources and jobs	Source(s)	Coal plants provide employment and domestic energy supplies	Increasing control over installed electricity generation capacity and jobs
	Example	Coal industry unions mobilizing against renewables and demanding continued use of coal (see Ref. [19] on Poland)	Increased inclusion of the community energy sector in government positions and in energy sector summits due to their importance for energy system operation (see Ref. [13] on Germany)
Media or social influence	Source(s)	Financial resources and existing public relations capabilities	Normative advantages stemming from a commitment to local economic development, community building, and, often, justice and climate action
	Example	Incumbent centralized utilities buying advertising to discredit rooftop solar (see Ref. [20] on the United States)	Ability to attract participation in, and support for, community energy through normative benefits (see Ref. [21] on New Zealand, Ref. [22] on The Netherlands)

I. Imaginaries

second and third columns provide examples of the sources and exercise of political power by incumbents and new actors, respectively.

3.1 Lobbying and elite access

Lobbying is one key coercive way that political power is shaped. Its effectiveness usually depends on access to both financial resources and decision-makers [2]. While any actor can lobby, most bigger companies pay professional lobbyists to pursue their interests. Lobbying encompasses all kinds of activities including meeting with politicians and bureaucrats, participating in official consultation processes, attending conferences and meetings, putting together information resources that support a specific policy position (e.g., on why a particular technology or system configuration makes the most sense), and forming political coalitions with like-minded companies or interests [23]. Because entrenched system actors are in regular contact with regulators and decision-makers, it is often also easier for them to have informal lobbying conversations (e.g., having direct phone access to key government or regulatory officials).

As new energy actors gain importance in the electricity sector, they increase the effectiveness of their lobbying. With increased capacity, they move beyond lobbying through submissions to official consultation platforms or seeking out impromptu conversations with decision-makers at industry conferences. Depending on the degree of market penetration, they may be invited (or, more frequently at this stage, demand) to feed into policy or regulatory development processes, or be invited to contribute to problem-solving or future systems planning [9]. These are opportunities for input that are regularly offered to centralized system actors [15]. This therefore offers the opportunity for new ideas and priorities to be introduced into political and regulatory decisions.

3.2 Information control

Control over relevant data and information also confers influence. Energy systems, and associated markets, require specific technical knowledge to ensure safe, efficient, and effective functioning (e.g., how much intermittent electricity can be accommodated on the grid). This means that decision-makers and regulators usually form close working relationships with energy companies, and these companies help to make the rules that govern their own operations. This allows for more efficient regulation and decision-making, and also puts these companies in a privileged position [15,24].

Technical capacity is another area where new energy actors are gaining expertise. The accessibility of renewables infrastructure (e.g., solar panels) means that many more people have the incentive to learn how electrical systems work. In many industrialized democracies, notably in the United States and Canada, companies used to the centralized system (i.e., generators, grid operators) sometimes try to control access to data by claiming that it is proprietary (e.g., Ref. [17]). Open data legislation, particularly in Europe, is becoming more common and requires disclosure of data on generation, loads, balancing, and transmission [25]. This lessens the political influence associated with information control. It also creates the conditions for a more efficient grid because people are able to collectively solve problems and innovate to solve system challenges.

3.3 Employment

The ability to provide stable, well-paying jobs also contributes to the ability to exert political influence. Political decision-makers usually hesitate to make policy decisions that will cause job losses [26], even when it is clear that industries must be phased out (e.g., coal) (e.g., Ref. [19]). Threats of industrial flight (i.e., moving to another jurisdiction), or job cuts, can be used to seek concessions from decision-makers. This confers structural political power on those who control jobs.

Employment issues are also complicated by the fact that, in many countries, labor unions have evolved to fight to ensure good working conditions in sectors associated with the centralized energy system. These unions negotiate directly with big companies and often have a great deal of influence with government decision-makers. While unions can secure better working conditions for workers, there are many examples where they have actively opposed transitions to renewable or decentralized energy [27].

Union opposition to renewables is compounded by the fact that workers in the renewables sector are often not unionized [28]. While distributed ownership and geography mean that the economic distribution of gains, and the physical distribution of jobs, are diffused, it also means that there is no single job site or employer around which workers can organize. Renewable jobs have therefore tended to be more precarious, lower paid, and lacking in worker's protections than jobs in sectors where there are well-established unions (e.g., oil and gas, nuclear) [28]. It is therefore unsurprising that workers use their political clout to ensure the maintenance of their industry, despite overarching needs to transition to different energy sources. However, there are emerging union movements in the renewables sector, as well as existing unions that do use their political power to lobby for transitions (e.g., Ref. [29]).

3.4 Media and social influence

Beyond direct political influence, actors with resources can influence political decision-making in more subtle ways. By shaping broader public perceptions of certain technologies, system configurations or policy proposals, it is possible to make it easier for decision-makers to support the interests of actors with these capacities [30]. For example, at key points in decision-making cycles in Ontario, Canada, the union that represents many centralized nuclear and oil and gas workers has taken out expensive ads in major newspapers arguing that renewables and decentralized generation are risky and unreliable [11]. This framing is intended to help shape public opinion so that there is less support for exploring new technologies and system architectures. Similarly, Shell's 2013[a] energy futures scenarios framed a decentralized world as politically unstable. These techniques help to shape public opinion regarding what is possible and desirable.

Many new, small-scale electricity actors do not have the resources required to run entire scenario planning exercises, produce television advertising, or take out full-page newspaper ads [9,17]. However, these actors often do have a significant amount of inherent social legitimacy in the eyes of the broader public. The prioritization of values beyond profit

[a] https://www.shell.com/energy-and-innovation/the-energy-future/scenarios.html.

maximization means that these actors can act very differently than traditional companies. While big companies often take discrete actions to balance their corporate social responsibility with profit-making, these smaller actors tend to actually be socially responsible. For example, local energy activities undertaken by cities or local authorities are often targeted at lower-income communities [31].

The local economic and social benefits attributed to community energy mean that these initiatives have significant normative legitimacy [32]. Indeed, big companies in the United Kingdom and Germany have developed highly publicized centrally-owned but geographically diverse "local" projects as a way to appeal to this public sentiment [33]. To address this, in order for an entity to be classified as a "Renewable Energy Community" and afforded market protections under the new EU Renewable Energy Directive (RED II), profit cannot take precedence over local social, environmental, and economic benefits, and governing control must be open and participatory [34]. This limits the extent to which incumbent actors can co-opt the values associated with small-scale energy in order to increase their legitimacy and social influence.

4 The state of shifting power

Nascent shifts in political power away from incumbent companies and toward nontraditional energy actors are visible in many, but certainly not all, industrialized democracies. In countries or regions that have specifically supported decentralized energy sectors, changes in the actors helping to shape political decisions are becoming visible. This is evident, for example, in the Netherlands and Germany [13,14]. Places like these have implemented policy instruments like feed-in tariffs or metering schemes that provide financial incentives for the uptake of distributed generation infrastructure by broad populations (e.g., rooftop solar panels). On the other hand, in places where energy systems are tightly locked in, incumbent centralized generators can apply pressure to limit the implementation or effectiveness of financial support schemes. This has happened, for example, in the parts of Canada, the United Kingdom, and the United States [9,35,36]. This makes it difficult for decentralized interests to gain the foothold needed to begin shifting some of the levers of political power described earlier.

There are also interesting dynamics emerging in places where cities are beginning to undertake activities related to energy or electricity. For example, Silvestri et al. [37], list examples of energy-related activities from Bristol in the United Kingdom, Ghent in Belgium, and several other Western cities. While cities are not always in regular communication with energy regulators, some do have associated utilities (i.e., companies that manage the actual electricity wires and flow through them) and associated relationships. At the very least, cities are generally used to lobbying higher jurisdictions to express their policy preferences. They, therefore, have more capacity in this area than other new energy actors. In places where cities are taking a proactive role in energy and electricity, and especially where cities partner with other new energy actors in expressing political preferences, there is greater potential for the successful exertion of political power at higher levels [38].

Some countries have also supported the political organization of new energy actors. This, for example, was the case in both the Netherlands and the United Kingdom, where the government supported the development and institutionalization of community energy lobby bodies [9]. New, decentralized electricity actors are, by definition, diffuse. This can make it hard for decision-makers to collect a coherent picture of the preferences of the sector. These dynamics can emerge even when those policymakers are generally supportive of an evolving energy system. By supporting the development of a specific lobby group, more progressive countries have made it easier for these interests to have their voices heard and accounted for.

5 Decentralization and democracy

A shift in political power away from interests that prioritize profit over social and environmental goals is potentially transformative. Democracies function as elected officials act in the interpreted interests of the people. However, it is increasingly clear that interpreting the interests of the people in purely economic terms has created an untenable social and environmental situation [39]. Most energy companies are legally bound to strive to make a profit. Their viability is usually determined based on the size of their economic returns. Even should they wish to make decisions that are environmentally sensible but economically tenuous, the business model they are locked into makes this impossible. As long as this is the case, and as long as these voices are loudest in the energy policy arena, we can expect policies that prioritize profits over other societal needs.

Within the democratic models established in most countries, gaining political power will depend on gaining at least some control over the levers, like those described earlier, that decision-makers must consider in order to ensure that society functions in a socially accepted manner. In the context of energy, this generally means keeping the lights on and keeping people employed. If there is a shift in control over those levers to new energy actors who lobby not just for favorable market conditions, but for market conditions that actively support social and environmental outcomes with equal strength, then there is the potential that we will begin to see different kinds of policies, driven by different priorities.

Traditional oil and gas companies have often lobbied aggressively and successfully to delay or diminish renewable energy and climate policy proposals [40]. Emerging energy actors like community, civic, and city-focused interests are much more likely to proactively support progressive social and environmental policies [32]. While there is evidence that many emerging energy actors are themselves economically and socially privileged (and often white), this is not an inevitability [41]. Progressive policy can be designed to ensure that access to energy ownership, as it becomes an increasingly viable option, is available to everyone.

6 Conclusion: What does this mean going forward?

The trajectory of the evolving energy transition is highly uncertain. For the foreseeable future, most jurisdictions will likely remain characterized by a jumble of centralized and decentralized infrastructures. Not every citizen wants to be an electricity producer, nor is decentralized energy always the most appropriate option. In some cases, it may make sense to wait to retire expensive centralized assets until the end of their lifespan (e.g., nuclear facilities). Some centralized resources may also be deemed appropriate for a low-carbon energy future. Hydroelectric dams, although associated with other environmental and social impacts, produce carbon-free energy and can be useful for storing electricity for later use. The best mix for any given area is likely to be highly contextual, depending on local resources and capacities [42].

While energy systems are slowly decentralizing, it is unclear what portion of decentralized infrastructure will be decentrally owned. It is entirely possible for large companies to own and operate infrastructure like rooftop solar panels in decentralized locations, paying roof owners a small fee to rent the space. This would still redistribute profits to some degree. However, it would be unlikely to create the broader shifts in control over political levers and the associated shifts in political power and democratic outcomes that accompany that control [33,43].

Despite an uncertain path forward, the decentralization of electricity systems has the potential to open a door to wider political and democratic transformation. Control over energy systems has historically translated into significant control over political systems [44]. If meaningful shifts in at least some fraction of electricity ownership from centralized to decentralized actors occur, this will change the range of voices that can exercise influence on political decisions on energy and related issues. This chapter has highlighted some of the ways that these new voices are engaging in political conversations: through their ability to lobby and access political elites; their ability to engage with relevant knowledge and technical information systems; their increased control over electricity systems and associated employment opportunities; and, their ability to engender public support through media and social influence. If widespread decentralization of ownership occurs, new voices could very well act in concert to hasten the adoption of policies that support bolder climate action, are more just, and prioritize social and environmental returns at least as highly as the return of profits to shareholders.

These outcomes remain uncertain not just because of structural patterns of incumbency. Meaningful decentralization of ownership will bring many new voices forward. From a democratic perspective, this is a net positive. Yet there remains no guarantee that all new energy actors will actually want to engage politically, or that they will indeed be focused on objectives outside of economic returns. Indeed, Judson et al. [43] find that new decentralized actors still tend to be conceptualized in economic terms. Bauwens [45] finds that many people are indeed motivated to participate in community energy by material incentives. However, there are ever-increasing examples of new energy actors engaging in political struggle (e.g., Ref. [46]). Regardless of the extent to which new energy actors effectively advocate for more progressive policy outcomes, there is a net democratic benefit in seeing a decrease in political control by those centralized actors who have thus far failed to meaningfully address current social and environmental crises.

There is certainly public support and ecological imperative for a new way of realizing energy systems and associated social and political infrastructures. A simmering political struggle is currently underway as new energy actors attempt to assert and insert themselves into previously locked-in systems. Achieving wider changes will depend on broad political organizing at multiple sites and scales around the globe. However, changes are happening within the Western world as evidenced by the new EU RED II regulations (themselves sites of struggle), and changing energy dynamics in multiple jurisdictions. Whether or not these struggles presage an overhaul of the priorities driving energy and environmental policy remains to be seen. However, the sociotechnical preconditions for these changes are in place.

References

[1] M.B. Lindberg, J. Markard, A.D. Andersen, Policies, actors and sustainability transition pathways: a study of the EU's energy policy mix, Res. Policy 48 (2019), 103668.

[2] D.A. Fuchs, Business Power in Global Governance, Lynne Rienner, Boulder, CO, 2007.

[3] M.J. Burke, J.C. Stephens, Political power and renewable energy futures: a critical review, Energy Res. Soc. Sci. (2017), https://doi.org/10.1016/j.erss.2017.10.018.

[4] International Energy Agency, World Energy Outlook 2020, 2020, Paris https://www.iea.org/reports/world-energy-outlook-2020.

[5] International Energy Agency, Power Systems in Transition, 2020. https://www.iea.org/reports/power-systems-in-transition.

[6] P. Joskow, Introduction to electricity sector liberalization: lessons learned from cross-country studies, in: Electricity Market Reform: An International Perspective, Elsevier Ltd, Oxford, 2006, pp. 1–32.

[7] N. Kelsey, J. Meckling, Who wins in renewable energy? Evidence from Europe and the United States, Energy Res. Soc. Sci. 37 (2018) 65–73.

[8] C. Burger, A. Froggatt, C. Mitchell, J. Weinmann, Decentralised Energy: A Global Game Changer, Ubiquity Press, London, 2020, pp. 1–19.

[9] M.C. Brisbois, Rescaling, reorganization and battles for influence: shifting political power in an era of electricity decentralization, Environ. Innov. Soc. Transit. 36 (2020) 49–69, https://doi.org/10.1016/j.eist.2020.04.007.

[10] S. Lukes, Power: A Radical View, Palgrave Macmillan, Hampshire, New York, 2005.

[11] D. Rosenbloom, H. Berton, J. Meadowcroft, Framing the sun: a discursive approach to understanding multi-dimensional interactions within socio-technical transitions through the case of solar electricity in Ontario, Canada, Res. Policy 45 (2016) 1275–1290.

[12] R. Lowes, B. Woodman, J. Speirs, Heating in Great Britain: an incumbent discourse coalition resists an electrifying future, Environ. Innov. Soc. Transit. 37 (2020) 1–17, https://doi.org/10.1016/j.eist.2020.07.007.

[13] C.H. Stefes, Opposing energy transitions: modeling the contested nature of energy transitions in the electricity sector, Rev. Policy Res. 37 (2020) 292–312.

[14] M.C. Brisbois, Shifting political power in an era of electricity decentralization: rescaling, reorganization and battles for influence, Environ. Innov. Soc. Transit. 36 (2020) 49–69.

[15] M. Lockwood, C. Mitchell, R. Hoggett, Governance of industry rules and energy system innovation: the case of codes in Great Britain, Util. Policy 47 (2017) 41–49.

[16] C. Hager, N. Hamagami, Local renewable energy initiatives in Germany and Japan in a changing national policy environment, Rev. Policy Res. 37 (2020) 386–411, https://doi.org/10.1111/ropr.12372.

[17] M.C. Brisbois, Powershifts: a framework for assessing the growing impact of decentralized ownership of energy transitions on political decision-making, Energy Res. Soc. Sci. 50 (2019) 151–161, https://doi.org/10.1016/j.erss.2018.12.003.

[18] H.-J. Kooij, M. Oteman, S. Veenman, K. Sperling, D. Magnusson, J. Palm, F. Hvelplund, Between grassroots and treetops: community power and institutional dependence in the renewable energy sector in Denmark, Sweden and the Netherlands, Energy Res. Soc. Sci. 37 (2018) 52–64, https://doi.org/10.1016/j.erss.2017.09.019.

[19] H. Brauers, P.-Y. Oei, The political economy of coal in Poland: drivers and barriers for a shift away from fossil fuels, Energy Policy 144 (2020), https://doi.org/10.1016/j.enpol.2020.111621, 111621.

[20] A. Peskoe, Unjust, unreasonable, and unduly discriminatory: electric utility rates and the campaign against rooftop solar, Tex. J. Oil Gas Energy L. 11 (2016) 211.

[21] A.L. Berka, J.L. MacArthur, C. Gonnelli, Explaining inclusivity in energy transitions: local and community energy in Aotearoa New Zealand, Environ. Innov. Soc. Transit. 34 (2020) 165–182, https://doi.org/10.1016/j.eist.2020.01.006.

[22] D. Sloot, L. Jans, L. Steg, In it for the money, the environment, or the community? Motives for being involved in community energy initiatives, Glob. Environ. Change 57 (2019), https://doi.org/10.1016/j.gloenvcha.2019.101936, 101936.

[23] A. Dür, D. De Bièvre, The question of interest group influence, J. Public Policy 27 (1) (2007) 1–12.

[24] R. Astoria, On the radicality of New York's Reforming the Energy Vision, Electr. J. 30 (2017) 54–58, https://doi.org/10.1016/j.tej.2017.04.018.

[25] European Network of Transmission System Operators, ENTSO-E Transparency Platform, 2021.

[26] C. Lindblom, Politics and Markets, Basic Books, New York, 1977.

[27] A. Thomas, N. Doerflinger, Trade union strategies on climate change mitigation: between opposition, hedging and support, Eur. J. Ind. Relat. 26 (2020) 383–399.

[28] D. Stevis, Labour Unions and Green Transitions in the USA: Contestations and Explanations, 2019, Toronto https://adaptingcanadianwork.ca/wp-content/uploads/2019/02/108_Stevis-Dimitris_Labor-Unions-and-Green-Transitions-in-the-US.pdf.

[29] Prospect, GMB Union, Unison, UNITE, Demanding a Just Transition for Energy Workers, 2019. https://library.prospect.org.uk/download/2018/02124.

[30] R. Falkner, Business Power and Conflict in International Environmental Politics, Palgrave Macmillan, 2008.

[31] M. Lacey-Barnacle, C.M. Bird, Intermediating energy justice? The role of intermediaries in the civic energy sector in a time of austerity, Appl. Energy 226 (2018) 71–81.

[32] M.J. Burke, J.C. Stephens, Energy democracy: goals and policy instruments for sociotechnical transitions, Energy Res. Soc. Sci. 33 (2017) 35–48.

[33] P. Devine-Wright, Community versus local energy in a context of climate emergency, Nat. Energy 4 (2019) 894–896.

[34] European Parliament, Directive (EU) 2018/2001 of the European Parliament and of the Council of 11 December 2018 on the promotion of the use of energy from renewable sources, 2018.

[35] D. Lee, D.J. Hess, Incumbent resistance and the solar transition: changing opportunity structures and framing strategies, Environ. Innov. Soc. Transit. 33 (2019) 183–195, https://doi.org/10.1016/j.eist.2019.05.005.

[36] M. Lockwood, C. Mitchell, R. Hoggett, Incumbent lobbying as a barrier to forward-looking regulation: the case of demand-side response in the GB capacity market for electricity, Energy Policy 140 (2020), https://doi.org/10.1016/j.enpol.2020.111426, 111426.

[37] G. Silvestri, I. Sanchez Cecelia, T. de Geus, J.M. Wittmayer, Factsheets on innovative energy practices in light-house cities, 2019.

[38] M. Betsill, H. Bulkeley, Cities and Climate Change, Routledge, 2003.

[39] K. Raworth, A Doughnut for the Anthropocene: humanity's compass in the 21st century, Lancet Planet. Heal. 1 (2017) e48–e49.

[40] InfluenceMap, Big Oil's Real Agenda on Climate Change, 2019, London https://influencemap.org/report/How-Big-Oil-Continues-to-Oppose-the-Paris-Agreement-38212275958aa21196dae3b76220bddc.

[41] C. Fraune, Gender matters: women, renewable energy, and citizen participation in Germany, Energy Res. Soc. Sci. 7 (2015) 55–65.

[42] J. Lowitzsch, C.E. Hoicka, F.J. Van Tulder, Renewable energy communities under the 2019 European Clean Energy Package–Governance model for the energy clusters of the future? Renew. Sust. Energy Rev. 122 (2020), 109489.

[43] E. Judson, O. Fitch-Roy, T. Pownall, R. Bray, H. Poulter, I. Soutar, R. Lowes, P.M. Connor, J. Britton, B. Woodman, C. Mitchell, The centre cannot (always) hold: examining pathways towards energy system de-centralisation, Renew. Sust. Energy Rev. 118 (2020), https://doi.org/10.1016/j.rser.2019.109499, 109499.

[44] T. Mitchell, Carbon Democracy: Political Power in the Age of Oil, Verso Books, 2011.

[45] T. Bauwens, Explaining the diversity of motivations behind community renewable energy, Energy Policy 93 (2016) 278–290, https://doi.org/10.1016/j.enpol.2016.03.017.

[46] E. Greenberg, C. McKendry, Contested power: energy democracy and the repoliticization of electricity in the western US, Energy Res. Soc. Sci. 73 (2021), 101942.

10

Democratic divergence and the landscape of community solar in the United States

Thomas Ptak[a] and Steven M. Radil[b],*

[a]Department of Geography and Environmental Studies, Texas State University, San Marcos, TX, United States [b]Department of Economics and Geosciences, United States Air Force Academy, Colorado Springs, CO, United States

1 Introduction

Community solar installations are broadly understood as relatively small solar energy projects that encompass various forms of involvement by a community. It must be noted, however, the notion of community in relation to community solar has largely and problematically remained undefined [1]. Respective communities either own, partially own, purchase electricity from, or otherwise participate (in a range of ways) in the development of these solar installations [2–4]. As a consequence, several different typologies of community solar governance have developed in parallel. When combined with a spatially fragmented policy landscape, significant challenges for advancing truly democratic participation and processes in developing community solar projects are evident. In short, there is a growing but often overlooked democratic divergence between the rhetoric and reality of community solar.

Community solar has been promoted by solar developers, industry groups, and others as a framework designed to "increase access to solar energy and to reduce upfront costs for participants" [5] and one which "expands access to solar for all, including in particular low-to-moderate income customers" [6]. Community solar installations have experienced varying degrees of success in expanding access as they can—but do not always—allow customers to enjoy the advantages of solar energy without having to install their own solar system [7]. There is, however, little evidence supporting the industry's arguments that community

---

* The views expressed are those of the author(s) and do not reflect the official policy or position of the US Air Force, Department of Defense or the US Government.

solar development increases the participation of low-to-moderate income customers [1], or as this short essay will detail, advance democratic practices and processes involved in energy transitions.

By the end of 2019, 40 states had at least one community solar installation in operation, totaling 2056 Megawatts (MW) [6]. An additional 3400 MW are anticipated over the next 5 years. If this is realized, the United States will host approximately 5.5 Gigawatts (GW) of installed community solar installations by 2025. When considering that the first community solar installation began development in 2006 and by 2016 installed community solar capacity was only 275 MW, growth rates for this particular model of renewable energy have been explosive. However, one consequence of the rapid proliferation of has been the absence of a national policy framework specific to community solar [8]. Furthermore, while 16 states have specific policies related to community solar, vast differences are inherent in most aspects of the policies [9], including how to define and regulate participation for low-to-moderate income [10,11]. A nationally integrated policy framework would start to level the playing field, increase the potential for an equitable distribution of benefits, one which reaches the intended—and indeed named—constituents.

2 Defining community solar

There are three distinct typologies or models of community solar installations in the United States. The first is the utility-sponsored model. In this model, community solar installations are defined as projects that are owned and operated by a utility that is open to voluntary customer subscription [12, p. 6]. The second type is a special purpose entity, where the projects are owned and controlled by investors, who are often not individual people or entities but rather consortiums or subsidiaries of energy and/or solar industry firms [ibid]. The third type, called the non-profit model, covers projects owned by charitable non-profit organizations (commonly municipal or cooperative utilities) and are typically sited at a community-owned area [ibid]. Within these three typologies, there is significant variability in the way projects are developed and operated. As Feldman et al. [13] detailed, individual program designs often vary widely in features such as ownership, financing, or participation structures. As such, the details of any given community solar program are highly place-contingent given the fragmented energy policy and legal frameworks across the United States [14]. Even projects that are structured almost identically often yield vastly different participation rates depending on the local contexts in which they are deployed [13, p. 4].

3 Energy democracy and community solar

As described by Angel [15, p. 557], the concept of energy democracy is increasingly "used to frame struggles that seek to keep fossil fuels in the ground, while developing alternative ideas and practices of low-carbon energy provision." Szulecki and Overland [16] argue that energy democracy has quickly become a buzzword while asserting that it still lacks a widely agreed-upon definition or common framework. They posit, however, that energy democracy commonly manifests as a concern about who controls the means of energy production and consumption, while the literature is often developed based on case studies with a specific national or local community focus. Moreover, they argue that the concept of energy democracy can be understood through three distinct categories: energy democracy as a decision-making process, energy democracy as a possible outcome of decarbonization of energy systems, and energy democracy as a goal or objective [ibid, p. 2]. We draw on this typology as it provides a framework to evaluate both individual energy projects and/or specific models for the extent to which they advance democracy. Here we focus on the first issue, energy democracy as a decision-making process, to consider the potential for public participation across different community solar models.

There are a number of assumptions bound up in contemporary energy transitions [17]. One assumption that is especially pertinent for this discussion is the idea that community energy can advance participation in democratic practices or even lead to broader democratization of energy systems and governance. While there is evidence to suggest particular forms of democratization at distinct spatial scales can occur through community solar development [18], democracy is complex and contested, involving a range of often competing actors, and does not advance in a linear or deterministic fashion. Further, much of the political decision-making process around energy is dominated by the influence of technical experts rather than by the public at large [19], making it difficult for non-experts to participate, a condition often referred to as post-political. Post-political decision-making is widespread in many spheres of governance and is often highly undemocratic in practice. It is especially prevalent in deliberations about highly technical issues [20].

The degree to which different models of community solar provide genuine opportunities for stakeholder participation informs the challenges of energy governance as a decision-making process. For instance, one possible consequence of the differences between the types of community solar models and the fragmented policy landscape in the United States is a divergence in the degree of stakeholder participation and processes that could be considered democratic. We argue that the special purpose entity model provides the lowest scope for any form of democratic engagement by stakeholders. Special purpose entity models are commonly developed by corporate organizers who structure the project as a discrete business entity in order to take advantage of tax incentives or to fulfill energy mix mandates. As such, there is nothing remotely democratic about the development of these models.

The utility-sponsored model can provide limited opportunities to advance stakeholder participation and democratic processes. The degree to which this occurs, however, is largely determined by the nature of the utility. In the United States, many utilities are large "investor-owned" corporate entities. As such, stakeholder engagement is largely reduced to customers residing inside a utility service area paying a subscription fee for figurative access to electricity generated by a community solar installation. In this model, these customers do not have any ownership stake in the systems, and all means of energy production and consumption are controlled by the utility. As a consequence, customers have little to no choice in where solar installations are sited nor how they might be leveraged to provide benefits for a formal community. In a previous paper, we named this phenomenon *community washing* [1] to connote the disjuncture between the usage of the term community and the actual decision-making practices in this model. Hence, there is little evidence to suggest that a utility-sponsored project developed by a large, "investor-owned" utility will support democratic participation or processes. Utility-sponsored models that are developed by a public utility district, electric cooperative, or municipal utility may have a higher potential to be driven by stakeholders and advance democratic processes. Genuine stakeholder engagement and closer adherence to democratic practices and processes can result from the customer-owned nature and governance structure of these utilities [12].

Finally, the non-profit model offers the most potential to genuinely engage stakeholders and advance democratic participation. In this model, non-profit entities such as schools and churches often partner with local citizens to develop community solar installations. This model provides opportunities for genuine stakeholder engagement and deliberation, as the non-profit entity and local community members can engage democratic processes to decide what type of system is developed, where it is sited, and how benefits can be distributed equitably to a formal community—not individual customers in a utility service area, a consortium of investors, or subsidiaries of energy and/or solar industry firms.

The first community solar installation developed in the United States is in Ellensburg, Washington as part of the city's "Renewable Energy Park" [21]. Launched in 2006 and expanded in four phases through 2013, the installation embodies more robust notions of community participation. Creation of the project was only possible through participation between a range of stakeholders, including community members, the director of the town's municipal utility, a non-profit entity owned by the community, along with the towns chamber of commerce and city council [22]. This community solar project is viewed as a benchmark model for engendering participation, and the renewable energy park contains multiple solar arrays as well as a range of wind turbines, drawing visitors from K-12 schools along with faculty and students from Central Washington University. Ellensburg's project demonstrates how community solar projects can support both community and democratic forms of participation [23].

4 Conclusions

In this short essay, we detail how fundamental differences across community solar models combined with a fragmented policy landscape present challenges for advancing democratic participation and processes. We argue there are two forms of *democratic divergence* evident in the US community solar landscape. First, there is a divergent relationship between specific community solar models and the degree to which stakeholders can participate and engage in democratic processes through their development to access a range of benefits beyond a reduction in the cost of electricity. Consequently, if expanded access to the benefits of solar energy along with increased forms of democratic participation is to develop, there is an obvious need to rethink the special purpose entity model. Currently, the utility-sponsored and non-profit models stand apart from the special purpose entity in their ability to engender stakeholder participation while advancing democratic principles and processes.

Second, there is a divergence between broadly agreed-upon objectives of community solar development and those of energy democracy. Community solar has been posited as a strategy for increasing access for individuals or organizations who do not possess the space or capital necessary to develop solar energy installations. Although as we have noted, energy democracy still lacks a widely agreed-upon definition, it is largely concerned about who controls the means of energy production and consumption. Clearly, there is a divergence between goals of increasing access to a specific form of renewable energy and broader control over means of production and consumption.

Given these divergences and the spatially fragmented energy governance landscape in the United States, there is a need to develop a nationally integrated policy framework rather than the current state-based model. Doing so would be an important measure to support continuing community solar development which, in some forms, can create pathways for increasing stakeholder participation while challenging orthodox power structures. However, this rapidly emerging type of community energy must be understood as a point of departure along a longer path toward decarbonized energy systems, one that will continue to face significant challenges in implementation. Like all issues, energy democracy is a complex, contested, and dynamic concept and participation involves much more than paying a utility bill or even casting a ballot on an energy initiative. Energy democracy must be understood also as a process and set of practices that can be used to press for—while not guaranteeing—more equitable and just energy outcomes.

References

[1] T. Ptak, A. Nagel, S. Radil, D. Phayre, Rethinking community: analyzing the landscape of community solar through the community-place nexus, Electr. J. 31 (2018) 46–51.

[2] E. Artale, H. Dobos, Community solar presents rewards and risks, Nat. Gas Electr. 32 (4) (2015) 19–24.

[3] P. Augustine, The time is right for utilities to develop community shared solar programs, Electr. J. 28 (10) (2015) 107–108.

[4] M. Peters, S. Fudge, A. High-Pippert, V. Carragher, S. Hoffman, Community solar initiatives in the United States of America: comparisons with – and lessons for – the UK and other European Countries, Energy Policy 121 (2018) 355–364.

[5] Bonneville Environmental Foundation, The Northwest Community Solar Guide, 2013. https://sparknorthwest.org/wp-content/uploads/2013/05/NW-Community-Solar-Guide.pdf.

[6] Solar Energy Industry Association, Community Solar, 2020. https://www.seia.org/initiatives/community-solar.

[7] National Renewable Energy Laboratory, Community Solar, 2019. https://www.nrel.gov/state-local-tribal/community-solar.html.

[8] G. Michaud, Perspectives on community solar policy adoption across the United States, Renew. Energy Focus 33 (2020), https://doi.org/10.1016/j.ref.2020.01.001.

[9] T. Stanton, K. Kline, The Ecology of Community Solar Gardening: A 'Companion Planting' Guide, 2016. www.nrri.org.

[10] J. Cook, M. Shah, Focusing the Sun: State Considerations for Designing Community Solar Policy, National Renewable Energy Laboratory, 2018. www.nrel.gov/publications.

[11] Interstate Renewable Energy Council, Shared Renewable Energy for Low- to Moderate-Income Consumers: Policy Guidelines and Model Provisions, 2016. https://irecusa.org/.

[12] J. Coughlin, J. Grove, L. Irvin, J. Jacobs, S. Johnson Phillips, L. Moynihan, J. Wiedman, A Guide to Community Solar: Utility, Private, and Non-Profit Project Development, National Renewable Energy Laboratory, 2010. https://www.nrel.gov/docs/fy11osti/49930.pdf.

[13] D. Feldman, A. Brockway, E. Ulrich, R. Margolis, Shared Solar: Current Landscape, Market Potential, and the Impact of Federal Securities Regulation, National Renewable Energy Laboratory, 2015. http://www.nrel.gov/docs/fy15osti/63892.pdf.

[14] B. Sovacool, R. Sidorstov, Energy governance in the United States, in: The Handbook of Global Energy Policy, John Wiley & Sons, UK, 2013, pp. 435–456.

[15] J. Angel, Towards an energy politics in-against-and-beyond the state: Berlin's struggle for energy democracy, Antipode 49 (3) (2017) 557–576.

[16] K. Szulecki, I. Overland, Energy democracy as a process, an outcome and a goal: a conceptual review, Energy Res. Soc. Sci. 69 (2020), https://doi.org/10.1016/j.erss.2020.101768.

[17] T. Trainer, Can renewables etc. Solve the greenhouse problem? The negative case, Energy Policy 38 (2010) 4107–4144.
[18] S. Becker, C. Kunze, Transcending community energy: collective and politically motivated projects in renewable energy (CPE) across Europe, People Place Policy 8 (3) (2014) 180–191.
[19] K. Szulecki, Conceptualizing energy democracy, Environ. Polit. 27 (1) (2018) 21–41.
[20] S.M. Radil, M.B. Anderson, Rethinking PGIS: participatory or (post) political GIS? Prog. Hum. Geogr. 43 (2) (2019) 195–213.
[21] City of Ellensburg, Renewable Energy, 2020. https://ci.ellensburg.wa.us/1031/Renewable-Energy.
[22] G. Rissman, Shared solar, in: C. Krosinsky, T. Cort (Eds.), Sustainable Innovation and Impact, Routledge, London, 2018, https://doi.org/10.4324/9781351174824.
[23] J. Farrell, Community Solar Power Obstacles and Opportunities, Institute for Local Self-Reliance, 2011. https://www.ilsr.org/wp-content/uploads/files/commsolarpower3.pdf.

C H A P T E R

11

The emerging energy future(s) of renewable power and electrochemistry: Advancing or undermining energy democracy?

Christian Brannstrom

Department of Geography, Texas A&M University, College Station, TX, United States

1 Reading energy futures through electrochemical interventions

Imagine this energy future: wind turbines and photovoltaic solar arrays power electrolyzers that split water into hydrogen and oxygen to produce green hydrogen or industrial chemicals, which could be commercialized using infrastructure adapted from the fossil fuel industry. As Lourdes Alonso-Serna and Edgar Talledos-Sánchez argue in their chapter (Part III, Theme 1), renewables in southern Mexico have become "fossilized" owing to their power supply to large industrial and commercial end users. Electrochemistry provides another route for "fossilization" as excess wind and solar power would produce renewable fuels such as green hydrogen. Investment capital for the electrochemical path to decarbonization could originate from the oil and gas firms that for decades plundered the "subterranean forest" [1]. In this view, "bottling renewables" [2] means that the renewable power sources that have struggled to gain entry into power grids will create the chemicals and fuels to continue an affluent existence for some of the Earth's human inhabitants and offer governments and industries a pathway to deep decarbonization. Wind and solar farms could become export factories that supply renewable fuels to countries trying to meet their goals for reducing carbon emissions and drive major changes to landscapes and sociospatial relations in sites of power production.

Should we be invigorated or appalled, inspired, or frightened, by this possible energy future? Predictably, the technical-scientific work in this field is highly optimistic, offering electrochemistry (and especially hydrogen) as the only sure path to deep decarbonization. Critical analysis, however, is lacking. Electrochemistry has not received similar critical scrutiny as the sustainability challenges of metals needed for decarbonization [3] or climate engineering [4,5].

This chapter first offers brief and non-technical definitions of electrochemistry interventions, then synthesizes three core aspects relevant to energy democracy. Next, the chapter evaluates four emerging concerns and concludes with observations on future work, which should include rigorous application of ethical and theoretical frameworks to evaluate this rapidly evolving technology that promises to have deep effects on energy democracies. This chapter does not aim to offer a comprehensive review of the literature or offers a particular ethical or theoretical framework; rather, it aims to outline some opportunities for dialogue among humanists, social scientists, scientists, and engineers on this emerging energy future. Recent work on hydrogen into the home [6] and geopolitical implications of the hydrogen trade [7] is indicative of emerging social science contributions that analyze electrochemistry futures.

2 Defining electrochemical interventions for decarbonization

Electrochemical interventions for decarbonization use hydro-, solar-, or wind-generated electricity to power water electrolysis (electrolyzers) that splits water into hydrogen and oxygen, providing "energy storage in chemical bonds" [8]. Reviews have described the different types of water electrolysis, which rely on different materials and configurations that have correspondingly different efficiencies, challenges, and costs [9,10]. According to one review, industrial scale-up of one electrolysis technology is "already happening at a rapid pace" [11].

Electrofuels or renewable fuels are liquid fuels produced from hydro, wind, or solar for the transportation sector. Electrochemistry powered by renewables may also produce green commodity chemicals. Power-to-gas (P2G) is the production of hydrogen (P2H), methane, and other complex hydrocarbons through electrolysis [12]. Solar or wind power could produce hydrogen and methane to power thermal electricity plants or to use in transport as a gas [13]. A "green, decarbonized industrial revolution" is possible through renewable hydrogen as an end-use fuel or "energy-carrying intermediate" [9]. Renewables supplying electrolyzers could produce ethylene oxide, heavily used in plastics manufacturing. This technology would "reduce net greenhouse gas emissions associated with plastics manufacture" [14, p. 1228].

Hydrogen (P2H) produced through electrolyzers from wind and solar power is critical to P2G and other electrofuels [15–17]. Methane would be produced from hydrogen reacting with carbon dioxide; dozens of plants in operation in Europe produce methane in this way through chemical and biological means [18]. Renewable power supply for this process would produce "green" methane. For example, in 2016, the Swiss firm Hitachi Zosen Inova purchased PTG technology known as ETOGAS that converts "volatile" renewable power into a synthetic natural gas [19]. A German steel manufacturer uses an electrolyzer connected to renewable power to produce "green hydrogen" that helps reduce carbon dioxide emissions [20]. Green hydrogen projects are also underway in Texas at a proposed 1200 MW Orange Power Station, where Mitsubishi turbines will burn green hydrogen mixed with natural gas, and in Utah, where hydrogen produced from electrolyzers powered from solar and wind will generate power in a 840 MW power plant [21]. Power-to-methane is a "promising option to absorb and exploit surplus renewable energy," especially in places with existing natural gas infrastructure; methane produced in this way would be used for power generation, heat generation, fuel, and raw material in industrial applications [22]. However, a critical view is that electrofuels "may contribute to a prolonged era of fossil fuels" if used as a "drop-in" alternative [23, p. 1902].

3 Core assumptions of electrochemical interventions for decarbonization

Arguments supporting electrochemical paths to decarbonization make several assumptions. One is that cheap or excess renewable power will exceed demand over extended periods. Scientists and engineers describe "waste," "curtailed," or "cheap" wind and solar [24]; "very cheap electricity" [25]; "curtailed, large-scale excess energy" from renewables [26, p. 170]; the "mismatch" between renewable power generation and electricity demand [22, p. 443]; the "increasing abundance of low-cost, renewable electrons" [9]; a need to store "excess energy" from renewables [10]; and "low-cost intermittent renewable energy" [27, p. 1]. The "precipitous decline in the cost of renewable power" will make renewable hydrogen "competitive" with industrial hydrogen by 2030 [28].

Evidence that wind and solar produce peaks in electricity supply that exceed load, resulting in "increasing amounts of cheap or even free excess energy," supports this view [13]. Electrochemistry is a way to solve the "variability (nondispatchability) challenge" that limits "widespread, terawatt-scale adoption of low-carbon energy sources" [8]. If wind power could produce hydrogen, it would "improve wind power investments" and "optimize the business case" for wind farms [15, p. 918].

A second assumption supporting electrochemical interventions is that battery technology is a less attractive way to store electricity. Authors emphasize the negative byproducts of battery manufacture and the narrow mineral base upon which batteries rely [15]. Batteries may only deliver short-term storage [29], providing storage "on the scale of hours or even days" [8]. By one estimate, the capacity equivalent of approximately 50 billion Tesla Model 3 batteries would be required [11]. Large-scale battery storage "is certainly not the solution to massive energy storage today" [30]. The temporal mismatch between renewable power production and electricity demand requires long-term storage options that batteries cannot provide [13]. Simulations from a 100% renewable California electricity grid show periods when "massive and seasonal energy storage is required," beyond the capacity of batteries. This situation favors the "power and energy capacity of hydrogen energy storage in current gas infrastructure (pipelines and storage facilities)," seen as "the only option that can technically balance renewable power and energy with the load on an annual basis" [26]. A prominent example is the widely reported practice of California paying Arizona and Nevada to take their excess solar power during certain times, described as "[dumping] its unused solar electricity" [31].

However, not all scientists and engineers see competition between electrolyzers and battery storage. A net zero-emission energy system has batteries, P2G, and P2H as complementary components [32]. The patented Renewstable system marketed by Hydrogène de France, scheduled to be deployed in French Guiana from a 50 MW solar farm [33] and in Barbados in 2022, uses electrolyzers to produce hydrogen for storage and includes battery arrays to store power generated by solar photovoltaic arrays.

The third claim embraces exportable green commodity chemicals, renewable fuels, and hydrogen as a way to avoid stranded assets in the form of oil and gas pipelines and storage capacity. P2H could replace diesel in commercial long-distance vehicles [34]. P2G is justified because "capacity of gas pipelines and gas storage is much higher than that of electricity transmission lines" [13, p. 1] and therefore compatible with oil and gas infrastructure. Renewable fuels "would allow for the re-use of existing trade infrastructure" [35, p. 2022]; moreover, countries or regions that produce power-to-fuel relatively cheaply should specialize to supply countries that face constraints, making "long-distance trade in renewable fuels economically viable" [35, p. 2027]. Wind farms in southern Argentina could produce Japan's hydrogen, linking "regions of renewable energy surplus" to "regions with high energy demands" [36]. The chemical industry could be revolutionized. Commodity chemicals, such as ethylene and ethanol, "when powered by renewable electricity… can be made with net negative carbon emissions footprint," therefore achieving a "carbon emissions-free means of chemical production," releasing the chemical industry from fossil fuels [8].

Several announcements during 2020 and early 2021 provided evidence that this third claim, exportable P2H branded as "green hydrogen," entered a new phase in terms of investments leading to commercial viability to support decarbonization:

- The hydrogen strategy released by the European Commission hydrogen strategy in July 2020 prioritized the development of renewable hydrogen from wind and solar. Green hydrogen is "the most compatible option with the EU's climate neutrality and zero pollution goal and most coherent with an integrated energy system" [37, p. 5].
- Former executives with Royal Dutch Shell and Exxon Mobil launched HydrogenOne Capital Ltd., a $315 million venture as the first hydrogen-dedicated investment firm [38].
- In Australia, Queensland's Stanwell Corp. announced that it will partner with Japan's Iwatami Corporation to develop a green hydrogen export facility powered by wind and solar farms [39].
- The $5 billion green hydrogen Helios Green Fuels project announced in Saudia Arabia includes the Saudi government, Acwa Power (partly owned by the Saudi sovereign wealth fund), and the Pennsylvania-based Air Products and Chemicals [40], while Thyssenkrup will build the 20 MW electrolysis plant that will produce hydrogen from wind and solar power to supply part of the NEOM project [41].
- Enegix Energy and Brazil's Ceará state government announced a $5.4 billion proposal for a green hydrogen facility powered by wind and solar farms and connected to the deepwater Pecém port facility. The Enegix Base One facility would be the largest in the world with 3.4 GW to be fed by ~8 GW of distributed wind and solar [42].

The Base One proposal offers opportunities to explore claims made in support of large investments in green hydrogen. Announced in February 2021, Base One would use Ceará wind farms to produce electricity for electrolyzers that would generate 600 million kg per year of hydrogen. Enegix made the export argument clearly, noting the proximity of Ceará to Europe and the desirability of the Pecém port next to Base One. Some of the wind power would originate in the proposed Caucaia offshore wind farm near Base One, discussed in Chapter 21 regarding Brazilian offshore wind farms, but other locations supplying wind and solar power were notably absent from Enegix and local media. Instead, local media emphasized that Base

One would create "hundreds of high-paying jobs" and quoted the Enegix CEO as promising job training and claiming that "we are creating rocket scientists in Ceará" [43]. Shortly after the investment was announced, the Ceará state government created a working group comprised of diverse state agencies (not civil society representatives) to guide policies for a green hydrogen hub. The decree described green hydrogen as a "vector that will permit the import of clean energy from regions favored by nature and with potential exceeding its needs," a reference to the excess renewable power idea noted earlier in this chapter. Referring to the "export factory" idea, the state's governor noted that the state "is in the vanguard...because we have favorable conditions to produce and export green hydrogen" [44].

4 Emerging energy futures

What sorts of energy futures might result from electrochemical interventions that would create fuels, hydrogen, ammonia, and other commodity chemicals from surplus wind and solar power? Geographical processes in low-carbon transitions include absolute and relative location, landscape, territoriality (political control over space), uneven development, scaling, and path dependency as critical variables [45]. Critical perspectives on capitalism's spatial fixes under climate change suggest that renewable power may offer attractive new sites for investment and accumulation, offering the means for capitalism to overcome the climate change crisis [46,47].

The electrochemistry literature reviewed here makes no mention of the territorial aspects of producing the assumed "excess" and "cheap" renewable power. Yet decarbonization through electrochemistry offers new possibilities for capital accumulation and territorial control, both of which are much more intensive than recent observations have suggested [46–48]. "Profound" changes to landscapes and sociospatial relations would occur if renewable energy regimes replaced the "subterranean energy regime" [47, p. 12], perhaps leading to "extractive peripheries" and "sacrifice zones" from renewable energy regimes that parallel fossil fuel regimes. One possibility is the perpetuation of the "decarbonization divide" and a new round of dispossession [49,50]. At a minimum, we should recognize that nearly all studies of host community acceptance and rejection assume that renewable power feeds regional or national grids. How will host communities respond when nearby wind and solar farms become export factories to support the carbon goals of affluent countries?

It is still unclear whether electrochemistry is a decarbonization solution that oil and gas will embrace, but opportunities for investment are enormous. Renewable electrochemical production could reduce "the financial burden of shutting down expensive existing assets" [8, p. 7]. Electrochemistry represents a deepening and intensifying of the renewable energy regime, saving us from burning through subterranean resources but intensifying the demand for wind and solar spaces. According to a recent report, the P2H industry is moving to mergers and acquisitions [51]. These investments raise the prospect that "centralized large-scale collective storage solutions and associated infrastructures, such as for hydrogen as a storage vector," may create "challenges for equity" [52].

This reading of electrochemical interventions suggests new forms of territorialization, or the spatial control of space [45]. International trade of electrofuels would follow principles of comparative advantage, represented visually as the flow of renewable fuels from South

America to Europe [35]. More critical observers will see new patterns of unequal exchange through renewable power. Instead of supplying regional or national power grids, wind and solar farms would export renewable power. This is destabilizing to our understanding of host community responses to wind and solar, which normally presumes this moral economy: wind and solar power may occupy large areas of land, but they supply regional or national grids. The emerging possibility of trade in renewable fuels and green hydrogen would dramatically increase the spatial distance between places of energy affluence and renewable power landscapes, contrary to the idea that renewable energy landscapes are morally good because they put consumers in close proximity to the power sources on which they depend [53]. International trade in renewable fuels or green hydrogen would upset this often unstated moral economy. Trade in hydrogen promises to "redraw the geography of global energy trade, create a new class of energy exporters, and reshape geopolitical relations and alliances between countries" [7, p. 1].

No less important are the numerous patents that will be required to protect winning intellectual property in the race to make electrolyzers economically viable. We do not yet know who might hold these patents, nor where they would be held; we also do not know the human capital requirements, nor the magnitude of the infrastructure that will sustain the electrolyzers. Would this technology could create demand for skilled human capital near solar and wind farms, benefiting places that host renewable power? The labor dimension is not yet well understood in the literature [54].

Do "renewables in a bottle" [2] threaten or undermine energy democracy? On the positive side, emerging work applies electrochemical interventions to solving problems of marginalized rural communities, suggesting that not all electrochemistry needs to be large scale and compatible with existing oil and gas infrastructure. For example, "distributed ammonia production" systems fed by renewable power could offer ammonia fertilizers to farmers in developing countries [18]. Other authors link PV with hydrogen storage, taking advantage of brackish groundwater, to supply power to an isolated rural area in Paraguay [55]. Promising results have been reported for a wind-to-hydrogen facility on the Orkney Islands of Scotland [56]. Perhaps wind and solar can supply national or sub-national electricity grids and also produce renewable hydrogen or other chemicals, enhancing electricity access for all, in new energy democracy.

But it is also possible that electrochemical interventions would lead to an energy future in which the affluent live far away from the wind turbines and solar PV arrays. This renewable power infrastructure, which covers increasingly large areas, would be connected to electrolyzers that produce exportable renewable fuels, hydrogen, and commodity chemicals for the benefit of the already affluent energy consumers, possibly to the detriment of places that host wind and solar. Markets for renewable hydrogen and chemicals produced through electrolyzers would make wind and solar lucrative mainstream investments, possibly allowing oil and gas firms to fully embrace renewables as a way to remain profitable in a carbon-constrained era, in a less tumultuous transition. A rather dystopian and undemocratic version of this decarbonized energy world would offer parallels to Earl Cook's description of the four worlds of energy [57, p. 257]: some countries host the firms that own the patents and employ the well-paid workforce that controls the electrochemical pathway to decarbonization, while other countries have the land and sea for wind and solar export factories that power electrolyzers producing "renewable" electrofuels and chemicals that maintain distant and affluent lifestyles. With wind and solar farms poised to become export factories critical to decarbonization, we need to renew our concern for the emerging justice issues that may guide just and equitable outcomes and processes for the people who are enmeshed in the landscapes and sociospatial relations that electrochemical interventions may inspire.

References

[1] R.P. Sieferle, The Subterranean Forest: Energy Systems and the Industrial Revolution, The White Horse Press, Cambridge, 2001.

[2] Nature Energy, Bottling renewables, Nat. Energy 4 (2019) 721.

[3] B.K. Sovacool, A. Hook, M. Martiskainen, L. Baker, The whole systems of energy injustice of four European low-carbon transitions, Glob. Environ. Chang. 58 (2019) 101958.

[4] R. Bellamy, A sociotechnical framework for governing climate engineering, Sci. Technol. Hum. Values 41 (2016) 135–162.

[5] J.A. Flegal, A.-M. Hubert, D.R. Morrow, J.B. Moreno-Cruz, Solar geoengineering: social science, legal, ethical, and economic frameworks, Annu. Rev. Environ. Resour. 44 (1) (2019) 399–423.

[6] M. Scott, G. Powells, Towards a new social science research agenda for hydrogen transitions: social practices, energy justice, and place attachment, Energy Res. Soc. Sci. 61 (2020) 101346.

[7] T. Van de Graaf, I. Overland, D. Scholten, K. Westphal, The new oil? The geopolitics and international governance of hydrogen, Energy Res. Soc. Sci. 70 (2020) 101667.

[8] C.H.P. Luna, et al., What would it take for renewably powered electrosynthesis to displace petrochemical processes? Science 364 (2019) eaav3506.

[9] K. Ayers, N. Danilovic, R. Ouimet, M. Carmo, B. Pivovar, M. Bornstein, Perspectives on low-temperature electrolysis and potential for renewable hydrogen at scale, Annu. Rev. Chem. Biomol. Eng. 10 (2019) 219–239.

[10] A. Buttler, H. Spliethoff, Current status of water electrolysis for energy storage, grid balancing and sector coupling via power-to-gas and power-to-liquids: a review, Renew. Sust. Energ. Rev. 82 (2018) 2440–2454.

[11] A. Hauch, et al., Recent advances in solid oxide cell technology for electrolysis, Science 270 (2020) eaba6118.

[12] N.M. Haegel, et al., Terawatt-scale photovoltaics: transform global energy, Science 364 (2019) 836–838.

[13] A. Ajanovic, R. Haas, On the long-term prospects of power-to-gas technologies, WIREs Energy Environ. 8 (2019) e318.

[14] W.R. Leow, et al., Chlorine-mediated selective electrosynthesis of ethylene and propylene oxides at high current density, Science 368 (2020) 1228–1233.

[15] D. Apostolou, P. Enevoldsen, The past, present and potential of hydrogen as a multifunctional storage application for wind power, Renew. Sust. Energ. Rev. 112 (2019) 917–929.

[16] F. Dawood, M. Anda, G.M. Shafiullah, Hydrogen production for energy: an overview, Int. J. Hydrog. Energy 45 (2020) 3847–3869.

[17] M. Götz, J. Lefebvre, F. Mörs, A.M. Koch, F. Graf, S. Bajohr, R. Reimert, T. Kolb, Renewable power-to-gas: a technological and economic review, Renew. Energy 85 (2016) 1371–1390.

[18] D. Hidalgo, J.M. Martín-Marroquín, Power-to-methane, coupling CO2 capture with fuel production: an overview, Renew. Sust. Energ. Rev. 132 (2020) 110057.

[19] Hitatchi Zosen Inova, Hitachi Zosen Inova acquires business of ETOGAS GmbH, Press Release (2016). 24 October.

[20] Salzgitter AG, GrinHy2.0: sunfire delivers the world's largest high-temperatur [sic] electrolyzer to Salzgitter Flachstahl, Press Release (2020). 25 August.

[21] S&P Global Market Intelligence, Hydrogen No Longer a Distant Mirage, 30 November 2020.

[22] K. Ghaib, F.-Z. Ben-Fares, Power-to-methane: a state-of-the-art review, Renew. Sust. Energ. Rev. 81 (2018) 433–446.

[23] S. Brynolf, M. Taljegard, M. Grahn, J. Hansson, Electrofuels for the transport sector: a review of production costs, Renew. Sust. Energ. Rev. 81 (2018) 1887–1905.

[24] Z. Yan, J.L. Hitt, J.A. Turner, T.E. Mallouk, Renewable electricity storage using electrolysis, Proc. Natl. Acad. Sci. 117 (2020) 12558–12563.

[25] W.C. Nadaleti, G.B. Santos, V.A. Lourenço, The potential and economic viability of hydrogen production from the use of hydroelectric and wind farms surplus energy in Brazil: a national and pioneering analysis, Int. J. Hydrog. Energy 45 (2020) 1373–1384.

[26] A. Saeedmanesh, M.A. MacKinnon, J. Brouwer, Hydrogen is essential for sustainability, Curr. Opin. Electrochem. 12 (2018) 166–181.

[27] M.J. Orella, S.M. Brown, M.E. Leonard, Y. Román-Leshkov, F.R. Brushett, A general technoeconomic model for evaluating emerging electrolytic processes, Energ. Technol. 8 (2020) 1900994.

[28] G. Glenk, S. Reichelstein, Economics of converting renewable power to hydrogen, Nat. Energy 4 (2019) 216–222.

[29] C. Smith, A.K. Hill, L. Torrente-Murciano, Current and future role of Haber–Bosch ammonia in a carbon-free energy landscape, Energy Environ. Sci. 13 (2020) 331–344.

[30] J.L. Aprea, J.C. Bolcich, The energy transition towards hydrogen utilization for green life and sustainable human development in Patagonia, Int. J. Hydrog. Energy 45 (2020) 25627–25645.

[31] I. Penn, California invested heavily in solar power, Los Angeles Times (2017). 22 June.

[32] S.J. Davis, et al., Net-zero emissions energy systems, Science 360 (2018) eaas9793.

[33] HDF, French Guiana region to install world's largest power station with 140 MWh renewable energy storage, Press Release, 28 May 2018.

[34] S. McDonagh, P. Deane, K. Rajendran, J.D. Murphy, Are electrofuels a sustainable transport fuel? Analysis of the effect of controls on carbon, curtailment, and cost of hydrogen, Appl. Energy 247 (2019) 716–730.

[35] J. Schmidt, K. Gruber, M. Klingler, C. Klöckl, L. Ramirez Camargo, P. Regner, O. Turkovska, S. Wehrle, E. Wetterlund, A new perspective on global renewable energy systems: why trade in energy carriers matters, Energy Environ. Sci. 12 (2019) 2022–2029.

[36] P.-M. Heuser, T. Grube, M. Robinius, D. Stolten, Techno-economic analysis of a potential energy trading link between Patagonia and Japan based on CO2 free hydrogen, Int. J. Hydrog. Energy 44 (2019) 12733–12747.

[37] European Commission, A Hydrogen Strategy for a Climate-Neutral Europe, Brussels, 2020.

[38] Reuters, Energy Industry Veterans to Launch Hydrogen Investment Fund, 9 July 2020.

[39] Stanwell Corporation, Central Queensland: A Future Hydrogen Export Powerhouse, 27 November 2020.

[40] Bloomberg, Saudia Arabia's Bold Plan to Rule the $700 Billion Hydrogen Market, 6 March 2021.

[41] Reuters, Germany to Contribute 1.5 Million Euros to the Thyssenkrupp's Saudi Hydrogen Plant, 15 December 2020.

[42] J. Ennes, Enegix Energy to build green hydrogen plant in Brazil, Power Finance & Risk, 2 March 2021.

[43] S. Quintela, A empresa australiana que investirá cerca US$5,4 bilhões no projeto espera ter o empreendimento 100% operacional em 2025, Diário do Nordeste (3 March 2021).

[44] J. Cruz, Governo do Ceará e instituições parceiras lançam HUB de Hidrogênio Verde, Fortaleza, 19 February 2021.

[45] G. Bridge, S. Bouzarovski, M. Bradshaw, N. Eyre, Geographies of energy transition: space, place and the low-carbon economy, Energy Policy 53 (2013) 331–340.

[46] J. McCarthy, A socioecological fix to capitalist crisis and climate change? The possibilities and limits of renewable energy, Environ. Plan. A 47 (2015) 2485–2502.

[47] M.T. Huber, J. McCarthy, Beyond the subterranean energy regime? Fuel, land use and the production of space, Trans. Inst. Br. Geogr. 42 (2017) 655–668.

[48] A. Scheidel, A.H. Sorman, Energy transitions and the global land rush: ultimate drivers and persistent consequences, Glob. Environ. Chang. 22 (2012) 588–595.

[49] B.K. Sovacool, Who are the victims of low-carbon transitions? Towards a political ecology of climate change mitigation, Energy Res. Soc. Sci. 73 (2021) 101916.

[50] T. Kramarz, S. Park, C. Johnson, Governing the dark side of renewable energy: a typology of global displacements, Energy Res. Soc. Sci. 74 (2021) 101902.

[51] S&P Global Market Intelligence, Hydrogen Finance Maturing From Day Trading to M&A, 3 December 2020.

[52] N. Eyre, S.J. Darby, P. Grünewald, E. McKenna, R. Ford, Reaching a 1.5°C target: socio-technical challenges for a rapid transition to low-carbon electricity systems, Phil. Trans. R. Soc. A 376 (2018). 37620160462.

[53] M.J. Pasqualetti, Morality, space, and the power of wind-energy landscapes, Geogr. Rev. 90 (2000) 381–394.

[54] R.P. Dicce, M.C. Ewers, Solar labor market transitions in the United Arab Emirates, Geoforum 124 (2021) 54–64, https://doi.org/10.1016/j.geoforum.2021.05.013.

[55] D.R.F. León, C.K.N. Cavaliero, E.P. Silva, Technical and economical design of PV system and hydrogen storage including a sodium hypochlorite plant in a small community: case of study of Paraguay, Int. J. Hydrog. Energy 45 (2020) 5474–5480.

[56] M. Westrom, Winds of change: legitimacy, withdrawal, and interdependency from a decentralized wind-to-hydrogen regime in Orkney, Scotland, Energy Res. Soc. Sci. 60 (2020) 101332.

[57] E. Cook, Man, Energy, Society, W. H. Freeman and Company, San Francisco, 1976.

12

The future of energy ownership

Clark A. Miller

School for the Future of Innovation in Society, Arizona State University, Tempe, AZ, United States

1 Introduction

Suppose for a moment that it is 2050. Suppose, too, as increasingly seems to be the expectation in the energy sector, roughly half of the world's total energy consumption is supplied by solar energy. That would require something like 100 TW of peak solar capacity [1], or 227 billion solar panels, roughly 29 panels per person on the planet.[a]

Who owns these 227 billion panels? It is literally a $100 trillion question, and one of the most consequential design choices confronting humanity in the coming energy transition [2,3]. At one extreme, solar ownership might be spread broadly across the world's nearly 8 billion people. At the other extreme, a small number of owners—whether public or private—could control most of the world's solar energy supply. In between, are hundreds of different ways of configuring answers to who will own the world's future solar panels.

At stake in the future of energy ownership, I propose in this chapter, is not merely a vast economic resource but the character of political economy and democracy in the 21st century and beyond. Energy is constitutional of social, political, and economic order [4]. As humans, the ways in which we work, move around, and exercise power are all tightly bound up with how we produce, organize, and use energy [5]. As such, energy transitions are never just shifts in technology or fuel source but always also transformations of society and markets [6].

The nature of these transformations is not predetermined, however. New technologies are interpretively flexible, capable of being integrated into human affairs in diverse ways [7]. The choice of which ways to design new technologies—and to design new forms of human life and orchestration around them—is thus a powerful moment of opportunity for social innovation [8]. As new technologies, industries, and organizations take shape, however, that opportunity closes and the forms of life and regimes of power they help constitute become much more obdurate and difficult to change [9].

---

[a] Assuming 440W Series 6 First Solar panels.

For the past two centuries, energy ownership has been the locus of some of the most complex and contentious political and economic fights, and the settlements achieved in those contests have created some of the most durable and powerful institutional, market, and technological arrangements in human history, including the oil and gas, coal, and automobile industries, and electric utilities [10]. Because solar energy, especially, is radically distinct from carbon-based energy, not only as a form of energy but also in its geographies, temporalities, technologies, businesses, and imaginaries, yesterday's energy arrangements are today poised for disruption, although disruption is by no means guaranteed [11].

At the crux of the question of disruption, I suggest, is the future of energy ownership. Energy, today, is narrowly owned. Tomorrow's energy systems may concentrate ownership even further. Or they may go in other directions. Which choices get made matter.

2 Why ownership matters to democracy

In his book *Carbon Democracy*, the historian Timothy Mitchell argues that 19th- and 20th-century democracies owe both their vibrancy and their limits to the regimes of ownership that were created and nurtured to unleash vast quantities of carbon-based energy for economic and societal use [12]. These regimes forged a stark division of ownership between technologies of energy generation and supply, on the one hand, which were centrally owned, and technologies of energy use, on the other hand, which were distributed. In both the electricity and fuel sectors, large, often quasi-monopolistic organizations mobilized efficient and reliable supplies of energy as the basis for industrial growth and, with that growth, prosperous societies that helped contribute to the success of democratic ideologies and institutions. At the same time, individuals, households, and businesses exercised extensive control over how they used energy to create the kinds of lives they wanted to live, e.g., through the ownership of automobiles, appliances, telephones, televisions, and computers, all of which helped fuel the imagination of liberty, embodying not only freedom from want but also freedoms of movement, identity, association, choice, information, and more.

These regimes are now in crisis, however [13]. At the center of the crisis is the need to end the massive burning of carbon-based energy that today fuels the global economy but also pours carbon dioxide emissions into the atmosphere, driving the Earth's climate system off the edge of a precipice and into potential destructive disaster for human civilization. That crisis is magnified by the extreme concentrations of wealth and inequality that have been created, among other drivers, by centralized energy systems, and by the corruption of politics and democracy that those centralized institutions have wrought in the service of perpetuating carbon-based forms of energy and the wealth they generate [14]. The very fact that these systems are now "too big to fail" complicates the prospect of energy reform—the reliable operation of energy systems must now be maintained to provide power and energy to forms of economy and society taken for granted by democratic publics—even as it makes clear that energy ownership is critical to the future of democracy.

To be sure, questions of energy ownership and its relationship to democracy are not new. Energy is already subject to diverse forms of ownership, including by individuals, not-for-profit collectives, private corporations, municipal governments, and national states

[15]. This diversity stretches across the entire energy supply chain, the result of a century of highly political debates in democratic societies over the status and control of energy, including, e.g., historical debates over public power vs private power, rural electrification, oil crises, and electricity markets, as well as more contemporary debates over renewable energy standards, net metering, distributed generation, and community solar [16].

What ultimately matters, however, is not just the diversity of forms of energy ownership but the distributional patterns of that ownership and the ways those patterns inflect, intersect, and integrate into larger social, economic, and political dynamics, geographies, and structures [17]. This is why, for many people involved in energy transitions, the prospect of reorganizing energy ownership has the potential to be revolutionary [18].

3 The future of solar ownership

If it were not for solar energy, it is not clear that there would be much of a debate about energy ownership in contemporary discussions of energy transitions. Solar energy has three key characteristics that mark it as highly distinct from past, carbon-based forms of energy and that deeply influence the possibilities of solar ownership.

The first is its ubiquity. Solar energy is everywhere, every day, in quantities sufficient to be meaningful [19]. Unlike carbon-based forms of energy, therefore, there is no need to find concentrated sources of solar energy and undertake extensive investments to locate, find, and extract the basic resource. It is all around us and potentially available to anyone with a claim on space.

The second is its flexibility. Solar energy has already proven capable of being effectively integrated into diverse human practices, technological arrangements, and business models that vary dramatically in scale, purpose, and organization [20]. Just in terms of scale, solar applications range over nine orders of magnitude, from a few Watts to a few Gigawatts. Solar is also already organized into diverse devices, into so-called solar home systems that power a few small devices, into microgrids, all the way up into national and even supranational electricity grids.

The third is its cost. Today, the cost of photovoltaic panels to transform sunlight into electricity has dropped to unimaginably low prices, and the cost for a full system is now price competitive with grid electricity at a wide variety of scales, in a wide variety of locations.[b] Expectations are that the competitiveness of solar energy will continue to expand in the future. McKinsey recently forecast, for example, that it will be cheaper in 2030 in the United Kingdom to build a new solar power plant, despite Britain's relatively high levels of cloudiness, than to continue to operate, maintain, and fuel a natural gas power plant that was already built [21].

Together, these three characteristics combine to fire diverse solar imaginaries [22]: myriad visionary ideas about how to configure the future of solar energy technologies with possibilities for a better human future, such as electrifying the lives of a billion people in rural and

[b] https://www.lazard.com/perspective/levelized-cost-of-energy-and-levelized-cost-of-storage-2020/.

remote parts of the world; ending colonial and post-colonial dependencies on oil; dismantling the power of monopoly electric utilities through distributed energy generation, taking individual houses, businesses, and even communities off-grid; empowering the flourishing of local energy cooperatives; and reenergizing the future of manufacturing, meaningful labor and work, and sagging industrial economies [23–25].

4 The intricacies and implications of energy ownership

To bring alternative visions of solar-powered futures into reality will require careful attention to the intricacies of ownership design in the energy sector. Today, in a world permeated by the ideas and institutions of capitalism, the idea of ownership is often viewed as an aspect of markets. Early Enlightenment philosophers, however, like Adam Smith, who laid the intellectual foundations of capitalism, were clear that ownership is, first and foremost, a legal and political institution [26].

So it is for ownership of solar panels. The right to own solar panels and to dispense the energy they produce requires a legal foundation, in all jurisdictions, and, even where allowed, is subject to a wide range of conditions [27]. At stake in those rules are a range of diverse questions: who is allowed or will have the opportunity, as individuals or groups, to own solar panels, to generate and use electricity with them, or to distribute or sell that electricity to others. At stake as well, are the rules that govern ownership conditions, such as where the solar panels can be placed, how many can be purchased and operated, what rates will be offered for the electricity, and how the solar panels will be financed, acquired, and installed [28]. Already, different jurisdictions within the United States and around the world have remarkably diverse rules covering these and other aspects of solar energy ownership. The result is a bewildering array of solar diversity, in terms not only of what solar markets look like from place to place but also the forms and arrays of socio-technical configurations entering into the world through solar energy systems.

To date, this diversity is largely a curiosity. In 2019, solar energy, worldwide, amounted to just over 3% of global electricity consumption (and less than 1% of global energy consumption). Yet, the world is now adding solar power plants significantly faster than any other form of electricity generation (115 GW of new solar capacity was added in 2019, according to the International Energy Agency) [29]. And, as coal power plants retire and vehicle fleets shift from gasoline and diesel to electricity, solar additions can be expected to accelerate even more rapidly, providing cheap solar electrons to fuel the next generation of transportation and electricity grids. By 2050, if in fact the world has achieved a carbon-neutral energy system, with 50% powered by solar energy, ownership of the resulting solar panels will generate considerable value for humanity and for the global economy.

At that point, the intricacies of solar ownership will no longer be a curiosity. How that value is distributed among the world's inhabitants will have deep consequences for the distribution of power and wealth. If individuals and communities around the globe are ultimately empowered through widespread, distributed ownership of solar energy, combined with the capacity to use that energy to generate social and economic value for themselves, their households, and their residents, the result could usher in a massive leveling of global inequalities, helping to bring billions of people out of poverty. By contrast, if some significant fraction of the world's solar value is captured by narrow groups of owners, either through ownership of solar panels (e.g., via utility-scale solar power plants or companies who own vast tracts of rooftop solar) or of their manufacturing and distribution (e.g., with solar manufacturing concentrated among one or two suppliers or countries), the resulting energy ownership patterns could be even more concentrated than the present.

Solar ownership is not merely a question of the financial worth of energy. In the United States alone, to be sure, energy sales are an almost $1 trillion-per-year business, approximately 5% of the US economy. Yet when you add together the full extent of the work that goes into the energy sector—the mining and refining of fuels, the operating of power plants, the construction and maintenance of pipelines and electricity grids, the financing of energy infrastructures, and the manufacturing of equipment and hiring of workers for all of this—the scope of impact is enormous. And, of course, the energy sector influences not only itself but every form of human activity that depends on it.

The shape of solar ownership will thus determine not only the distribution of energy wealth but also the forms of energy work generated in the future, how that work is compensated, the degree of energy abundance or scarcity in the global economy, and how that abundance or scarcity is distributed across diverse energy users and uses [30]. A great deal is at stake in how the future of solar energy—or whatever form of green energy dominates the future—is owned, controlled, and governed.

5 The future of energy and democracy

We do not often think about the design of ownership as a choice, but it is. For the past century or so, most of the world has approached the accessibility and reliability of modern forms of energy and energy services as the primary consideration, prioritizing and privileging ownership regimes that could provide vast quantities of relatively low-cost energy to the economy and the public. That effort fueled a global industrial revolution, as well as an array of environmental and social crises that now threaten to undermine planetary civilization. Democracy is sacrificed on the altar of prosperity.

The question, now, is how to transform the future of energy to save humanity—and democracy—not just the climate. With what should we replace carbon-based energy systems in the future? Carbon-neutral energy, for sure, but owned by whom, in what ways, aligned with what imagined and real relationships between energy technologies the idea of the good life?

Our choices vary markedly: public ownership vs private; centralized ownership vs decentralized; equal ownership vs variable; ownership retained in the hands of the existing energy sector vs ownership redistributed, e.g., to those left behind by the current economy. While there is no space in this chapter to explore all of these options in-depth, a few examples may be helpful. The debate over whether energy resources and infrastructures should be publicly or privately owned has a long history in the United States, dating back to the beginnings of the industry. There are those, today, who believe that public ownership of energy is the only path to a green future [30]. Yet there are also many instances of centralized, state-owned energy systems around the globe that have worsened inequalities, invited corruption, and undermined democracy. Many believe that decentralized ownership of solar energy offers a clear path to a more democratized energy system. But even if a deeply decentralized energy system can be made to work, there are many potential designs for what it might look like. The model in which individual homeowners own their own rooftop solar systems is only one of many different kinds of models for more distributed ownership, others of which (e.g., cooperatives, municipally owned systems, or community solar) might be, for example, better options for strengthening economic security and opportunities for low-income communities or for re-cultivating a new kind of collaborative spirit amongst local communities.

6 So how do we decide?

Just now, the future of energy ownership is largely being left in the hands of markets. Solar panels are largely bought and sold in global markets, installed by private contractors, and the electricity is sold through diversely formulated electricity markets. The net result is that private ownership of solar panels is far, far ahead of public ownership, with almost all solar panels owned by and acquired through the private sector (including individual households). Already, however, solar markets are disrupting traditional energy ownership patterns, with roughly one-third of global solar additions happening via distributed systems (although, as with the example of Elon Musk's Tesla, not all distributed solar systems are owned in a de-centralized fashion). On the other hand, because solar ownership is currently dominated by markets, its ownership remains largely in the hands of the world's wealthy, whether through investments in rooftop solar systems, solar energy manufacturers, solar energy companies like NextEra Energy, or electric utilities. The fascinating exception, of course, is solar home systems, owned by households in some of the world's poorest communities. This is not currently a large segment of the world's energy, but it is perhaps the one with the biggest opportunity to rapidly and significantly impact the global distribution of wealth and inequality.

If we want to make a different choice about how to arrange the future of energy ownership, the time is now, while alternative energy systems and arrangements remain modest, and while the vast majority of investments in solar and other carbon-neutral energy sources remain small. Now is when the world's financial sector—led by companies like Blackrock, the world's largest investment firm, with $7 trillion under management—is gearing up to determine how money flows into the future of the energy sector. Now is when energy customers—companies like Apple, Google, and Delta Airlines—are determining how to ensure that they are powered by green energy in perpetuity. Now is when the automotive sector—General Motors, Ford, Honda, Toyota, and a host of prospective new entrants like Tesla, Rivian, and Nio—is building the future of transportation technologies to run on green electrons. Now is when the world's electric utilities are wrapping their heads around how to run the world's electricity grids without emitting carbon into the atmosphere. Now is when the oil majors—BP, ExxonMobil, Shell—are figuring out how to stay relevant for decades into the future.

All of these companies have made major announcements—and, behind the scenes, major financial and economic maneuvers—in 2020. These maneuvers are likely to set the design parameters for energy ownership for the next century unless alternatives emerge and take shape quickly. These entities understand that energy is constitutional to today's societies and to the future of global markets. They understand how deeply the shape and distribution of energy ownership will shape the future of the political economy. Their interests are narrow, however, and focused on securing their own positions—and those of their investors—in that future.

Democratic publics have much broader and deeper interests in the future of the political economy and should therefore be paying very careful attention to how the future of solar energy ownership is being structured by decisions being made today. I do not have a recommendation for how energy ownership should be designed for 2050 and beyond. I honestly do not know what forms of energy ownership will be best for the future of democracy. The history of industrial democracies suggests that different places will arrive at different answers to that question, as they often do to questions of how best to entwine technology and society [31]. Whichever choices do get made in the next few decades, however, about how to enable or constrain different forms of energy ownership and control, will have vast consequences not only for the Earth's climate but also for how Earth's inhabitants live, work, and play.

The question of energy ownership, therefore, deserves to be at the heart of public policy debates between now and mid-century. It is crucial to ask not only how fast we can deploy solar energy but also how to distribute its ownership among the world's citizens.

References

[1] S.R. Kurtz, et al., Revisiting the Terawatt Challenge, MRS Bull. 45 (3) (2020) 159–164.

[2] J. Eschrich, C.A. Miller (Eds.), The Weight of Light, Center for Science and the Imagination, Tempe, 2019. https://csi.asu.edu/books/weight/.

[3] J. Eschrich, C.A. Miller (Eds.), Cities of Light, Center for Science and the Imagination, Tempe, 2021. https://csi.asu.edu/books/cities-of-light/.

[4] G. Hecht, The Radiance of France: Nuclear Power and National Identity After World War II, MIT Press, Cambridge, 1998.

[5] E. Shove, M. Pantzar, M. Watson, The Dynamics of Social Practice: Everyday Life and How It Changes, Sage, Thousand Oaks, 2012.

[6] C.F. Jones, Routes of Power, Harvard University Press, Cambridge, 2014.

[7] T.J. Pinch, W.E. Bijker, The social construction of facts and artefacts: or how the sociology of science and the sociology of technology might benefit each other, Soc. Stud. Sci. 14 (3) (1984) 399–441.

[8] D. Nye, Electrifying America: Social Meanings of a New Technology, MIT Press, Cambridge, 1992.

[9] T. Hughes, Networks of Power: Electrification in Western Society, 1880–1930, JHU Press, Baltimore, 1993.

[10] J. Urry, The problem of energy, Theory Cult. Soc. 31 (5) (2014) 3–20.

[11] A. Lovins, Reinventing Fire: Bold Business Solutions for the New Energy Era, Chelsea Green Publishing, Hartford, 2013.

[12] T. Mitchell, Carbon Democracy: Political Power in the Age of Oil, Verso Books, Brooklyn, 2011.

[13] J.-M. Chevalier, The New Energy Crisis, Palgrave Macmillan, London, 2009.

[14] B. Sovacool, R.V. Sidortsov, B.R. Jones, Energy Security, Equality and Justice, Routledge, London, 2013.

[15] E.M. Gui, I. MacGill, Typology of future clean energy communities: an exploratory structure, opportunities, and challenges, Energy Res. Soc. Sci. 35 (2018) 94–107.

[16] J.D. Lambert, The Power Brokers: The Struggle to Shape and Control the Electric Power Industry, MIT Press, Cambridge, 2015.

[17] C.A. Miller, J. Richter, J. O'Leary, Socio-energy systems design: a policy framework for energy transitions, Energy Res. Soc. Sci. 6 (2015) 29–40.

[18] K. Abramsky, Sparking a Worldwide Energy Revolution: Social Struggles in the Transition to a Post-Petrol World, AK Press, Chico, 2010.

[19] W. Herche, Policy, Geospatial, and Market Factors in Solar Energy: A Gestalt Approach, Arizona State University, PhD Dissertation, 2017.

[20] J.S. Lacerda, J. Van Den Bergh, Diversity in solar photovoltaic energy: Implications for innovation and policy, Renew. Sust. Energy Rev. 54 (2016) 331–340.

[21] McKinsey Energy Insights, Global Energy Perspective: Reference Case 2018, December 2017.

[22] S. Jasanoff, S.-H. Kim (Eds.), Dreamscapes of Modernity: Sociotechnical Imaginaries and the Fabrication of Power, University of Chicago Press, Chicago, 2015.

[23] J. Cloke, A. Mohr, E. Brown, Imagining renewable energy: towards a Social Energy Systems approach to community renewable energy projects in the Global South, Energy Res. Soc. Sci. 31 (2017) 263–272.

[24] A. Martin, Community Solar Imaginaries: framing distributed solar energy developments and just transitions in California and New York, in: AGU Fall Meeting 2019, 2019. PA11B-0945.

[25] R. Williams, 'This Shining Confluence of Magic and Technology': solarpunk, energy imaginaries, and the infrastructures of solarity, Open Libr. Humanit. 5 (1) (2019).

[26] R. Islam, et al., World Development Report 2002: Building Institutions for Markets, The World Bank, Washington, 2001.

[27] S. Moore, Sustainable Energy Transformations, Power and Politics: Morocco and the Mediterranean, Routledge, London, 2018.

[28] K.H. Solangi, et al., A review on global solar energy policy, Renew. Sust. Energy Rev. 15 (4) (2011) 2149–2163.

[29] International Energy Agency, World Energy Outlook, IEA/OECD, Paris, 2019.

[30] K. Aronoff, et al., A Planet to Win: Why We Need a Green New Deal, Verso Books, Brooklyn, 2019.

[31] S. Jasanoff, Designs on Nature: Science and Democracy in Europe and the United States, Princeton University Press, Princeton, 2005.

Transitions

Introduction to Part II: Energy Futures

Martin J. Pasqualetti

School of Geographical Sciences and Urban Planning, Arizona State University, Tempe, AZ, United States

This section considers the empirical application of energy alternatives as an effect—and as motivation—to transition to a future that rests firmly on a foundation of increased energy democracies around the world. That would seem enough of a goal, although the sequence remains unclear. That is, will a transition to greater energy democracy be a tagalong to concerns such as global warming and the risks of long-lived radioactive wastes? Or, instead, will we overcome existing inertia by promoting greater energy democracy as a critical component of a fairer and more just future world? In truth, both will likely take place somewhat simultaneously. They will be two strands in a thread of energy that wrap around others including ethics, affordability, access, mobility, culture, communication, history, landscape integrity, and—perhaps dominantly—geopolitics, the least predictable strand of them all.

Part II examines the combination of influences that will influence how successful we will be in achieving greater energy democracies. It questions whether energy democracy movements and investments can deliver sustainable energy transitions locally, nationally, and globally. As we have been slow to learn, the transition to states of energy democracy will have little to do with scientific discoveries or technological advances. Yes, we are likely soon to solve the problem of long-term storage of electricity. And, yes, we are developing safer commercial nuclear reactors. However, there are many problems we are not likely to solve before greater energy democracy needs to be in place around the world as the norm rather than the exception.

One of the forces that propel transitioning to energy democracy is the unconscionable weight of injustice and inequity in the Global South. It is here where live the majority of those without electricity, whose cookstove emissions kill more than malaria, who have little to no hope for living at more than a subsistence existence for the rest of their abbreviated lives, who are often exploited by graft and corruption from those who control whatever energy resources might be otherwise available, who might have more children than they can feed, whose women and girls risk physical assault as they collect fuels at the expense of time in school, who sometimes feel no choice but to resort to crime and terrorist activities to survive.

As the contributors to Part II explain, the transition to energy democracy requires not just wistful aspiration but empirical analysis into a wide assortment of topics, such as (1) the range of energy democracy investments occurring around the world, (2) energy democracy responses to energy in/securities and risks, (3) the structural, relational, and cultural dynamics of energy democracy movements and investments, (4) social, economic, political, and economic intersections informing sustainable and democratic energy transitions, (5) the challenges involved in decentralizing energy ownership and production, (6) the capacities needed to bring stakeholders into energy governance, and (7) ecological implication, such as whether energy democracy initiatives will balance human interests with those of non-human entities, such as other species and ecologies.

While discussions of energy democracy stretch across the Global South, many issues arise elsewhere as well. McEvoy, for example, looks at energy imaginaries in Ontario where past inequities pepper the development of hydropower and where reside some of the largest nuclear power plant installations in the world. Across the Atlantic, Lacey-Barnacle and Nicholls address matters in England where questions include whether the collective voices of those living there are yet audible above the confusing din of support for non-democratic options such as the proposed Sizewell C reactor on the Sufolk coast.

Greater independence and security are integral parts of most conceptions of greater energy democracy. So it is the goal of the US Commonwealth of Puerto Rico, whose grid was destroyed by Hurricane Maria in 2017. As Sokol points out, rebuilding reliability faces opposing opinions from those calling for a return to the status quo ante and those who purport that renewables offer wiser choices, as well as improvements in energy democracy.

Far to the west, across the Pacific Ocean, Matsukawa explains that Japan faces a particularly daunting challenge as his country tries to adjust to the political repercussions from the destruction of three boiling water reactors at Fukushima on the northeast coast of Honshu in 2011. Until that time, Japan was relying on 54 reactors to supply 30% of its electricity. As of March 2021, only five plants with nine reactors have gained the agreement of residents to resume operations, all of them on the western side of the country and all of them pressurized water reactor designs. Meanwhile, Japan has been redirecting its attention to renewables such as wind and solar, which inherently promise more democratic forms of generation.

Indeed, part of the needed adjustments in any effort toward energy democracy must rely at least to some degree on an increase in energy literacy. While the industrial countries are making some headway in this direction, billions of people hold only the most basic understanding of the human price of commercial energy sources. This minimal knowledge is especially true regarding the fundamental question of knowing how best to gain access to whatever energy resources might be available, as Heynen and his colleagues discuss for Papua New Guinea.

Taking the wind energy path promises to lead us to a future of greater energy democracy, and its rise in global installed capacity started with on-shore installations. However, off-shore wind resources attract increasing attention because winds over open water tend to be more consistent and speedier. Even better, they do not interfere with existing land uses. Opportunities for these installations to bolster energy democracy include places like Brazil as Xavier and his colleagues discuss.

For obvious reasons, cooperatives often hold great promise for energy democracy. More direct public involvement is typical, for example, as is likely with independent power producers or government entities. Caroline Wright discusses the challenges and opportunities of cooperatives, while Stephanie Lenhart focuses on their evolution.

The massive—and growing—populations of India and Pakistan present some of the most intense challenges, both for their citizens and for those who would supply them. What should be the best energy sources to satisfy energy demand? Does either country possess sufficient infrastructure, and if not, who can afford to pay for improvements? Do the impoverished people in either country have any chance of affording power that might be on offer? Will everyone understand or have a say in what is available? These are some of the questions Naqvi addresses regarding the Karachi electric sector and Bedi tackles in India.

The two groups of papers that make up Part II of this book provide a sample of empirical work that must accompany—and frequently precede—achieving greater energy democracy. They illustrate the complexities of shifting from fossil-, nuclear-, and hydro-based supply systems that have evolved over the past 125 years. They remind us of the trillions of dollars of infrastructure scattered over landscapes that continue to create massive inertia against change. However, it is not just the tangible elements of energy that create resistance to change. Instead, many policies, jobs, laws, attitudes, access, and literacy must be considered. These elements and many others reside at a substantial remove from the days of greater energy democracy. In truth, they are nearer the opposite end of the broad spectrum of energy options. It will take us some time to move the needle from where it is now to where we would like it to be. The chapters in Part II are examples of what is needed.

Organizing

13

Energies of resistance? Conceptualizing resistance in and through energy democratization

Joshua K. McEvoy

Department of Political Studies, Queen's University, Kingston, ON, Canada

1 Introduction

What does it mean to democratize energy? How should we interpret the multiple discourses, imaginaries, and practices that animate the complex processes of energy transition and democratization? These fundamental questions have led to productive theorizing in recent years [1–3]. A prominent strain of research postulates that energy democratization can be understood in terms of *resisting, reclaiming*, and *restructuring* dominant energy systems [1,4–7]. From this view, energy democratization is conceptualized as a process through which "dispersed grassroots initiatives and a transnational social movement, is challenging energy incumbents" [8]. This understanding centers resistance in its vision of energy democratization, which is most often interpreted as opposition to "the fossil-fuel-dominant energy agenda" [1].

In this brief chapter, I explore the concept of resistance in and through energy democratization and attempt to open a dialogue between resistance and energy literature. I argue that this dialogue helps make more explicit both the object of resistance and the potential for transformative change. In part, I do so by drawing on an understanding of democratization that centers the capacity for "effective intransigence" [9]. This understanding of resistance in and through energy democratization also takes a cue from activists and groups engaged in resistance struggles. The Tiny House Warriors on Secwepemc territory in British Columbia and the Water Protectors at Standing Rock have repeatedly asserted that their resistance is directed not only toward a specific energy project but to the undemocratic systems of domination of which those projects are part and represent—patriarchy, settler colonialism, white supremacy, and capitalism [10–13]. This conceptualization of resistance explicitly recognizes the entanglement of energy infrastructure with socio-political structures. This understanding helps to reconcile opposition to non-fossil energy infrastructures, such as resistance to wind projects in the Isthmus of Tehuantepec in Mexico [14–16] or biofuels in Kenya's Tana Delta [17], with resistance in and through energy democratization more broadly. By refusing to disaggregate the material infrastructure of energy from the socio-political context in which it is embedded, the object of resistance is made clear as the configurations of energy—in whatever form it takes—and power that produce and uphold systems of domination.

The chapter is organized into three sections. First, I briefly sketch a critical understanding of energy democratization as a struggle against prevailing neoliberal discourses and practices. I argue that if we recognize energy (especially but not exclusively fossil energy) as being deeply implicated in the technocratic and undemocratic social relations of neoliberalism, then we must also acknowledge that democratizing energy always entails not only resistance *in* but also *through* energy. In other words, if we understand energy not only as an instrument of power but as being deeply enmeshed in and co-constitutive with a depoliticizing and undemocratic project, we must develop a concept of resistance that reflects this entanglement. Second, I bring energy scholarship into conversation with the language of *economic diversity* developed by J.K. Gibson-Graham [18–21] and critical resistance literature more broadly. I show how reading the processes of energy democratization through the language of resistance helps make legible the stakes involved and the transformative potential (or lack thereof) of particular manifestations of energy democracy. Finally, I conclude with a brief discussion of what this understanding of resistance might mean for our understanding of the processes that lead to co-optation and capture.

2 Energy/democracy

In energy democratization research and activism, energy systems are conceptualized as part of the broader socio-political, cultural, and economic processes that shape modernity [9,22–26]. For instance, in *Carbon Democracy*, Timothy Mitchell argues that the centralization and technocracy of energy systems in the "age of oil" are intricately tied to, and co-constitutive with, the rise and reproduction of a liberal order "that employs popular consent as a means of limiting claims for greater equality and justice by dividing up the common world" [9]. From this view, the liberal project of the postwar years, which Mitchell contends was animated by Keynesian imaginaries of unlimited energy supplies powering perpetual economic growth, should be understood as "an engineering project, concerned with the manufacture of new political subjects and with subjecting people to new ways of being governed" [9]. In other words, the sociotechnical assemblage of fossil energy and liberal ideology enabled the rise of a particular governmentality and related subjectivity—a particular mode of governing premised on economization and the atomized subject of *homo economicus*. Similarly, Wendy Brown argues that the prior, alternative subjectivity of *homo politicus*, characterized by popular sovereignty, was "slimmed" or restricted with the rise of liberal democracy [27]. Brown contends that this slimming eventually turned to erasure with the rise of neoliberalism [27,28]. It is through the "ascendancy of neoliberal reason that the citizen-subject converts from a political to an economic being and that the state is remade from one founded in juridical sovereignty to one modeled on a firm" [27]. Read together, Brown and Mitchell offer profound insight into the sociotechnical processes that gave rise to an undemocratic socio-economic order that has not only accelerated and expanded environmental despoilation and inequality but simultaneously circumscribed the capacity for meaningful political intervention to address these crises.

There are, at least, three important implications of the interpolation of energy in the rise and reproduction of an undemocratic neoliberal order. First, there is an obvious connection to the longstanding critique in green political theory of the inability or unwillingness of liberal democratic politics to address the related crises of environmental despoliation and inequality [29–33]. Related to this critique is what Mitchell describes as the phenomenon of "democracy without democratization" [9]. By this, Mitchell is referring to the way that processes of democratization, in fact, often entail a struggle to broaden or redefine liberal understandings and institutions of democracy so as to both extend effective political agency to those excluded or otherwise marginalized by them and to expand what is subject to democratic processes (e.g., "economic democracy") [9]. Finally, as energy democracy and transition research more broadly has demonstrated, the interpolation of energy in the neoliberal order means that we need to take both the discursivity and the materiality of energy seriously if we are to transform energy systems. Taken together, then, it is clear, as many others have pointed to, that the democratization of energy entails much more than a technical switch or change in sectoral governance and design. Democratization *in* energy necessarily entails a process of resistance *through* energy.

3 Democratization/resistance

Taking this understanding of energy democratization seriously requires a reading of its discourses and practices that are attuned to processes of resistance. Resistance, here, is understood as the possibility of opposition to *constitutive exclusion* [34]. As Kramer explains, "constitutive exclusion describes the phenomenon of *internal* exclusion, or those exclusions that occur *within* a philosophical system or a political body. Constitutive exclusions occur when a system of thought or a political body defines itself by excluding some difference which is intolerable to it" [34]. Echoing the language of constitutive exclusions, Gibson-Graham describes the neoliberal notion of the economy as establishing "the bottom line for action" making "us perform in certain ways" which directs and limits "politics to certain channels, blinding us without realizing it to the possibility of other options" [20]. In a similar vein, Simon Dalby argues that under neoliberalism, "citizens are increasingly turned into consumers, and the functions of state agencies rearticulated in terms of clients and customers, not democratic political subjects ... Collective action to reduce pollution or prevent environmental degradation has been replaced by people looking to private consumption, rather than acting as political subject collectively determining the conditions of our lives" [35]. Dalby's description of life in the neoliberal era is consistent with what Gibson-Graham calls *capitalocentrism*, in which economization and particular understandings of the economy act as constitutive exclusions that restrict the subject's ability to recognize or imagine noncapitalist economic and social relations [20]. In this way, the economy becomes a powerful heuristic that makes capitalism, and by extension neoliberalism, seem natural and inevitable [20,36,37].

To dislodge capitalocentrism and "create the conditions for the emergence of noncapitalist modes of economic subjectivity," Gibson-Graham argues for a language and practice of economic diversity [20]. Drawing on feminist political economy, Gibson-Graham's discourse of economic diversity is first meant to draw our attention to existing non-commodity production, non-market exchange, and other economic activities that diverge from traditional capitalist relations. By centering heterogeneity and divergence in our thinking about economic relations, Gibson-Graham contends that we can better resist the totalizing tendency and economization of neoliberalism while also identifying spaces to nurture and strengthen efforts to build toward a post-neoliberal capitalist world. Dislodging a totalizing concept like the singular "economy" is also meant to assist in opening space for heterodox subjectivities and future imaginaries. In this way, practicing and imagining resistance to neoliberalism through economic diversity echoes Foucault's understanding of critique in What is Enlightenment [38] as both "the analytical practice of unsettling the present and the experimental practice of living differently" [39].

In the context of energy democratization, the language of difference has significant purchase. Resistance in and through energy democratization can be understood fundamentally as the act or manifestation of difference—the articulation of discourses and the production of energy configurations that diverge from the technocratic and exploitative relations of neoliberalism. Although the constraints of this chapter preclude in-depth empirical study, we can take the example of cooperatives, which Gibson-Graham raises as a potential site of economic divergence and are a frequent topic in energy democratization scholarship. The deliberative decision-making of renewable energy cooperatives (RECs) and other common characteristics of community energy initiatives, including the control over surpluses, localized ownership, and the construction of member communities, are significant markers of non-conformity with neoliberal capitalism [40,41] and in this way, RECs "play a symbolic role in shaping public perception of the possible" [42].

However, in some cases, RECs, and cooperatives more broadly, are difficult to distinguish in a meaningful sense from other profit-seeking endeavors [42,43]. RECs often operate as investment cooperatives through which capital is raised via membership and capital markets to construct solar panels or wind turbines, which then generate electricity that is sold to grid operators with the surplus distributed to members, used for future energy projects, circulated back into capital markets, and/or directed to community development projects. Members, then, are often cast as investors, with one Canadian-based REC advertising "local, sustainable, and profitable clean energy investments" for "individuals, organizations and businesses" [44]. Additionally, in study conducted in Germany, the demographics of REC membership have been found to be disproportionately male, highly educated, and affluent [45,46]. While different from typical capitalist firms in significant ways and perhaps even generative of alternative subjectivities or imaginaries, it is not necessarily the case that all RECs can be understood as resistant to neoliberalism and, in fact, they may in some cases serve to entrench it further. Kacper Szulecki identifies this phenomenon in energy democracy as the emergence of a liberal *prosumer* subjectivity (the subject as both consumer and producer of energy) which iterates but does not challenge neoliberal orthodoxy [2].

We must, therefore, possess an analytic of resistance able to distinguish oppositional or resistant divergence from that which presupposes and maintains existing social relations. David Hoy's distinction between resistance with and without critique is helpful here. Hoy argues that without critique, "utopian imaginings of freedom may not be aware of the extent to which they presuppose the patterns of oppression that they are resisting" [47]. Hoy argues that we must, therefore, embed critique in our resistance, as "[c]ritique is what makes it possible to distinguish emancipatory resistance from resistance that has been co-opted by the oppressive forces" [47]. This is similar to Brown's position that resistance requires an "alternative horizon" as resistance is deeply entangled with, and co-constitutive of, restructuring processes [48,49]. As Lars Cornelissen puts it, "resistance is not secondary to the elaboration of alternatives; rather, moments of refusal must guide the formulation of alternative analyses" [39]. In this way, no specific configuration of energy, such as a REC, is inherently resistant to undemocratic systems of domination like neoliberalism. Instead, their resistance and, in turn, their transformative potential is made contingent.

4 Resistance/materiality

Here, however, it is essential to emphasize that resistance is always situated and contextual and that introducing a concept of resistance to energy necessitates an *energizing* of resistance. Most notably, we must take account of the materiality of energy. This means dislodging the anthropocentrism at the core of much resistance literature. Of course, this is not meant to suggest that the agentic capacity of materiality is absent from resistance literature [50], but rather that energy research has developed sophisticated analytics specifically sensitized to energy. In particular, energy research has drawn on actor-network to develop this line of thinking.

For instance, in their ethnographic research on wind energy development in the Isthmus of Tehuantepec, Cymene Howe and Dominic Boyer use a "networks of enablement" approach in which heterogeneous elements (e.g., discourses, infrastructures, technologies, ecologies) interact to potentially open space and perform resistance [15,16]. Boyer, focusing specifically on the energetic aspects of these networks, posits an *energopower* analytic that iterates on Foucault's biopower [15,51]. Motivating his conceptual iteration, Boyer argues that the advent of the Anthropocene and the related emergent energy transition has "shaken the foundations of the contemporary biopolitical regime," as modern life is "unstably intertwined with infrastructures, magnitudes, and habits of using electricity and fuel" [51]. Boyer has no intention of displacing biopolitics but instead, advocating a "both/and" approach, views energopower as "an alternative genealogy of modern power, as an analytic method" [51]. Thus, Boyer suggests that just as Foucault's biopower provides an analytic for exploring the management and control of populations, energopower offers a complementary analytic oriented to the management of climate change and energy transition. They are, in this sense, mutually reinforcing. Energopower, should be understood as always being "shaped by particular forms and politics of life," and likewise, biopower is "always plugged in" [51]. In other words, Boyer argues that the edifices of modern power, such as schools, prisons, and myriad surveillance technologies, are today enabled by electrification and "energy-intensive building materials like cement" (i.e., they are "plugged-in") [51]. Critically, then, Boyer's concept of energopower posits an understanding of power and, in turn, resistance, that is always relational and located in the varied configurations of energy systems across space and time.

Thinking about resistance through the analytic of energopower, or an alternative networks of enablement schema, further allows us to think in less reductive ways about co-optation and capture. For instance, Pel argues in his invitation to think dialectically about co-optation in "sustainability transitions" that energy infrastructure, flows, and the institutions that enable them may possess "latent transformative contents" [52]. From this view, although discursive critique might be absent, entities like RECs may, to an extent, enact or enable transformative resistance through the production of material relations that open new possibilities for resistant power. As Kramer argues, we must be sensitive to the materiality of constitutive exclusions [34], such as the dispossession of land or the inequity of pay, which inversely means we must also be cognizant of the potential for material processes in generating or enabling resistance to these exclusions. Such an understanding of the relational nature of resistance in many mirrors Mitchell's description of the processes that led to a prior episode of energy democratization in the late-19th century. In that case, the materiality of coal, the technomaterial artifact of the steam engine, egalitarian discourses, and the institutions of unions acted in concert to enable a more democratic politics that saw the successful contestation of exploitative labor practices [9]. Today, resistance in and through energy that serves as a democratizing force is desperately needed to politicize and address the vast inequality, mass extinction, and environmental and climatic despoliation that fossil-fueled neoliberal social relations have produced.

5 Conclusions

The language of resistance offers a promising conceptual lens through which to interpret the processes of, and struggles for, energy democratization. It helps us to make sense of the stakes involved and the transformative potential of energy democratizing efforts by orienting our analyses to the ways in which democratizing discourses and practices relate to dominant structures of power. Although further conceptualization is required, this brief foray into energizing resistance also suggests we may gain a more nuanced and less reductive understanding of co-optation and capture. This chapter, it is hoped, will act as an invitation for further engagement with the concept of resistance in and through energy democratization. Developing an understanding of resistance that acknowledges the profound ways energy has been, and continues to be, implicated in systems of domination and exploitation is an essential step toward identifying viable pathways to an emancipatory and democratic energy future.

References

[1]   M.J. Burke, J.C. Stephens, Energy democracy: goals and policy instruments for sociotechnical transitions, Energy Res. Soc. Sci. 33 (2017) 35–48.
[2]   K. Szulecki, Conceptualizing energy democracy, Environ. Polit. 27 (1) (2018) 21–41.
[3]   S. Becker, M. Naumann, Energy democracy: mapping the debate on energy alternatives, Geogr. Compass 11 (8) (2017), e12321.
[4]   E. Allen, H. Lyons, J.C. Stephens, Women's leadership in renewable transformation, energy justice and energy democracy: redistributing power, Energy Res. Soc. Sci. 57 (2019), 101233.
[5]   J.C. Stephens, Energy democracy: redistributing power to the people through renewable transformation, Environ. Sci. Policy Sustain. Dev. 61 (2) (2019) 4–13.
[6]   J.C. Stephens, M.J. Burke, R. Gibian, E. Jordi, R. Watts, Operationalizing energy democracy: challenges and opportunities in Vermont's renewable energy transformation, Front. Commun. 3 (2018) 391.

[7] S. Sweeney, Resist, Reclaim, Restructure: Unions and the Struggle for Energy Democracy, Trade Unions for Energy Democracy, Rosa Luxemburg Stiftung, 2013. http://unionsforenergydemocracy.org/resist-reclaim-restructure-unions-and-the-struggle-for-energy-democracy/. (Accessed 05.06.20).

[8] K. Szulecki, I. Overland, Energy democracy as a process, an outcome and a goal: a conceptual review, Energy Res. Soc. Sci. 69 (2020), 101768.

[9] T. Mitchell, Carbon Democracy: Political Power in the Age of Oil, Verso, New York, 2011.

[10] J.C. Stephens, Diversifying Power: Why We Need Antiracist, Feminist Leadership on Climate and Energy, Island Press, Washington, 2020.

[11] N. Estes, Our History Is the Future: Standing Rock Versus the Dakota Access Pipeline, and the Long Tradition of Indigenous Resistance, Verso, New York, 2019.

[12] Tiny House Warriors, Our Land Is Home. http://www.tinyhousewarriors.com/. (Accessed 12.01.21).

[13] R. Minutaglio, How Tiny Houses Became a Symbol of Resistance for Indigenous Women, Elle, 2019. https://www.elle.com/culture/career-politics/a29738953/tiny-house-warriors-trans-mountain-pipeline-canada/. (Accessed 12.02.20).

[14] S.H. Baker, Revolutionary Power: An Activist's Guide to the Energy Transition, Island Press, Washington, 2021.

[15] D. Boyer, Energopolitics: Wind and Power in the Anthropocene, Duke University Press, Durham, 2019.

[16] C. Howe, Ecologics: Wind and Power in the Anthropocene, Duke University Press, 2019.

[17] K. Neville, Fueling Resistance: The Contentious Political Economy of Biofuels and Fracking, Oxford University Press, New York, 2021.

[18] G. Roelvink, K. St. Martin, J.K. Gibson-Graham (Eds.), Making Other Worlds Possible: Performing Diverse Economies, University of Minnesota Press, Minneapolis, 2015.

[19] J.K. Gibson-Graham, J. Cameron, S. Healy, Take Back the Economy: An Ethical Guide for Transforming Our Communities, University of Minnesota Press, Minneapolis, 2013.

[20] J.K. Gibson-Graham, A Postcapitalist Politics, University of Minnesota Press, Minneapolis, 2006.

[21] J.K. Gibson-Graham, The End of Capitalism (as We Knew It), University of Minnesota Press, Minneapolis, 1996.

[22] O. Barak, Powering Empire: How Coal Made the Middle East and Sparked Global Carbonization, University of California Press, Oakland, 2020.

[23] C.D. Daggett, The Birth of Energy: Fossil Fuels, Thermodynamics and the Politics of Work, Duke University Press, Durham, 2019.

[24] C. Hoffmann, Beyond the resource curse and pipeline conspiracies: energy as a social relation in the Middle East, Energy Res. Soc. Sci. 41 (2018) 39–47.

[25] A. Malm, Fossil Capital: The Rise of Steam Power and the Roots of Global Warming, Verso, New York, 2016.

[26] E. Shove, G. Walker, What is energy for? Social practice and energy demand, Theory Cult. Soc. 31 (5) (2014) 41–58.

[27] W. Brown, Undoing the Demos: Neoliberalism's Stealth Revolution, Zone Books, New York, 2015.

[28] W. Brown, In the Ruins of Neoliberalism: The Rise of Antidemocratic Politics in the West, Columbia University Press, New York, 2019.

[29] J. Wainwright, G. Mann, Climate Leviathan: A Political Theory of Our Planetary Future, Verso, New York, 2018.

[30] B. Latour, Down to Earth: Politics in the New Climatic Regime, Polity Press, Cambridge, 2017/2018.

[31] B. Doherty, M. de Geus (Eds.), Democracy and Green Political Thought: Sustainability, Rights, and Citizenship, Routledge, London, 1996.

[32] R. Eckersley, Liberal democracy and the environment: the rights siscourse and the struggle for recognition, Environ. Polit. 4 (4) (1995) 169–198.

[33] J.S. Dryzek, Ecology and discursive democracy: beyond liberal capitalism and the administrative state, Capital. Nat. Social. 3 (2) (1992) 18–42.

[34] S. Kramer, Excluded Within: The (Un)intelligibility of Radical Political Actors, Oxford University Press, Oxford, 2017.

[35] S. Dalby, Anthropocene Geopolitics: Globalization, Security, Sustainability, University of Ottawa Press, Ottawa, 2020.

[36] T. Mitchell, Rethinking economy, Geoforum 39 (2008) 1116–1121.

[37] T. Mitchell, Fixing the economy, Cult. Stud. 12 (1) (1998) 82–101.

[38] M. Foucault, What is enlighenment? in: P. Rabinow (Ed.), The Foucault Reader, Pantheon Books, New York, 1984, pp. 32–50.

II. Transitions

[39] L. Cornelissen, On the subject of neoliberalism: rethinking resistance in the critique of neoliberal rationality, Constellations 25 (2018) 133–146.

[40] J.K. Gibson-Graham, J. Cameron, S. Healy, Commoning as a postcapitalist politics, in: A. Amin, P. Howell (Eds.), The Shrinking Commons? Rethinking the Futures of the Common, Routledge, New York, 2016, pp. 192–212.

[41] J. Cameron, J. Hicks, Performative research for a climate politics of hope: rethinking geographic scale, 'impact' scale and markets, Antipode 46 (1) (2014) 53–71.

[42] J.L. MacArthur, Empowering Electricity: Co-operatives, Sustainability, and Power Sector Reform in Canada, University of British Columbia Press, Vancouver, 2016.

[43] M.D. Tarhan, Renewable energy co-operatives and energy democracy: a critical perspective, in: Presented at the Canadian Association for Studies in Co-operation, 2017. https://www.researchgate.net/publication/317369738_Renewable_Energy_Co-operatives_and_Energy_Democracy_A_Critical_Perspective. (Accessed 05.06.20).

[44] Solar Power Investment Cooperative of Edmonton, Invest in Your Future. https://joinspice.ca/. (Accessed 03.03.21).

[45] J. Rommel, J. Radtke, G. von Jorck, F. Mey, Ö. Yildiz, Community renewable energy at a crossroads: a think piece on degrowth, technology, and the democratization of the German energy system, J. Clean. Prod. 197 (2018) 1746–1753.

[46] Ö. Yildiz, J. Rommel, S. Debor, L. Holstenkamp, F. Mey, J.R. Müller, J. Rognli, Renewable energy cooperatives as gatekeepers or facilitators? Recent developments in Germany and a multidisciplinary research agenda, Energy Res. Soc. Sci. 6 (2015) 59–73.

[47] D.C. Hoy, Critical Resistance: From Poststructuralism to Post-critique, MIT Press, Cambridge, 2014.

[48] W. Brown, Edgework: Critical Essays on Knowledge and Politics, Princeton University Press, Princeton, 2005.

[49] W. Brown, Resisting left melancholy, boundary 2 26 (3) (1999) 19–27.

[50] M. Callon, Y. Millo, F. Muniesa (Eds.), Market Devices, Blackwell Publishing Ltd., Oxford, 2007.

[51] D. Boyer, Energopower: an introduction, Anthropol. Q. 87 (2) (2014) 309–333.

[52] B. Pel, Trojan Horses in transitions: a dialectical perspective on innovation 'capture', J. Environ. Policy Plan. 18 (5) (2016) 673–691.

C H A P T E R

14

The role of ownership and governance in democratizing energy: Comparing public, private, and civil society initiatives in England

M. Lacey-Barnacle[a] and J. Nicholls[b,c]

[a]Science Policy Research Unit, University of Sussex, Brighton, United Kingdom [b]Bristol Law School, University of Bristol, Bristol, United Kingdom [c]School of Sociology, Politics and International Studies, University of Bristol, Bristol, United Kingdom

1 Introduction

In this chapter, we seek to highlight the critical importance of ownership and governance in shaping energy democracy. In doing so, we assert that the extent to which energy infrastructures can be said to be "democratic," is tied to both the ownership type and model of governance used to facilitate the deployment of new energy technologies. We also highlight that the relationship between ownership and governance is empirically underexplored in energy democracy literature. For example, Szulecki [1] offers a compelling vision of how governance should relate to ownership to support energy democracy:

> Governance in energy democracy should be characterized by wide participation of informed, aware, and responsible political subjects, in an inclusive and transparent decision making process relating to energy choices, with the public good as its goal. To create and safeguard civic empowerment and autonomy, high levels of ownership of energy generation and transmission infrastructure through private, cooperative or communal/public means are necessary. *Szulecki [1, p. 35]*

Here, Szulecki [1] is clear that governance arrangements should champion the public good, while ownership should advance civic empowerment. Explorations of how this happens in everyday practice, however, are lacking. Accordingly, theoretical articulations of energy democracy are common, yet limited empirical research exists that draws on real world

Energy Democracies for Sustainable Futures
https://doi.org/10.1016/B978-0-12-822796-1.00014-0

examples of these models in practice, let alone engages in critical comparisons between public, private, and civil society initiatives. Such analysis is required to bring greater clarity to the relationship between governance and ownership in the pursuit of energy democracy. Close empirical examination also begins to substantiate some of the broader claims made by energy democracy scholars, particularly with regards to how different types of ownership and models of governance affect developments in practice.

Drawing on PhD research data derived from qualitative research in England (2015–19) [2,3], we have compared and synthesized findings on the relationship between ownership and governance according to our shared research interests and methods. The novel insights considered in this chapter are largely drawn from in-depth interviews with directors of local energy projects, local politicians, and members of the public involved in the development of or living near to renewable energy projects. One authors PhD was focused on energy justice insights in local energy systems [2], and the other, on energy democracy more specifically [3]. In the energy justice-oriented PhD [2], the procedural justice tenet of energy justice was used, among other energy justice tenets, to analyze primary data. As procedural justice concerns itself with the participation of people in decision-making procedures around energy systems, sufficient insights were generated on the democratic aspects of local engagement. In the energy democracy-oriented PhD [3], the theoretical lens of deliberative democracy was employed to analyze three solar farm developments through detailed ethnographic case studies. The three solar farm cases correspond to different governance models, including civil society, public/municipal, and corporate governance.

Our chapter draws on quotes from our respective research participants to highlight the differences between various ownership and governance structures, making clear the implications this has for energy democracy generally. Anonymized details of our research participants and their associated identifiers are included as an Appendix at the end of the chapter.

We look first at different ownership types in relation to renewable energy developments, broadly distinguishing between two categories of ownership: civic ownership and private ownership. Importantly, we note that ownership, on its own, does not reflect a key distinction within the energy sector associated with governance structures, of which we work with a tripartite distinction between public, private, and civil society governance structures.

Drawing on prominent English models of energy ownership, this chapter focuses on the private Limited company model and civic models, such as the Co-operative and the Community Benefit Society (BenCom) models. We also engage briefly with the public/municipal energy governance model as several UK municipal energy companies use the Limited company model to facilitate engagement in energy markets, but are governed by local government authorities.

We then touch on governance issues around voting rights and involvement in Annual General Meetings (AGMs), reflecting on how such issues are variously configured and understood through governance structures.

Our findings draw on our data to show both the importance and complexity of different ownership types and how questions of democratic involvement and participation are intimately tied to specific organizational structures and forms of governance. In our conclusion, we note that further interdisciplinary and comparative research into both ownership and governance structures used to democratize energy infrastructures is vital. This is particularly pertinent if we seek to understand energy democracy in practice.

2 Energy democracy: Conceptualizing ownership and engagement

While research into energy democracy has grown significantly over the last decade, little direct attention has been paid to the governance models and ownership types used to facilitate energy democracy. Rather, energy democracy literature has tended to engage with different types of ownership and governance structures theoretically. Van Veelen [4] notes that civil society models of ownership are emphasized in energy democracy literature, while Burke and Stephens [5] provide a high-level outline of the core focus of ownership in energy democracy:

> Central to an energy democracy agenda is a shift of power through democratic public and social ownership of the energy sector and a reversal of privatization and corporate control [...] diverse forms of ownership are needed [...] that respect the political, economic and social requirements, diversity, and challenges of specific locations or communities. *Burke and Stephens [5, p. 38]*

We support Burke and Stephen's [5] emphasis on tailoring local ownership types to the specifics of local and regulatory contexts. However, Burke and Stephens do not elaborate on the specific types of ownership they have in mind. Thinking about how energy democracy broadly approaches key ownership questions, Becker and Naumann [6] note that:

> Energy democracy focuses on variegated forms of collective organization and ownership [...] the main question here is who owns and controls what kind of energy infrastructure and with what kind of consequences. *Becker and Naumann [6, pp. 4–5]*

While Becker and Naumann [6] ask "who" owns and controls "what kind" of energy infrastructures, they do not ask "how"? Through examining different ownership types and governance structures via several local renewable energy projects in England, we seek to begin to address how exactly energy democracy is being realized through different organizational forms and the challenges these forms present.

3 Ownership types in England: Private and civic

This section briefly outlines the regulatory context for different legal models used in England, introducing two key pieces of legislation that underpin the ownership types under analysis in this chapter: The Co-operative and Community Benefit Societies Act 2014 [7] and the Companies Act 2006 [8].

For our analysis of local energy initiatives in England, the key structures associated with these are the Limited company, Co-op and BenCom, with the addition of the local government-led municipal energy model, which largely utilizes the Limited company model. While we focus on the municipal energy use of the Limited company model, it is important to acknowledge that some notable examples employ BenCom ownership forms in the United Kingdom. Importantly, the specific structures that we outline in this section are commonly used across England to facilitate the development of renewable energy or the purchase of existing renewable energy projects.

Before we outline the critical aspects of each legal structure, it is important to differentiate between ownership types and their associated legal structures and governance models—which refer to the internal processes of an organization. It is also critical to note that governance models are not necessarily associated with legal structures. This distinction helps readers to understand different sectoral governance logics in relation to varying legal structures. In the following sections, we show that both ownership and governance have implications for energy democracy and that these should be considered in combination when seeking to understand energy democracy in practice.

For the purposes of simplicity, we have created a table (Table 14.1) to illustrate how we have aligned different English legal structures to different ownership types and governance models. In the next section, we briefly summarize the relevant pieces of legislation that have created the legal framework for local energy initiatives to own renewable energy infrastructures in England.

3.1 Civic energy developments—Co-ops and BenComs

Focusing first on civic ownership types for local energy, the primary piece of UK legislation is the Co-operative and Community Benefits Societies Act 2014. This act helped to consolidate what are known as Registered Societies in the United Kingdom, in the form of Co-ops and the newly introduced Community Benefit Societies (CBS) or BenComs.

Co-ops are member-oriented organizations that exist for the benefit of members, which often adhere to governance rules and core principles outlined by the International Co-operative Alliance [9]. BenComs, in contrast, are established more decidedly for the benefit of the wider community, with a legal mandate to operate beyond the interest of their members and to serve the local community. These differences in model rules are attested to by a director of a BenCom which owns a 2.7 MW solar farm in the East Midlands, England:

> The main difference between a CBS and a Co-op [...] is that with a CBS you are primarily set up to serve the community, so therefore instead of paying your shareholders first, you pay your community benefit fund first. You pay your shareholders – or the members – second. That is a big difference between the two. [CB1]

Thus, BenComs can be seen as more community oriented, while Co-ops more member oriented. While this clearly has implications for energy democracy in practice, with some of our participants suggesting Co-ops are more insular in nature, the ICA Co-operative Principles focus on ensuring that Co-operatives serve communities around them, with "concern for community" being one of the core seven Principles [9]. In addition, one of the member-shareholders interviewed sought to alter this interpretation of Co-ops:

> Even a Co-operative society, when compared to a BenCom, will have clear social goals and in fact cannot be a Co-op unless it has those clear social goals [...] for the benefit of members Co-ops still do a huge amount of wider social things [...] A member's Co-op is very different from a private investors club. [MS1]

TABLE 14.1 Overview of ownership type, legal structures, and governance models.

Ownership type	Associated legal structures	Governance models
Private	Limited Company	Private—Shareholder governance
Civic	Co-operative, Community Benefit Society (BenCom)	Public—Council-led governance Civil society—Co-operative governance

These differences in legal structure, tied to the ownership type also have implications for the provision of benefits. Benefits can take a variety of forms. While they are often financial, across different ownership models, they can also involve educational activities, wildlife enhancement, civic engagement, or local workshops. In the community/civic energy sector, such benefits tend to be a financial payment in the form of a Community Benefit Fund for a designated community or to another community organization. Both of our data sets identified that the BenCom ownership examples drew on Co-operative governance structures, despite having a different legal model under the aforementioned UK legislation. Accordingly, both Co-ops and BenComs typically allow members to vote on dividend interest (in the case of Co-ops), what happens with Community Benefit Funds (in both BenComs and Co-ops), and the election of new directors (in both cases). While this reflects an important distinction in model rules, the two ownership structures are often merged in practice in civic energy projects.

3.2 Private energy developments—Limited companies and municipal developments

The second piece of important UK legislation is the Companies Act 2006. This act has helped to update rules on the Limited company model, which is used by both private organizations and public authoritiesto facilitate the development and purchase of existing renewable energy generating technologies.

In terms of governance structure, Limited companies are governed by the interests of their shareholders, with shareholder control being married to the size of shareholdings. Those with a larger holding of shares have a greater say in decision-making. As is apparent from above, this contrasts with Co-operative governance models typically used by Co-ops and BenComs, whereby each member has an equal say in how the organization is governed.

In practice, the private ownership type and company structure used for developing renewable energy infrastructures are varied. What is common practice, however, is for several Limited companies to be used to separate out the different development functions and their associated risks. This can, for example, mean that a separate developer company is subcontracted to manage the development by another company that ultimately owns the asset, which is itself separate from a holding company. Corporate company structures used in this way are not only employed to limit risk, but also to limit the disclosure of commercially sensitive information for competitive advantage. A complex arrangement of several companies that are legally separate but financially interconnected can, for example, create opaque company accounts. Such intransparent accounting reduces public oversight and therefore the possibilities of democratic participation.

These opaque corporate information practices were also apparent from the research findings in relation to disparate ownership arrangements. For example, in one case concerning a 5 MW solar farm in South West England, where a Limited company structure was adopted, the developer company (a Special Purpose Vehicle—"SPV") was removed from the parent company (who retained financial control). The SPV company directors were both employees of the parent company and thus the SPV was in practice being run by the parent company. Complex ownership structures made it challenging to trace back ownership through company records and even harder to understand the governance arrangements and model. Residents living near to the solar farm reported having limited knowledge of who owned the solar farm, and the lead developer (from a separate developer company to the parent company) commented on this complex ownership arrangement:

> It's a really bad habit in our sector that these types of projects are getting sold every two to three years. At the end, nobody knows who owns the solar farm, and if something happens, you even don't know who to call … It creates uncertainty, and nobody likes uncertainty. [LC1]

Complex ownership (and governance) arrangements, therefore, prevented detailed economic information from coming to light, which in turn prevented public involvement in decision-making. As the development decisions were made by separate companies, they were legally disconnected and removed. In practice, critical decisions were being taken within (and across) the private companies involved. However, such decisions were not open for public scrutiny as decision-making was legally fragmented across separate company structures. This paucity of publicly available information made it challenging to form a basic assessment of the ownership arrangements and severely limited public participation, thereby undermining any potential for energy democracy.

Similar limits on opportunities for public participation were also present in the case of a municipal energy governance model. While municipal energy developments are likely to be publicly funded, meaning that they are more transparent than corporate governance models, a municipal energy example demonstrated little involvement of the public in its decision-making procedures as recorded in the primary data collected. As attested to by a local councillor, governance issues were central to concerns around democratic legitimacy:

> The elected mayor is the principal shareholder […] For me, we're getting back to who else should be sitting on that shareholder thing […] where is democracy? You know, who else gets to sit on the shareholder group and make those decisions about what they do. So there is an issue of governance. [LA1]

In addition, one manager working for a Limited company model set up to compete in UK energy markets on behalf of an English city council, highlighted the exclusive nature of the governance of the municipal energy model, in which key financial decisions remained solely the preserve of the local council:

> When we're due to make a profit […] the money will go back to the council and it will go through the democratic process of the councillors and the mayor at the time will decide how that money is spent. We're here as a vehicle to create that profit. [MEC1]

Thus, the Limited company structure used for both private and public developments in the cases in question exhibited a tendency to prevent public engagement, in contrast to the Co-operatively owned developments that sought greater levels of public engagement. This limitation on public participation presented by the Limited company structure poses serious challenges to the realization of energy democracy in practice, alongside raising further questions around the use of "private" organizational structures to facilitate energy democracy.

3.3 Comparison

While it is common for several companies to be used across all forms of development, the critical distinction between private and civic led developments identified from the research data was that operational decisions in the private cases were remote and lacked transparency. This lack of transparency was traced back to a void in local representation in the developer organization.

Comparatively, in the civic cases, the developer organization established a vital link to the public through encouraging local participation. The civic ownership developments that were examined in both of our empirical research were all premised on a place-based community of interest, where membership tended to be encouraged from the local area in which the solar farms were sited. For example, member-shareholding in a large renewable energy Co-op in South West England was encouraged by both a low share offer and generating benefits for the local community:

> [the] Co-operative funded the installation of solar panels on [the] community centre. I bought some shares [...] you can spend £50, which isn't very much, and be an owner, which I think is quite cool to think that you can be part of something like that. I wouldn't invest money in the stock market or anything like that [...] it's reduced the cost of the bills and the running cost of the community centre. [MS2]

In contrast, in the case of a private development of a solar farm in England, ownership and investment was remote and distant, with decision-makers having limited interest in the public living near to the solar farm.

> To be honest, we don't really care what happens with the solar farms. [LC2]

Lastly, a BenCom solar farm case exhibited demonstrable differences to the private developments in our data, employing a Co-op governance structure but with a pronounced civic ethos and genuine concern for benefits to be retained locally. This generated a high level of public participation that supported a democratic decision-making process. Substantive public participation was evident from the significant level of support registered in the local planning application (69 letters of support and 5 letters of objection).

4 Governance models, voting rights, and AGMs

4.1 Co-operative governance: Shareholder voting and AGMs

While Co-operative (including BenComs) models of governance do support a more democratic decision-making process than private Limited companies, allowing for one-member-one vote, they alone do not guarantee energy democracy. It is our contention that for an assessment of energy democracy to be made, the governance and detailed micro-politics of on the ground decision-making and operations needs to be closely analyzed. In this section, we consider the micro-politics of voting rights and AGMs, treating these as similar across the case of BenComs and Co-operatives, and therefore group them under "Co-operative governance models." Members of Co-operatives are usually investors; however, several organizations studied also had a class of non-investor members, where members of the public can take part but have no say over certain financial decisions which remain the purview of investor members only.

Investor members have a right to vote on decisions that are raised through meetings, which in practice involves voting at the Annual General Meeting (AGM). This voting supports a mutualist link where shareholders (who own some capital) directly benefit from dividend payments.

Directors of Co-ops are voted in by members of the organization at the AGM, or they can be Co-opted directly by other directors if the organizations' rules allow. Where members stand as directors, a vote by members takes place which enhances the internal democratic processes of the organization. Voting rights are based on one-member-one vote, and accordingly, voting is unaffected by the size of shareholdings—in contrast to private developments. This is of vital importance as it prevents monetary power from influencing decision-making. However, conversely, one director of a BenCom that owned 2.8 MW of solar PV in England noted that lowering the minimum share offer to facilitate wider ownership and engagement in governance could result in member manipulation:

> What would be the motivation for buying a £10 share? Why would you do that? [...] Just to have your say – why would you want to? There is a concern that you open it up to malicious share ownership. 20 people or 30 people can buy their £10 share and come to shareholders meetings and start to manipulate us. You are less likely to get that with a £500 minimum. [CB3]

While in theory this could take place; this also limits members (owners) to relatively affluent individuals and communities, posing important social equity and justice questions for energy democracy in local energy initiatives. However, when deciding on how surplus funds should be allocated as community benefits, a director of one of the UK's biggest energy Co-ops, which at the time of research owned over 9 MW of solar PV assets in England, stated that:

> Every AGM, there will be motions on how much community benefit and how it will be allocated at a gross level. The tiny detail will [...] be devolved. But we need to retain confidence that the process that we're part of is meeting our needs, it fits in with our intentions with community benefit. [CB2]

As mentioned above, new Co-operative directors can also be "co-opted" by existing directors, and this is a fairly common practice of Co-ops working in the energy sector. Whilst a co-opted director will likely then have a confirmatory vote by members at the next AGM, this practice diminishes the voting process as a director's position is already established. Hence the practice moves voting to more of a confirmatory process as the co-opted director already has a legitimatized position, established by current directors. This practice may allow for the effective operations of the Co-operative or BenCom, but it may also diminish the potential of energy democracy in practice.

Importantly, the AGM is the core decision-making forum for Co-operatives. AGMs bring members together to (in theory) deliberate and then vote on organizational issues. They therefore represent the core organizational decision-making forum that holds the greatest potential for realizing energy democracy in practice. However, it is important to recognize that AGMs will not generate energy democracy just by taking place. The AGM itself is a political space of competing interests and different perspectives. We suggest that it is necessary to carefully analyze the operations of AGMs in each case to determine the potential for energy democracy in practice.

What happens in the run-up to a vote is of vital importance for transforming the notion of energy democracy from an academic theoretical conceptualization into a substantive real-world process. Voting can, for example, be based on sustained deliberative exchanges

[3,10], or voting can be acclamatory, simply ushering in decisions that have been made by directors before the AGM.

It is important to bear in mind that AGMs, by their nature, happen once a year, and for participation to be substantive it is necessary to have sustained engagement from members. Research findings comparing a civil society development with a municipal development, which both drew upon a BenCom legal structure, found that the AGMs in both cases largely did not function as deliberative democratic forums. A key issue identified was that while it may have been possible in principle for members to influence the agenda of the AGMs, this did not appear in practice. The directors of the BenComs set the agenda and steered the AGM discussion—particularly in the case of the municipal development. This left the directors to largely set the course and operations of the BenComs in both cases.

4.2 Corporate governance: Shareholder voting and AGMs

Similar to Co-operatives, corporate company decision-making tends to be made through AGMs where shareholders vote on key decisions, including voting on directors and agreeing to dividend payments. Although an AGM for a corporate and a Co-operative organization may appear similar, a vital difference is that shareholders in a corporate company have varying levels of control dependent on the size of their shareholdings. This crucial difference shifts the power of shareholders completely. By allowing material wealth and voting rights to be combined, the equality of members is undermined by differences in wealth and money. This essentially reduces or removes the say of shareholders without a controlling share in the company.

While anyone (with sufficient wealth) can invest in the parent company of a renewable energy development to have a say on decisions via their shareholder voting rights, the localist and place-based link (compared to the Co-operative model) is lost; particularly when companies are remote and have no association to the local area. The motives of a company are also important. It was found that there was significant interest for purchasing pre-existing solar farms in England by international investment companies, which meant that decision-making was transferred to international shareholders driven largely by commercial interests, rather than any interest in involving local citizens in decision-making around the deployment of solar farms.

5 Conclusion: The role of ownership and governance for energy democracy

In this chapter, we have demonstrated the critical role of different ownership and governance structures for both energy projects on the ground and a conceptual understanding of energy democracy. Exploring the different models of ownership and governance of renewable energy developments, alongside sectoral interactions between private, public, and civil society spheres [11], we have shown that both locally rooted ownership and transparent and inclusive governance are of critical concern to the realization of energy democracy. We have also shown that the complex connections between ownership and governance demand detailed analysis of particular civic initiatives and their operations to make sense of how we can achieve energy democracy in practice.

As noted earlier at the start of our chapter, we drew upon Szulecki's [1] theoretical distinction between governance and ownership for energy democracy, where broad claims were made around public, private, and Co-operative types of ownership being used to safeguard civic empowerment, alongside governance models ensuring wide participation. Our chapter has shown that there are significant interlinkages and overlaps between such civic ownership types in practice, with both BenCom and Co-op legal models frequently utilizing Co-operative governance structures. We also drew upon Burke and Stephens [4] and Becker and Naumann's [5] understandings of ownership within energy democracy as largely comprised of public and civil society models, with the aim of supporting new models that seek to democratize energy and reverse privatization. While we felt this was a more apt description of energy democracy ownership types, we have also shown the potential differences in civic and private ownership and differences between public, private, and civil society models of governance.

Our chapter demonstrates, through drawing on our respective findings, that Co-operative ownership types are more supportive of democratic processes, but these do not guarantee energy democracy alone. The private Limited company ownership structure is often complex and intransparent due to fragmented ownership arrangements, which significantly impairs local decision-making and undermines democratic potential. In addition, we found that, in our empirical cases, municipal energy governance models allowed for little democratic involvement in decision-making procedures. Despite being publicly funded, the "democratic process" was the purview of local government executives, offering limited opportunities for wider public input into decision-making. This has implications for how public models of energy ownership may reconsider their approach to governance and the broader "remunicipalization" agenda [12], as it is local government executives that engage in governance processes, rather than local citizens able to purchase shares or participate in local democratic forums, such as AGMs.

Nevertheless, we also need to think about how the micro-politics of decision-making and operations of organization, with careful and detailed analysis of what makes for inclusive and participatory decision-making in practice. As one of our research participants noted, shareholding is inherently political and does not guarantee democratic outcomes. For example, if local citizens partially fund municipal ventures through their taxes, there is a strong case that they should automatically be owners and/or members. Therefore, unless publicly owned and funded energy companies provide genuine and substantive means for public participation, they cannot be said to be deeply compatible with energy democracy.

One of our chapter's key findings for energy democracy scholars is that in our empirical cases, private Limited company developers demonstrate limited democratic engagement with anyone beyond their shareholders. Furthermore, they also exhibited little direct involvement with the communities around where energy infrastructures are sited. This is supported by further research [10] that considers certain cases considered in this chapter. In contrast, the civic ownership and civil society governance models featured in our analysis demonstrated significant involvement with the local communities near to where they were situated. These developments also took an active interest in supporting local economies and embedding local community benefits into their business models, which garnered further local support.

In our view, energy democracy is achieved when ownership structures and governance models work together to facilitate regular and meaningful citizen engagement in energy projects; ideally to support open participation in the critical decision-making procedures that direct the energy development and to reaffirm place-based ownership that is rooted in the locality. This makes apparent the vital link between ownership types and governance models and the reality of local decision-making in practice. In addition, opportunities for widening engagement among different socio-economic demographics, through both lowering the threshold for minimum investment and allowing for non-member participation, may contribute toward the realization of energy democracy and have positive implications for social justice in energy transitions [13]. Lastly, we urge scholars to continue critical and interdisciplinary comparative research into the ownership types, legal structures and governance models used to democratize energy infrastructures. These are, after all, the principal vehicles through which energy democracy may, or may not, be realized.

Appendix

See Table 14.2.

TABLE 14.2 Anonymized identifier system for research participants.

Participant positions	Associated organization (if applicable)	Assigned identifier
Corporate Developer	Limited company	LC1, LC2
Manager	Municipal energy company	MEC1
Director Member-Shareholder	Co-operative/BenCom	CB1, CB2, CB3 MS1, MS2
Local councillor	Local Authority	LA1
Resident	N/A	R1

Acknowledgments

Both authors thank the many research participants from across England that took part in their PhD research, whose respective insights helped to enhance and inform our thinking around the realities and practicalities of energy democracy on the ground. Max's PhD research was supported by an EPSRC studentship from 2014 to 2017, Grant number BV25012101/3. Jack's PhD research was supported by an ESRC studentship from 2014 to 2018, Grant number ES/J50015X/1.

References

[1] K. Szulecki, Conceptualizing energy democracy, Environ. Polit. 27 (1) (2018) 21–41.
[2] M. Lacey-Barnacle, Exploring Local Energy Justice in Times of Austerity: Civic Energy Sector Low-Carbon Transitions in Bristol City (Doctoral dissertation), Cardiff University, 2019.
[3] J. Nicholls, Owning the Sun: Energy Democracy and Public Participation in Solar Farm Developments in England (Doctoral dissertation), University of Bristol, 2020.
[4] B. Van Veelen, Negotiating energy democracy in practice: governance processes in community energy projects, Environ. Polit. 27 (4) (2018) 644–665.

[5] M.J. Burke, J.C. Stephens, Energy democracy: goals and policy instruments for sociotechnical transitions, Energy Res. Soc. Sci. 33 (2017) 35–48.

[6] S. Becker, M. Naumann, Energy democracy: mapping the debate on energy alternatives, Geogr. Compass 11 (8) (2017), e12321.

[7] Legislation, U. K, Co-operative and Community Benefit Societies Act 2014, 2014.

[8] Parliament, U. K, Companies Act 2006, Cabinet Office, 2006. https://www.legislation.gov.uk/ukpga/2006/46/contents.

[9] A.Z. Kimberly, R. Cropp, Cooperatives: Principles and Practices in the 21st Century, University of Wisconsin, USA, 2004.

[10] J. Nicholls, Technological intrusion and communicative renewal: the case of two rural solar farm developments in the UK, Energy Policy 139 (2020), 111287.

[11] E. Creamer, W. Eadson, B. van Veelen, A. Pinker, M. Tingey, T. Braunholtz-Speight, et al., Community energy: entanglements of community, state, and private sector, Geogr. Compass 12 (7) (2018), e12378.

[12] A. Cumbers, Remunicipalization, the low-carbon transition, and energy democracy, in: State of the World, Island Press, Washington, DC, 2016, pp. 275–289.

[13] M. Lacey-Barnacle, Proximities of energy justice: contesting community energy and austerity in England, Energy Res. Soc. Sci. 69 (2020), 101713.

Lessons from electric cooperatives: Evolving participatory governance practices

Stephanie Lenhart

Energy Policy Institute – Center for Advanced Energy Studies and School of Public Service, Boise State University, Boise, ID, United States

1 Introduction

Cooperatives are often viewed as an organizational form with the potential to promote more decentralized, sustainable, and just societies. The energy democracy movement connects the renewable energy transition to a redistribution of political and economic power, and cooperatives are mechanisms that can reclaim energy decision-making to prioritize the public interest [1,2]. Founded on well-established ideals and principles, cooperative institutions provide a model for pluralistic governance [3,4]. Electric cooperatives were first formed in the early 20th century to provide an essential service to a community as a whole, rather than a specific group of users [5]. In recent decades, electric cooperatives have emerged as prominent sites of experimentation in energy democracy [6,7]. Yet, the ability for cooperatives to further the ideals of energy democracy varies across local contexts and a diversity of nested institutional relationships.

This chapter explores how electric cooperatives, new and old, enact participatory governance. As sustainability and social justice concerns, demographic shifts, and technology innovations change opportunities and expectations for communication and participation, cooperatives are sites of contestation. Many established cooperatives are experiencing disruption in traditional participation mechanisms, and more recent cooperatives are conceiving of new participation mechanisms within the context of increasing interdependencies and restructured markets.

The next section describes the evolution of the cooperative institutional model. The third section draws on existing literature to explain the implications of the cooperative institutional design for institutional robustness, instrumental values, and civil society. The fourth section examines how cooperatives frame relationships with their members and the symbolic and material mechanisms of member participation. The chapter closes with an assessment of the challenges and opportunities cooperatives face in forming new structures for participation and democratic member control.

2 Cooperatives and member participation

The earliest cooperatives were established to address social justice failures during the industrial revolution [3]. These cooperatives were formed to benefit specific social groups that lacked access to economic or political power, such as small farmers or craftsmen. The benefits of cooperation demonstrated by these institutions also spurred the emergence of cooperatives formed to provide essential services to all members of a community. Electric cooperatives are an example of a cooperative form designed specifically to further social progress and based on the institutional design concept of *open membership*. These institutions providing an essential service of general interest were critical to rural electrification and economic development in Europe and North America. By the turn of the 20th century, a new cooperative form emerged in which *concern for community* is a central aim. In other words, this institutional form is explicitly designed to further the public interest [5].

In practice, a diversity of cooperative forms exists. In Europe, most of the early electric cooperatives established to provide a service of general interest were nationalized during the second half of the 20th century, and new community electric cooperatives have been established following the market liberalization that began in the late 1990s [7,8]. In contrast, many of the nearly 900 electric cooperatives in the United States have provided continuous service for more than a century [9]. These cooperatives operate with a focus on both member benefits and a concern for the community.

Today, electric cooperatives are an example of the potential and limits of providing an essential service to a community as a whole aimed specifically at furthering the public interest. In practice, electric cooperatives are either evolving or newly forming within existing institutional relationships and technical networks that complicate adherence to pluralistic ideals.

3 Institutional robustness, instrumental values, and civil society

A critical element of energy democracy is a demand for more accountability to the public interest in energy policymaking, which in turn contributes to a transformation of existing political and economic systems through the empowerment of individuals and communities [10]. By developing active participants, distributed energy can deepen civil society and reorder complex sociotechnical systems that reinforce existing political and economic inequities [11]. These ideals are promoted through individual and community ownership of energy infrastructure [1] and increased collective participation in decision-making [12]. Specific to governance, energy sector democratization seeks new institutional designs that are more inclusive, open to contestation, procedurally transparent, supportive of the resources and capacity to facilitate meaningful participation, and facilitated by the exchange of information, education, and new policy instruments [10–15].

Many of these features are embedded in cooperative principles and practices. Electric cooperatives are member-owned institutions that provide services by sharing the costs and financial risks of complex infrastructure and technology within a democratic governance framework [5]. Electric cooperatives are considered member-owned if they are supported by power purchases from residents within their service territory, as in the case of traditional electric distribution cooperatives, or by a subset of residents that choose to invest in the cooperatives as members, as in the case of relatively new renewable energy cooperatives. These examples reflect two basic categories of cooperatives: customer-owned or supplier-owned [8]. The principles that emerged from the cooperative movements over the past two centuries guide current day cooperatives even as the institutional form continues to evolve and diversify. The International Co-operative Alliance, which was founded in 1895, provides cooperative definitions and principles that span national boundaries and represents an estimated three million cooperatives worldwide [3]. Cooperatives are guided by seven principles (Table 15.1).

TABLE 15.1 Cooperative principles.

Voluntary and open membership	Cooperatives are voluntary organizations, open to all persons able to use their services and willing to accept the responsibilities of membership, without gender, social, racial, political, or religious discrimination.
Democratic member control	Cooperatives are democratic organizations controlled by their members, who actively participate in setting their policies and making decisions. Men and women serving as elected representatives are accountable to the membership. In primary cooperatives, members have equal voting rights (one member, one vote) and cooperatives at other levels are also organized in a democratic manner.
Member economic participation	Members contribute equitably to, and democratically control, the capital of their cooperative. At least part of that capital is usually the common property of the cooperative. Members usually receive limited compensation, if any, on capital subscribed as a condition of membership. Members allocate surpluses for any or all of the following purposes: developing their cooperative, possibly by setting up reserves, part of which at least would be indivisible; benefiting members in proportion to their transactions with the cooperative; and supporting other activities approved by the membership.
Autonomy and independence	Cooperatives are autonomous, self-help organizations controlled by their members. If they enter into agreements with other organizations, including governments, or raise capital from external sources, they do so on terms that ensure democratic control by their members and maintain their cooperative autonomy.
Education, training, and information	Cooperatives provide education and training for their members, elected representatives, managers, and employees so they can contribute effectively to the development of their cooperatives. They inform the general public—particularly young people and opinion leaders—about the nature and benefits of cooperation.
Cooperation among cooperatives	Cooperatives serve their members most effectively and strengthen the cooperative movement by working together through local, national, regional, and international structures.
Concern for community	Cooperatives work for the sustainable development of their communities through policies approved by their members.

From International Cooperative Alliance, Cooperative Values, Identity, and Principles, n.d. https://www.ica.coop/en/cooperatives/cooperative-identity (Accessed 2 February 2020).

The cooperative governance design asks member-owners to be actors that cooperate to meet their needs through self-governing communities. This element of participation is core to the institutional form and requires more than indirect participation through the election of governing boards or annual general meetings. Rather, the benefits of the cooperative form require active engagement through co-production, direct participation in governance, and concern for the community [5,16]. From a participatory governance perspective, engaging member-owners that have diverse perspectives and local knowledge provides normative value by deepening civil society and instrumental value by increasing legitimacy, effectiveness, and equity [17,18]. Cooperatives also provide an opportunity to enhance institutional robustness through experimentation, innovations, and limited reliance on formal statutory authority [16].

Ownership of the organization, not just the energy infrastructure, confers rights to governance, residual profits, and internal information. Furthermore, the cooperative principles promote participatory governance functions including information exchange (both to and from members), monitoring, transparency, and direct engagement in operations and decision-making [16]. Finally, member-owners having been granted authority to co-govern have a responsibility to ensure democratic accountability [19].

4 Changing opportunities and expectations for member participation

Across electric cooperatives, a variety of deviations from cooperative ideals and principles have been observed, and while the election of board members and annual member meetings are widespread mechanisms, meaningful member-owner participation in cooperative governance varies in practice [6,20,21]. Moreover, the organizational identity of electric cooperatives is in transition as increased attention to sustainability and social justice concerns, demographic shifts contributing to more heterogeneous community interests, and technology innovations are changing opportunities and expectations for communication and participation. Within this context, some electric cooperatives are facing challenges in maintaining connections with member-owners and differentiating themselves from more prevalent investor-owned utilities. By contrast, other electric cooperatives are creating opportunities for voluntary participation and many are enabling the expression of community values through material or symbolic participation. Based on an in-depth study of a subset of electric cooperatives in the United States, variation in member-owner participatory practices and communication can be viewed as a continuum (Table 15.2).

Many electric cooperatives are experiencing disruptions in traditional mechanisms of participation. Long-standing electric cooperatives centered member-owner participation in traditions of meeting with community members regularly in local shops, talking with members when they came to pay their bill, and the connection established by generations of families that lived and worked in communities and understood the importance of "when the lights came on." In recent decades, these practices have been limited or discontinued by changing lifestyles, demographics, or technology innovations that limit face-to-face interactions. Many current electric cooperative members do not know what a cooperative is or how it differs from an investor-owned utility, and this disconnect is exacerbated when infrastructure has been moved underground and communication has been automated. Today, many member-owners want to do business on the phone or through a website and electric cooperatives have limited

TABLE 15.2 Continuum of participatory governance practices among cooperatives.

Disruptions in traditional mechanisms for participation	Opportunities for voluntary participation	New structures for inclusive participation
Provide information or define procedures that limit individual agency Lack of adaptation to shifting demographics, new communication technologies, and other factors	Education and input into decision-making Responding to individual demands and presenting opportunities to individuals	Material and symbolic participation as expression of values Engage individuals through knowledge, transparency, and deliberation that enables individual agency

Modified from G. Chan, S. Lenhart, L. Forsberg, M. Grimley, Barriers and Opportunities for Distributed Energy Resources in Minnesota's Municipal Utilities and Electric Cooperatives, University of Minnesota, 2019.

interaction with or understanding of their member-owners. The traditional connections to the community are being disrupted and new forms of participatory governance are needed.

Despite these challenges, most electric cooperatives are working to provide opportunities for voluntary participation. Typically, cooperatives have open public meetings that are intended to allow community participation, and efforts to solicit input through surveys and other forms of voluntary participation are common. Some electric cooperatives actively focus on community education, engage volunteer groups, or provide opportunities for the community to interact with energy service providers. Building on a tradition of community service projects and annual community events, cooperatives are holding town hall meetings with door prizes, activities or meals, working with distributed generation vendors and members to assess design and investment options, and developing social media and interactive applications for information exchange. Yet, these traditional approaches to participatory governance vary in their ability to provide meaningful information exchange, monitoring, transparency in decision-making, or deliberation in policy direction.

Finally, in addition to the political power derived from material participation in ownership and control of energy system assets, many electric cooperatives recognize participation as an important means for expressing community values. Cooperatives in generation and transmission relationships have created "billboard" solar energy projects in every community they serve, and distribution cooperatives claim that the value of placing small-scale solar generation directly within communities are examples of efforts to increase the salience of energy use and the legitimacy of energy service providers. These material assets are seen to help with education by giving individuals a visible, physical manifestation of renewable energy. Furthermore, distributed renewable energy infrastructure is viewed as creating a sense of ownership, increasing member-owner involvement, and lending legitimacy to cooperative decisions. Renewable and distributed energy resources provide multiple attributes beyond de-carbonization. As explained by one practitioner, "People are trying to do something else. They are trying to affirm their values around renewables, they're trying to maybe have a little additional security, or they're making an economic play … ." Thus, how the resource is provided makes a difference. These new structures make cooperative power visible and can provide new participatory mechanisms to revive community engagement that has diminished as infrastructure is increasingly underground and distant. Visible on-site renewable power serves to increase the legitimacy of electric cooperatives as community members are losing a sense of how cooperative power differs from investor-owned power.

5 Challenges and opportunities

Member-owner participation in cooperative governance provides an alternative to top-down elected representative forms of accountability and also establishes responsibility for active co-governance. Abdication or obstruction of this responsibility limits the potential benefits of cooperatives by disconnecting the network of electricity system institutions from public interest accountability.

Electric cooperatives are seeking to develop new participatory mechanisms and maintain legitimacy as the foundational structures of participatory governance are shifting with evolving demographics and changing expectations for the use of communication technologies. Cooperatives that traditionally relied on face-to-face communication and deep community ties to ensure adherence to cooperative principles and democratic control are now challenged to maintain connections with members and differentiate themselves. The formal approaches to participatory governance such as surveys and open meetings vary in the ability to provide meaningful deliberation, particularly where informal participation is diminished. Yet, what is emerging in cooperatives reflects unique structures and policy instruments of governance.

Many cooperatives are embracing material and symbolic forms of participation through distributed generation, energy efficiency competitions, and other community engagement activities. Building on traditional cooperative principles, some cooperatives are re-envisioning their community identity around goals of economic development, social cohesion, or sustainability. They are providing voluntary programs, like community solar or demand management offerings that allow community investment and joint ownership or joint control of resources. Furthermore, they are engaging in cooperation, education, and information sharing through new information and communication technologies and by engaging with new actors including distributed generation vendors. These new structures make cooperative member-owned power visible and could provide new participatory mechanisms to revive community engagement and democracy.

References

[1] M.J. Burke, J.C. Stephens, Energy democracy: goals and policy instruments for sociotechnical transitions, Energy Res. Soc. Sci. 33 (2017) 35–48, https://doi.org/10.1016/j.erss.2017.09.024.

[2] J.C. Stephens, Diversifying Power: Why We Need Antiracist, Feminist Leadership on Climate and Energy, Island Press, 2020.

[3] International Cooperative Alliance, Cooperative Values, Identity, and Principles, n.d. https://www.ica.coop/en/cooperatives/cooperative-identity (Accessed 2 February 2020).

[4] K. Taylor, Governing the Wind Energy Commons: Renewable Energy and Community Development, West Virginia University Press, 2019.

[5] P.A. Mori, Community and cooperation: The evolution of cooperatives towards new models of citizens' democratic participation in public services provision, Ann. Public Coop. Econ. 85 (2014) 327–352, https://doi.org/10.1111/apce.12045.

[6] D. Fairchild, A. Weinrub, Energy Democracy: Advancing Equality in Clean Energy Solutions, Island Press, Washington, DC, 2017.

[7] Ö. Yildiz, J. Rommel, S. Debor, L. Holstenkamp, F. Mey, J.R. Müller, J. Radtke, J. Rognli, Renewable energy cooperatives as gatekeepers or facilitators? Recent developments in Germany and a multidisciplinary research agenda, Energy Res. Soc. Sci. 6 (2015) 59–73, https://doi.org/10.1016/j.erss.2014.12.001.

[8] P.A. Mori, Customer ownership of public utilities: new wine in old bottles, J. Entrep. Organ. Divers. 2 (2013) 54–74, https://doi.org/10.5947/jeod.2013.004.

[9]   B.J.M. Carmody, Rural electrification in the United States, Ann. Am. Acad. Political Soc. Sci. 201 (2016) 82–88. Ownership and Regulation of Public Utilities.

[10] K. Szulecki, Conceptualizing energy democracy, Environ. Polit. 27 (2018) 21–41, https://doi.org/10.1080/0964 4016.2017.1387294.

[11] M.J. Burke, J.C. Stephens, Political power and renewable energy futures: a critical review, Energy Res. Soc. Sci. 35 (2018) 78–93, https://doi.org/10.1016/j.erss.2017.10.018.

[12] B. van Veelen, D. van der Horst, What is energy democracy? Connecting social science energy research and political theory, Energy Res. Soc. Sci. 46 (2018) 19–28, https://doi.org/10.1016/j.erss.2018.06.010.

[13] B. Van Veelen, Negotiating energy democracy in practice: governance processes in community energy projects, Environ. Polit. 27 (2018) 644–665, https://doi.org/10.1080/09644016.2018.1427824.

[14] B. Cozen, D. Endres, T.R. Peterson, C. Horton, J.T. Barnett, Energy communication: theory and praxis towards a sustainable energy future, Environ. Commun. 12 (2018) 289–294, https://doi.org/10.1080/17524032.2017.13 98176.

[15] S. Welton, Grasping for energy democracy, Mich. Law Rev. 116 (2018) 581–644, https://doi.org/10.1533/97818 45699789.5.663.

[16] K. Taylor, Learning from the co-operative institutional model: how to enhance organizational robustness of third sector organizations with more pluralistic forms of governance, Adm. Sci. 5 (2015) 148–164, https://doi.org/10.3390/admsci5030148.

[17] A. Fung, Varieties of participation in complex governance, Public Adm. Rev. 66 (2006) 66–75, https://doi.org/10.1111/j.1540-6210.2006.00667.x.

[18] A. Fung, Continuous institutional innovation and the pragmatic conception of democracy, Polity 44 (2012) 609–624, https://doi.org/10.1057/pol.2012.17.

[19] A. Fung, Empowered Participation: Reinventing Urban Democracy, Princeton University Press, 2004.

[20] G. Chan, S. Lenhart, L. Forsberg, M. Grimley, Barriers and Opportunities for Distributed Energy Resources in Minnesota's Municipal Utilities and Electric Cooperatives, University of Minnesota, 2019.

[21] S. Lenhart, G. Chan, L. Forsberg, M. Grimley, E. Wilson, Municipal utilities and electric cooperatives in the United States: interpretive frames, strategic actions, and place-specific transitions, Environ. Innov. Soc. Transit. 36 (2020) 17–33.

16

Bringing democratic transparency to Karachi's electric sector

Ijlal Naqvi

Singapore Management University, Singapore

1 Introduction

Karachi's electricity sector has been transformed in the past 10 years due to the actions of KE (Karachi Electric), the privatized electrical utility in this metropolis of 15 million people on Pakistan's Arabian Sea coast. The indicator which best exemplifies this transformation is that of electricity losses, or the proportion of electricity that is supplied to the grid but never billed to any consumer. Although there are also engineering reasons for losses, losses primarily signify theft by consumers who avoid billing by illegally tapping into overhead power lines. There are financial implications for high losses which include higher prices for paying consumers, greater theft as higher prices make electricity unaffordable, and a broader detrimental impact on public sector finances from accumulated subsidies and arrears.

The inadequacy of electricity supply has been a national political issue of key importance in Pakistan; it is estimated to reduce GDP growth by some 2%, while putting a heavy burden on the national budget. Against this backdrop, KE reduced its losses from 35.9% in 2009 to 19.1% in 2019—a remarkable improvement. A key element of the campaign to reduce losses in Karachi has been the strategic decision to preferentially serve those areas of the city which do pay for electricity (known as "segmented loadshedding"). Additionally, KE has attempted to modify the behavior of its consumers by offering a reduction in rolling blackouts (loadshedding) if an area can reduce its losses and has initiated several engineering and social interventions aimed to achieve this change. Although the mechanisms whereby the reduction in losses was achieved requires closer investigation, KE presents its initiative of segmented loadshedding as a success [1], yet citizens of Karachi still have serious complaints about the performance of KE [2]. In this essay, I explore how the principles of democratic transparency can be used to guide a proposed intervention that serves both the human necessity of access to electricity as well as the utility's need to balance its books.

2 The problem of electricity supply in Pakistan

The problem of electricity supply in Karachi is relevant to Pakistan more broadly, to other countries of the Global South, and also to other aspects of human well-being for which public services are important. The United Nation's Sustainable Development Goals spell out the many different ways that public service delivery, including electricity supply, is crucial for human freedoms and capabilities, and thus for development goals in general [3,4]. Beyond simple access, the quality of service delivery is also important [5]. Although Karachi is unique in Pakistan for its privatized utility, electricity distribution is heavily regulated, and the problem of high losses is present to varying degrees throughout the country [6,7]. Electricity losses are a problem present in many countries across the Global South, and one that is seen as particularly difficult to address because of the political and social dimensions of regulating citizen conduct [8,9].

There are two competing explanations of KE's reduction in losses. The first explanation is that consumer behavior has not changed, but that KE simply reduced its aggregate losses by shifting its provision of electricity to low-loss areas rather than high-loss areas. In other words, high-loss areas remain high loss, but the reduced quantum of electricity supply means that there is less impact on KE's aggregate losses. The second explanation is that KE has indeed succeeded at altering consumer behavior, and that previously high-loss areas have reduced their percentage losses. Some evidence for both explanations exists, though the second is the one usually advanced by KE. In the absence of a careful empirical analysis using longitudinal data, it is difficult to say which explanation has been more central to the KE reduction in losses, and I proceed herewith on the basis that both mechanisms are likely to be at work to different degrees in different areas of Karachi. This distinction matters because governance interventions will need to be tailored to their locality accordingly.

The reduction in losses notwithstanding, grounds for considering an intervention aimed at introducing democratic transparency stem from the fact that problems persist with electricity supply in Karachi: the cost of electricity is still high; the KE solution to high losses appears to entail a disciplining of the citizenry, to compel them to pay for service; there is a constraining of *jugaad* (a South Asian term for improvised solutions that reflect resourcefulness and self-reliance), and possible inhibition of collective action; KE has been accused of rampant overbilling to boost recovery [10]; and the practice of allocating loadshedding based on feeder losses serves as a collective punishment mechanism that differentially impacts the poor.

3 Democratic transparency

The concept of democratic transparency developed by Archon Fung and collaborators is based on principles of putting information of appropriate richness and depth into the hands of ordinary citizens and NGOs so that they can take action which can improve access to and the quality of vital public services [11]. Despite their wide uptake and broad intellectual and normative appeal, the track record of transparency-related interventions is decidedly mixed, with a key factor being the social and political context in which these interventions toward democratic transparency are made [12]. In none of these interventions was it the case that a monopoly service provider was keen on reform.

However, for Karachi, and indeed this would be true for any electrical utility, the demand for high quality and reliable electricity supply by citizens is something that the utility has every incentive to fulfill—provided that it can cover its own costs of operations. Hence the crucial nature of losses: If a compact can be made with citizens that they will pay for a certain quality of electricity supply, then all parties gain something. Moreover, the political actor who could broker and sustain such an arrangement would gain tremendous popular and business support. How then could such a virtuous cycle be engaged?

KE's existing approach to transforming high-loss areas into low-loss areas already combines technological change with local buy-in through the involvement of elected local government officials and bazaar associations. What is not provided, however, is any information that allows the citizens, their representatives, or any civil society groups to monitor the performance of KE with regards to schedules for blackouts or to scrutinize the campaigns to collect past arrears (regarding which there have been complaints of excessive bills). KE's approach is therefore incomplete, and the feedback loop needs to be closed. The map of electricity consumption and losses in Fig. 16.1 could form the basis for such a feedback loop, and could be used by traditional news media as well as in web-based applications for public consumption. A spatial overlay which incorporates local landmarks and other infrastructure-based services would further facilitate citizen's understanding of urban governance in their neighborhoods and the city around them [13].

KE seeks to enable behavioral change but does not give citizens the information they need to monitor KE's operations and see whether changes in their neighborhood are taking hold. Fig. 16.1 could be the basis for providing this information to citizens, but it would have to be supported by social action which enriches two-way communications between citizens and the utility. In particular, local civil society groups could play a vital role in processing information on KE's operations—such as loadshedding schedules, losses, and bill collection campaigns—and sharing insights from them with citizens. Elected local government leaders can also play a key role as champions of this transformation, and thus claim credit for the improved service delivery environment.

4 Conclusions

The argument for an intervention modeled on the principles of democratic transparency is both utopian and practical [14]. While local power brokers will definitely have sensitivities to any changes in the status quo, and Karachi's history of violence must be taken into account here, there is also ample scope to benefit all parties through the provision of higher quality services. Karachiites continue to demand such change and express their dissatisfaction with KE. By enabling a virtuous cycle of democratic transparency, the provision of actionable information in partnership with local civic organizations and elected leaders can be the basis for securing that improved service delivery environment.

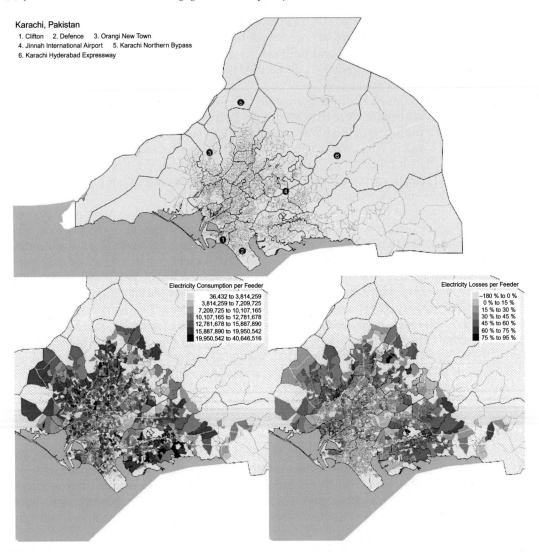

Karachi, Pakistan
1. Clifton   2. Defence   3. Orangi New Town
4. Jinnah International Airport   5. Karachi Northern Bypass
6. Karachi Hyderabad Expressway

Electricity Consumption per Feeder
36,432 to 3,814,259
3,814,259 to 7,209,725
7,209,725 to 10,107,165
10,107,165 to 12,781,678
12,781,678 to 15,887,890
15,887,890 to 19,950,542
19,950,542 to 40,646,516

Electricity Losses per Feeder
−180 % to 0 %
0 % to 15 %
15 % to 30 %
30 % to 45 %
45 % to 60 %
60 % to 75 %
75 % to 95 %

FIG. 16.1   Top: Karachi overview map for orientation. Bottom left: Energy consumption map. Bottom right: Energy losses map.

References

[1] Karachi Electric, Segmented Load-Shed in Line With National Power Policy: K-Electric, 2020, Available from: https://www.ke.com.pk/segmented-load-shed-in-line-with-national-power-policy-k-electric/.
[2] PTI protests over loadshedding, demands end to KE monopoly, Dawn (2020). July 6, 2020. Available from: https://www.dawn.com/news/1567299.
[3] United Nations General Assembly, United Nations Millennium Declaration, in 55/2, 2000, Available from: https://www.ohchr.org/EN/ProfessionalInterest/Pages/Millennium.aspx.

[4] A. Sen, Development as Freedom, Random House, 1999.

[5] P. Farmer, et al., Reimagining Global Health: An Introduction, University of California Press, 2013.

[6] I. Naqvi, Pathologies of development practice: higher order obstacles to governance reform in the Pakistani electrical power sector, J. Dev. Stud. 52 (7) (2016) 950–964.

[7] K. Munir, S. Khalid, Pakistan's power politics, Econ. Polit. Wkly. 47 (25) (2012) 24–27.

[8] D.G. Victor, T.C. Heller, The Political Economy of Power Sector Reform: The Experiences of Five Major Developing Countries, Cambridge University Press, 2007.

[9] T.B. Smith, Electricity theft: a comparative analysis, Energy Policy 32 (2004) 2067–2076.

[10] K. Hussain, K-Electric: fattened for sale? Dawn (2017). January 28, 2017. Available from: https://www.dawn.com/news/1310013.

[11] A. Fung, Infotopia: unleashing the democratic power of transparency, Polit. Soc. 41 (2) (2013) 183–212.

[12] S. Kosack, A. Fung, Does transparency improve governance? Annu. Rev. Polit. Sci. 17 (2014) 65–87.

[13] I. Naqvi, A. Poorthuis, A. Govind, Urban governance and electricity losses: an exploration of spatial unevenness in Karachi, Pakistan, Energy Res. Soc. Sci. 79 (2021), https://doi.org/10.1016/j.erss.2021.102166.

[14] E.O. Wright, Envisioning Real Utopias, Verso, 2010.

17

Energy literacy: Democratizing energy access initiatives in Papua New Guinea

Anthony P. Heynen[a], Matthew J. Herington[b],
Craig B. Jacobson[a], Lilly P. Sar[c], and Paul A. Lant[a]

[a]Energy & Poverty Research Group, School of Chemical Engineering, The University of Queensland, Brisbane, QLD, Australia [b]Centre for Communication and Social Change, School of Communication and Arts, The University of Queensland, Brisbane, QLD, Australia [c]Centre for Social and Creative Media, University of Goroka, Goroka, Papua New Guinea

1 Introduction

Energy literacy represents a foundational process in democratizing energy, particularly in efforts to tackle energy poverty, a term used to describe a lack of access to modern forms of energy. Evidence suggests that energy access on its own is insufficient to lead to sustained, positive, social impacts for those who are currently energy poor. We contend here that an engagement approach enabling such communities to articulate their energy needs and experiences is important to allow energy access transitions to achieve sustained development outcomes. This would ensure that consumers can more fully participate in new energy projects. This chapter provides an outline of such a program and its implementation in Papua New Guinea (PNG), where electricity access levels are among the world's lowest.

A pilot program was implemented in PNG via a train-the-trainer model involving local university students, who then conducted workshops with community youth leaders in selected rural locations in PNG. The workshops discussed energy poverty impacts, energy access opportunities and benefits, and also the participatory tools used to create community ownership and empowerment. Workshops concluded with a discussion of governance and the available resources to achieve sustainability and longevity of the future projects.

The approach may provide a first and important step toward deep community engagement and contribute to more sustainable energy access in PNG and the Global South. The chapter concludes with a reflection on the approach's importance in the broader discussion on democratizing energy.

2 Context

In the Global South, efforts to tackle energy poverty are aimed at improving many dimensions of human well-being and development. These include health (by reducing indoor air pollution, a cause of respiratory and cardio-pulmonary diseases); gender empowerment (by reducing the need for women and girls in particular to collect fuel wood); and education (through household lights extending student learning hours). Energy, whether modern or in more primitive forms, is used for basic necessities such as cooking, heating and lighting, as well as more contemporary uses like charging cellular phones. Energy can be considered a critical input into the functioning of communities with or without modern forms of energy access. However, prolonged use of traditional forms of energy can become a ubiquitous and static part of energy poor communities that can fundamentally limit development progress.

Governments and the private sector have undertaken multi-scale programs focused on providing access to clean, reliable forms of energy, commensurate with the United Nations' Sustainable Development Goal 7. Notably, India has connected electricity to millions of households through concerted government efforts [1], while the private sector continues to lead progress in sub-Saharan Africa [2].

A growing body of research on energy justice suggests that culturally appropriate, co-designed energy initiatives based on meaningful community engagement processes, integrated alongside a supportive ecosystem of financing and productive energy uses, are more likely to lead to positive social impacts and sustained community transitions [3–5].

Underpinning these integrated approaches are the principles of energy democracy, where progressive social change is engendered through a bottom-up approach [6]. In energy poor communities, which often face other systemic disadvantages, energy democracy is achieved through empowering individuals, households, and communities with the tools to articulate their energy goals and aspirations. This energy literacy represents a foundational change in the power dynamics of energy provision. Historically, energy access has been implemented via state-funded donor-gift paradigms (1970s and 1980s) and market creation paradigms (1990s and 2000s) [7]. As energy access paradigms shift to more sustainable approaches, shared-value multi-party approaches necessitate the democratic articulation of energy needs and aspirations by energy's intended beneficiaries.

Evidence across theory and practice shows that greater local participation is essential for achieving meaningful development outcomes [8,9]. Buzz words such as people-centered development, empowerment, and capacity building have become common lexicon in the design and implementation of development projects globally [10]. For any given transition in an energy poor context, it is likely that there have been previous experiences with energy projects in the communities, sometimes perceived to deliver positive outcomes and sometimes not.

However, not all local participation and engagement is equal. Arnstein's ladder of participation [11] describes a spectrum of approaches toward local engagement, from nonparticipation and token participation to citizen control and ownership over the development process. More meaningful and purposeful participation at higher levels of the ladder enables a true voice in decision-making, and creates local ownership and motivation to change from the status quo.

Applied to the energy sector, Herington et al. [7] argue that the "purposes for which planners undertake participatory processes in rural energy decision making fall into one of five broad categories: to exclude, legitimize, consult, partner or empower." As a result, the level of stakeholder participation in decision-making processes for energy access projects can range from non-existent and shallow engagement, such as in the form of information sessions or briefings, to deep engagement in the form of workshops and deliberate capacity building efforts.

Transitioning to energy democracy requires the development of more nuanced understandings of local participation and the interrogation of what purposes underpin those efforts and what outcomes result. Thus, a targeted program of energy awareness and literacy building, particularly before or during community energy projects, is a first and important step toward deep engagement in the energy sector and the democratization of energy.

3 Papua New Guinea

PNG is a nation occupying the eastern half of the island of New Guinea and surrounding islands, in the southwest Pacific Ocean region. Its population, approaching 9 million, is young (76% are under 35 years of age), and 80% of people live in rural areas [12]. The nation has extraordinary natural resources: it contains approximately 5% of the world's biodiversity [13] and has globally significant oil, gas, and minerals reserves, which underpin its export-orientated economy [14]. Despite this, approximately 40% of its population live below the poverty line and 75% of households depend on subsistence agriculture [12]. There are large differences in poverty levels between the nation's capital city, Port Moresby, and other towns and rural areas.

PNG has one of the lowest electricity access rates in the world, at just 13% [15]. With most electricity grid infrastructure located in major cities and towns, the electrification rate in rural areas may be as low as 8%. Access to electricity is inextricably linked to human development outcomes such as health, education, and employment. Therefore, the low levels of electricity access perpetuate inequalities and exacerbate vulnerabilities in rural communities. Unconnected communities are faced with inadequate levels of connectivity, access to markets, education, and medicinal storage. When these services are available, they are significantly more expensive: this represents a "poverty trap" for such locations [16].

The nation's electricity system is a monopoly, managed by the state-owned electricity company, PNG Power Limited (PPL), through three major grids and 29 diesel-operated mini-grids [17]. The system is generally viewed as lacking generation capacity and distribution coverage, with low levels of reliability [15]. The PNG Government has a National Energy Policy to increase electricity access to 70% of the population by 2030 [15]. This target was given added momentum in late 2018 through the PNG Electrification Partnership, with Australia, the United States, Japan, and New Zealand committing to assist PNG on this agenda [18].

For the substantial areas of the country away from electricity networks, pico systems such as solar lanterns provide light to approximately 60% of PNG's households [19]. These off-grid systems are generally provided to households by non-governmental organizations (NGOs). Private purchases of imported solar lantern products are also common [17].

As governments and NGOs rise to the challenge of providing electricity access, there are a number of barriers to overcome. In addition to PNG's mountainous topography and low population density, lack of governance in the electricity sector and lack of capacity and support given to rural communities are significant barriers to energy development [17]. Sovacool et al. [20] identified poor community coordination from government agencies when grid connectivity was being expanded.

Off-grid projects have also been largely unsuccessful in PNG, which experiences consistent issues pertaining to low levels of energy democracy. Communities generally exhibited low levels of energy literacy, with little understanding of the benefits of electricity. Without this understanding, the level of motivation to prioritize energy access is generally insufficient to shift stagnant practices in a village or household context. This manifests in communities having little involvement in the design and implementation of energy access projects [20]. Exacerbating this situation, many communities have also lacked the knowledge and skills to maintain the installed systems and have struggled with affordability issues [17,20]. Lack of buy-in from landowners also presents a significant barrier, with most land in PNG under customary ownership [21].

4 The process

Building on a similar program developed for market literacy [22], a youth-focused energy leadership and literacy program was designed and tested. The approach involved three stages: (1) building literacy around energy poverty impacts and opportunities; (2) participatory methods to help generate and instill community ownership and a shared understanding of energy sources and needs; and (3) a discussion of previous energy projects, resources available, appropriate governance structures, and capacities required to be built to achieve sustainability of the future projects.

The approach was implemented in selected communities in PNG under the moniker PNG Future Energy and Environment Leadership Strategy (PNG FEELS). The PNG FEELS approach focused on empowering the community, via a youth leadership strategy to improve energy knowledge and articulation. The community and youth leaders expressed a wish to focus on understanding local environmental resources, including renewable energy sources, and associated community values and uses. The program in PNG followed a "Train-the-Trainer" model, first through capacity building of 12 PNG university students as energy development trainers and community advocates. This was followed by participatory workshops in rural communities, facilitated by the youth development trainers and community advocates, to build energy literacy, create awareness on energy poverty issues, and the development of tangible knowledge-based outputs.

The overarching objective of PNG FEELS was to empower student leaders with both the knowledge (know-why) and process (know-how) capacity to enable beneficial, locally relevant discussions concerning energy and the environment with key community members. The ultimate vision is to use a broader rollout of PNG FEELS to effect positive change in rural PNG.

Key elements of the PNG FEELS program included:

- **Broad priorities and issues around energy poverty (know-why)**: Explanation of energy poverty as well as the benefits of energy access. To facilitate this aspect, student facilitators asked the community groups about their priorities, how energy can support these priorities, and the resources currently able to be accessed. Participants were also asked to reflect on the past experiences with respect to energy projects.
- **Key issues identified and solutions (know-how)**: Ownership and empowerment using problem/solution tree analyses, based on the development of an understanding of energy sources and needs. Participants were asked to analyze one problem to work out the causes, and then develop one corresponding solution to determine the effects.
- **Next steps**: Finally, community members were asked to consider governance within their community and how decisions are made. They were then required to reflect on the resources required for positive change.

The train-the-trainer workshop occurred in Madang, the capital of Madang Province in PNG in late August 2018. Four researchers from the Energy & Poverty Research Group at the University of Queensland shared their knowledge and research experience in energy poverty, facilitating interactive workshops with local stakeholders and students.

The student facilitators, now equipped with shared knowledge in energy literacy, then delivered a co-designed workshop program to youth and community members from five rural villages in the central coast of Madang Province, approximately 80 km north of Madang town, over 5 days (Fig. 17.1). The workshops were conducted in Tok Pisin, the most widely used language in PNG, and included a number of opportunities to communicate outcomes to wider audiences within the villages (Fig. 17.2).

FIG. 17.1 Youth and community members participated in small-group workshops, facilitated by university students. *Photo: A.P. Heynen.*

FIG. 17.2 Outcomes of the workshops were discussed with village audiences. *Photo: A.P. Heynen.*

Following the workshop, the student facilitators and researchers completed an Energy Needs and Opportunities Assessment report detailing the outputs of these participatory workshops, which were delivered back to the community via local partners.

5 Outcomes

Key achievements of the pilot program included:

- The engagement of over 30 community participants across five villages.
- Increased community understanding of energy poverty, energy uses, and opportunities, plus health impacts of indoor air pollution caused by biomass cooking practices, including pneumonia, stroke, and heart disease. Problem and solution articulations developed for each village, articulating priorities and local energy access challenges that can then be developed into action plans for local-level governments (LLG).
- Enduring partnerships established across universities, youth organizations, and community, along with an enthusiastic student cohort of energy transition advocates with knowledge and experience using participatory community engagement tools.
- Development of a methodology that can inform future program implementation.

Central to the project was the development of close and effective partnerships with key stakeholders in PNG, each bringing unique strengths. For example, an important academic partner, namely the University of Goroka, contributed significantly with intimate local connections to specific localities in PNG. The PNG National Agricultural Research Institute (NARI) provided critical resource support and local governance oversight and administration.

The program also partnered with The Voice, Inc. (TVI), a dynamic youth development organization that runs student leadership and empowerment programs at universities and high schools across PNG. TVI provided the framework through which the 12 student leaders from the University of Goroka and Divine Word University could participate in the pilot program as the key community facilitators of PNG FEELS.

6 Aiding the transition to energy democracy

Overall, the PNG FEELS program used energy knowledge and its articulation in a community context as the key vectors for empowerment. Thus, the approach is designed to achieve a more socially sustainable and supported model when compared to other programs that may rely on energy products or services exclusively. The development of local knowledge and support is considered crucial.

Research has indicated that providing new or improved energy access on its own is unlikely to deliver sustainable development outcomes. Approaching energy access as a part of a wider ecosystem of not only infrastructure, assets, and resources, but also energy literacy within local communities, may lead to the improved and sustained capabilities necessary to achieve the desired outcomes and social benefits across health, education, and livelihoods.

A significant social benefit rarely discussed, and less understood in the context of improved energy access efforts, is the potential to use energy access planning and implementation processes as an opportunity for strengthening democratic outcomes. That is, to use energy access as a means to empower local voices in decision-making in a sector that has traditionally been delivered through top-down processes, with local engagement often confined to lower rungs on Arnstein's ladder of participation [11]. That is, to exclude, to legitimize or at best, to consult on already-decided-upon objectives, approaches, and desired outcomes.

To have any hope of reducing energy poverty and achieving its full suite of promised development outcomes, energy access must be accompanied by meaningful community engagement and capacity building, initiated through processes of energy democracy for communities and citizens. The focus on local youth empowerment through the program helps equip the future leaders of PNG with the confidence and skills to create a positive and prosperous future for their people. While not tied to a specific energy intervention, the PNG FEELS pilot program could form the basis for wider energy democracy efforts as electrification programs are implemented across PNG and elsewhere.

Acknowledgments

The authors acknowledge Total and The University of Queensland (UQ) for funding the PNG FEELS program. The authors are grateful to the support provided by The Voice, Inc.; the University of Goroka; Divine Word University; Dr. Sim Sar (National Agricultural Research Institute); Anna-Claire Zanetti (UQ); participating students; and youth leaders of the villages of Aronis, Megiar, Gabagsal, Waliak, and Wasab.

References

[1] A.P. Heynen, P.A. Lant, S. Smart, S. Sridharan, C. Greig, Off-grid opportunities and threats in the wake of India's electrification push, Energy Sustain. Soc. 9 (1) (2019) 1.

[2] P. Yadav, A.P. Heynen, D. Palit, Pay-as-you-go financing: a model for viable and widespread deployment of solar home systems in rural India, Energy Sustain. Dev. 48 (2019) 139–153.

[3] P. Balachandra, Modern energy access to all in rural India: an integrated implementation strategy, Energy Policy 39 (12) (2011) 7803–7814.

[4] United Nations Development Program, Towards an 'Energy Plus' Approach for the Poor: A Review of Good Practices and Lessons Learned From Asia and the Pacific, UNDP, Bangkok, Thailand, 2011.

[5] United Nations Development Program, EnergyPlus Guidelines: Planning for Improved Energy Access and Productive Uses of Energy, UNDP, New York, USA, 2015.

[6] J.C. Stephens, Energy democracy: redistributing power to the people through renewable transformation, Environ. Sci. Policy Sustain. Dev. 61 (2) (2019) 4–13.

[7] M.J. Herington, E. Van de Fliert, S. Smart, C. Greig, P.A. Lant, Rural energy planning remains out-of-step with contemporary paradigms of energy access and development, Renew. Sust. Energ. Rev. 67 (2017) 1412–1419.

[8] J.E. Stiglitz, Towards a new paradigm for development: strategies, policies and processes, Appl. Econ. Int. Dev. 2 (2002) 116–122.

[9] G. Mansuri, V. Rao, Localizing Development: Does Participation Work?, World Bank Publications, 2013. Available from: http://hdl.handle.net/10986/11859.

[10] P.N. Thomas, E. Van de Fliert, Participation in theory and practice, in: P.N. Thomas, E. Van de Fliert (Eds.), Interrogating the Theory and Practice of Communication and Social Change: The Basis for a Renewal, Palgrave Macmillan, London, 2014, pp. 39–51.

[11] S. Arnstein, A ladder of citizen participation, J. Am. Plan. Assoc. 35 (4) (1969) 216–224.

[12] United Nations Development Program, About Papua New Guinea, 2020. https://www.pg.undp.org/content/papua_new_guinea/en/home/countryinfo.html. (Accessed 10 October 2020).

[13] D.P. Faith, H.A. Nix, C.R. Margules, M.F. Hutchinson, P.A. Walker, J.G. West, G. Natera, J. Stein, J.L. Kesteven, A. Allison, G. Natera, The BioRap biodiversity assessment and planning study for Papua New Guinea, Pac. Conserv. Biol. 6 (4) (2001) 279–288.

[14] A.L. D'Agostino, B.K. Sovacool, Unsold solar: a post-mortem of Papua New Guinea's Teacher's solar lighting project, J. Energy Dev. 36 (1/2) (2010) 1–21.

[15] O. Renagi, J.A. Babarinde, An appraisal of PNG National Energy Policy 2018-2028, in: Paper of the International Sustainable Energy Research Conference, Lae, Papua New Guinea, June, 2018. Available from: https://www.researchgate.net/profile/Jacob_Babarinde/publication/326079987_An_Appraisal_of_PNG_National_Energy_Policy_2018-2028/.

[16] M. Aklin, P. Bayer, S.P. Harish, J. Urpelainen, Escaping the Energy Poverty Trap: When and How Governments Power the Lives of the Poor, MIT Press, Cambridge, Massachusetts, 2018.

[17] M. Rawali, A. Bruce, A. Raturi, B. Spak, I. MacGill, Electricity access challenges and opportunities in Papua New Guinea (PNG), in: Proceedings of the Asia Pacific Solar Research Conference, 2019. Available from: http://apvi.org.au/solar-research-conference/wp-content/uploads/2020/02/Rawali-M-Electricity-Access-Challenges-and-Solar-Energy-Opportunities-in-PNG.pdf.

[18] Australian Infrastructure Financing Facility for the Pacific, Media Release: Papua New Guinea Electrification Partnership, 30 June 2020, Available from https://www.aiffp.gov.au/news/papua-new-guinea-electrification-partnership.

[19] T.F. Engelmeier, N.R. Gaihre, Going the Distance: Off-Grid Lighting Market Dynamics in Papua New Guinea, International Finance Corporation, 2019.

[20] B.K. Sovacool, A.L. D'Agostino, M.J. Bambawale, The socio-technical barriers to solar home systems (SHS) in Papua New Guinea: "choosing pigs, prostitutes, and poker chips over panels", Energy Policy 39 (3) (2011) 1532–1542.

[21] S. Chand, Registration and release of customary-land for private enterprise: lessons from Papua New Guinea, Land Use Policy 61 (2017) 413–419.

[22] M. Viswanathan, S. Sridharan, R. Gau, R. Ritchie, Designing marketplace literacy education in resource-constrained contexts: implications for public policy and marketing, J. Public Policy Mark. 28 (1) (2009) 85–94.

Communities

18

A just development energy transition in India?

Heather Plumridge Bedi

Environmental Studies Department, Dickinson College, Carlisle, PA, United States

1 Energy transition woes

India simultaneously faces rising energy demand, development challenges, and heightened climatic vulnerability. There are a range of challenges facing the nation as it attempts to provide electricity to over 300 million Indians without electricity and 500 million people who lack clean cooking fuel [1]. Concurrently, there is increasing energy demand from the growing number of middle- and upper class consumers. Since 2000, energy use among India's population of over a billion people has almost doubled [2]. National energy demand is predicted to increase 2.7–3.2 times between 2012 and 2040 [1].

Government officials promote efforts to transition the nation to renewable energy sources.

India's National Action Plan on Climate Change, introduced in 2008, promotes a transition to renewable energy, balancing economic growth with climate change mitigation [3]. Increasing from the 2017 levels of 50 gigawatts (GW) per year, India plans to escalate renewable grid power capacity to 175 GW by 2022 [4]. From 2015 to 2020, India's installed renewable energy capacity increased by 226% with renewable energy comprising 23% of this installed generation capacity [5]. This is a significant shift, as carbon-intensive coal provides 60% of the nation's electricity supply.

To fulfill energy transition ambitions, Indian states pursue large-scale private or public-private renewable energy initiatives, which require large swaths of land, a resource not readily available in most of India. The geographies of this infrastructure do not correlate with per capita state wealth, infrastructural electricity needs, or geographies amenable to national energy projects. For example, renewable efforts in the relatively impoverished states of Rajasthan and Madhya Pradesh in north and central India outpace the wealthy western coastal state of Maharashtra [6].

The profound social and development implications of energy infrastructure emanate from existing land and development injustices. Evidence from state-level renewable energy infrastructure projects reveals a repetition of historical patterns of marginalized populations.

These groups disproportionately bear the brunt of the environmental and social externalities associated with procuring electricity, whether non-renewable or renewable [7]. For example, Yenneti et al. [8] document how one of the world's largest solar plants heightened precarity for vulnerable social groups in the state of Gujarat through the transition of common grazing lands and agricultural land into non-agricultural use. In the southern state of Kerala, the Kasaragod solar park had to be scaled back from 200 to 50MW following a contentious land acquisition process. Making up 11% of Kerala's population, Indigenous *Adivasi* people and scheduled castes represent 30% of the poor in the state [7]. With land acquired for renewable infrastructure, the historical marginalization of *Adivasi* peoples endures. *Adivasi* residents without legal land titles lost their lands and livelihoods to make way for the solar park, which fueled local political opposition and halted the park expansion [7]. These examples demonstrate how the "deep-seated inequities" associated with energy access and other development benefits permeate renewable energy projects and emphasize the need to ensure energy justice [9]. This essay applies an energy justice framework to examine the potential for a "just development" energy transition in India.

2 A just development energy transition in India

An energy democracy perspective may provide potential governance lessons for decentralizing future energy infrastructure planning in India. The Indian government has developed an energy transition plan that focuses on "harnessing Solar Power for Prosperous Rural India" [5]. In practice, the emerging climate-centered renewable efforts reveal that the benefits of the transition do not extend to all Indians. As detailed in the previous section, renewable energy challenges emanate from planners and officials in India pursuing energy infrastructure projects from a narrow perspective not attuned to the land, development, and social justice implications of their interventions. A justice-centered energy democracy approach provides the opportunity to engage more Indians in energy decision-making processes.

The energy democracy movement calls for a transition of political power from a government and corporate centralized system to a decentralized system centered on the interests of workers, communities, and the general public [10]. Decentralized renewable energy initiatives that are controlled at the community level present an alternative to the dominant institutional control of energy systems [11]. Efforts to ensure India's energy transition must prioritize that it be a just energy transition. Without governance changes, gains achieved through the promotion of renewable energy will prove pyrrhic if all Indians can't access electricity and populations are displaced for the large-scale energy infrastructure.

As energy access is intimately connected to human development indicators, a democratic energy transition should prioritize social and environmental justice concerns. India's energy transition should adopt a just development framework that addresses energy and development considerations holistically. A key step to implementing a development-centered approach to an energy transition that is socially and environmentally responsible is to consider recognitional, procedural, and distributional justice concerns in all decisions regarding renewable energy in India. Table 18.1 details how the application of justice dimensions in planning, implementation, and access to energy will increase the capacity for all Indians to benefit from the nation's energy transition.

TABLE 18.1 Justice dimensions of India's energy transition.

Dimension	Definition	Application
Recognitional justice	Acknowledging historical and contemporary vulnerability of under-represented populations	Energy access and projects acknowledge impacts on vulnerable groups, including women, children, elderly, *Adivasis*, and scheduled castes
Procedural justice	Access to legal due process, planning process, and fair treatment for all	Inclusion of energy impacted communities throughout the proposal and life of energy projects. Affordable legal channels if injustices occur
Distributive justice	Equitable distribution of costs and benefits of services and projects	Access to affordable electricity for all Indians. Ecological costs of energy provision distributed equitably, and not disproportionately burdening historically marginalized communities

Data from B.K. Sovacool, Lecture presented at the University of Bergen, Norway (2019); B.K. Sovacool, J. Kester, L. Noel, G.Z. de Rubens, Energy injustice and Nordic electric mobility: inequality, elitism, and externalities in the electrification of vehicle-to-grid (V2G) transport, Ecol. Econ. 157 (2019) 205–217.

2.1 Recognitional justice

Recognitional justice focuses on the acknowledgment of the historical and contemporary vulnerability of under-represented populations [12]. Precarity related to renewable energy includes displacement, livelihood loss, energy poverty, and exposure to waste toxins. Renewable energy infrastructure planning must actively acknowledge these potentials and plan to mitigate current or future injustices.

The principles of recognitional justice should also be applied to non-renewable energy sources, as India will remain dependent on coal for the indefinite future to power the nation's growing energy demand. Further, legacy fossil fuel actors, including the state Coal India Limited, are increasingly key players in renewable energy. Recognition of the history of disproportionate marginalization of populations provides important context for India's energy transition. India's coal dependency has had grave impacts on people living in and around coal extraction, processing, and waste disposal sites [13,14]. Acknowledgment of enduring unevenness can inform planning and decrease the potential for repeating historical injustices.

2.2 Procedural justice

Procedural justice emphasizes the importance of access for all to legal and planning processes and to fair treatment. With energy processes, decision-making channels should include all stakeholders. This engagement would extend throughout all life cycles of the energy system. Stakeholders should have a voice in the siting and planning for energy production, processing, and waste projects. Before the start of projects, planning documents must be provided in local languages and stakeholders should have forums to have their concerns addressed by government officials and in project proposals. For large-scale renewable projects, there would be full compensation for required land acquisition, regardless of land title status. If they choose to do so, stakeholders should have the ability to access affordable energy through the projects.

If there are barriers to stakeholder access to information and planning channels, they must have access to free or reduced-cost legal processes and/or conflict mediation. This process could be a part of India's innovative National Green Tribunal, which fast-tracks environmental lawsuits in the nation's notoriously slow court system. The unaffordability of legal channels prohibits many constituents in India from using the courts as a forum to seek justice [15].

2.3 Distributional justice

Distributional justice centers on the principle that there should be equitable dissemination of both the costs and benefits of services and projects. Historically, the costs of coal extraction, processing, and waste have fallen disproportionately on India's Indigenous *Adivasi* population. The distributional benefits and costs must also be applied to non-renewable energy during the transition period. Although the geographical contours of coal are predetermined by where the resource is located, distributional justice is equally relevant to renewable and non-renewable sources of energy.

Emerging evidence points to *Adivasi* land precarity related to large-scale solar projects [7]. To address these injustices, there must be equitable distribution of energy initiatives across states and ecosystems. Energy development projects should not heighten social disconnection or land dispossession for marginalized populations. Given advancements in moving renewable energy to market, it is increasingly possible to concentrate large-scale solar energy in states, where population displacement would not be necessary. This approach would involve a change in the national approach to renewable energy infrastructure from state-level goals to national targets. Rather than setting state-level goals, the central government should concentrate renewable energy initiatives in states with vacant land, where large-scale energy infrastructure would not displace or further dispossess marginalized people.

3 Just energy democracy

Bulkeley and colleagues question if "interventions in the name of climate change serve to maintain the interests of an elite at the expense of a minority, and as such perpetuate patterns of inequality" [16]. India's renewable energy efforts are laudable, but there are examples of associated injustices. The potential for future inequalities underscores concerns that the energy transition will further vulnerability for the nation's poor and climate change impacted populations. Energy democracy, guided by principles of justice, provides tangible steps to make India's energy transition more just and inclusive for Indians. At a minimum, stakeholders should have access to procedural, recognitional, and distributional justice for energy infrastructure, processing projects, and energy access. The world's largest democracy has the potential to improve development standards for all Indians while transitioning from fossil fuels to renewable energy sources.

References

[1] NITI Aayog, Government of India, Draft National Energy Policy, New Delhi, 2017. http://niti.gov.in/writere-addata/files/new_initiatives/NEP-ID_27.06.2017.pdf. (Accessed 26 March 2019).
[2] International Energy Agency, World Energy Outlook 2015, 2015. https://www.iea.org/publications/freepub-lications/publication/WEO2015.pdf. (Accessed 25 March 2019).

[3] Government of India, National Action Plan on Climate Change, Prime Minister's Council on Climate Change, 2008. http://www.nicra-icar.in/nicrarevised/images/Mission%20Documents/National-Action-Plan-on-Climate-Change.pdf.

[4] International Energy Agency, Global Energy & CO2 Status Report 2018: The Latest Trends in Energy and Emissions in 2018, 2018. https://www.iea.org/geco/emissions/. (Accessed 15 July 2019).

[5] Ministry of New and Renewable Energy, Renewable Energy Website. https://mnre.gov.in/. (Accessed 25 September 2020).

[6] E. Chatterjee, The Asian anthropocene: electricity and fossil developmentalism, J. Asian Stud. 79 (1) (2020) 3–24.

[7] H.P. Bedi, "Lead the district into the light": solar energy infrastructure injustices in Kerala, India, Glob. Transit. 1 (2019) 181–189.

[8] K. Yenneti, R. Day, O. Golubchikov, Spatial justice and the land politics of renewables: dispossessing vulnerable communities through solar energy mega-projects, Geoforum 76 (2016) 90–99.

[9] S. Sareen, S.S. Kale, Solar 'power': socio-political dynamics of infrastructural development in two Western Indian states, Energy Res. Soc. Sci. 41 (2018) 270–278.

[10] M.J. Burke, J.C. Stephens, Political power and renewable energy futures: a critical review, Energy Res. Soc. Sci. 35 (2018) 78–93.

[11] D. Fairchild, A. Weinrub, Energy democracy, in: The Community Resilience Reader, Island Press, Washington, DC, 2017, pp. 195–206.

[12] S. Williams, A. Doyon, Justice in energy transitions, Environ. Innov. Soc. Transit. 31 (2019) 144–153.

[13] K. Lahiri-Dutt, The Coal Nation: Histories, Ecologies and Politics of Coal in India, Routledge, 2016.

[14] P. Oskarsson, H.P. Bedi, Extracting environmental justice: countering technical renditions of pollution in India's coal industry, Extr. Ind. Soc. 5 (3) (2018) 340–347.

[15] H.P. Bedi, Judicial justice for special economic zone land resistance, J. Contemp. Asia 45 (4) (2015) 596–617.

[16] H. Bulkeley, V. Castán Broto, A. Maassen, Low-carbon transitions and the reconfiguration of urban infrastructure, Urban Stud. 51 (7) (2014) 1471–1486.

Community adaptation to microgrid alternative energy sources: The case of Puerto Rico

Nicholas J. Sokol

Found Spatial, Knoxville, TN, United States

1 Introduction

Puerto Rico's energy infrastructure has taken a new form in the 21st century. To meet the challenges of modern demand on the island's grid, many Puerto Ricans have taken direct control of the production, distribution, and storage of their energy. Indeed, when examining Puerto Rico's current predicament, many of the island's problems can be resolved by returning to the island itself. Natural hazards, economic recession, lack of resources, loss of skilled labor to the continental United States, poor energy policy, limited land for infrastructure, and government corruption make it difficult for the island to maintain an efficient energy infrastructure. Much like the protests against former governor Ricardo Roselló, many Puerto Ricans are conducting a small-scale revolution for the sake of energy reliability, affordability, and availability. Should they succeed, they could become a global model for regions that struggle with decrepit energy infrastructure.

2 An overview of Puerto Rico's power grid and natural challenges

To understand Puerto Rico's current movement toward microgrids, it is important to know the history of the island's energy and power infrastructure. Puerto Rico first experienced electricity in 1897 when 120-V generators were built in the town of Utuado. Much of this energy was used to power the town square and a few houses nearby [1]. In 1898, the United States acquired Puerto Rico as a territory through its victory in the Spanish-American war and the signing of the Treaty of Paris. Much of the energy infrastructure from 1898 until 1940 was controlled by individual municipalities [1]. With the advent of the World War II, the governor

Energy Democracies for Sustainable Futures
https://doi.org/10.1016/B978-0-12-822796-1.00019-X

of Puerto Rico, Rexford Tugwell, determined that a centralized grid was necessary to enhance island-wide defenses. This led to the creation of the Puerto Rico Water Resources Authority (PRWRA), which centralized the island's grid in an attempt to modernize and equalize distribution of energy. Today, the PRWRA is known as the Puerto Rico Electric Power Authority (PREPA) and is the island's sole provider of electricity.

To its credit, the centralization of the electric grid was beneficial for many Puerto Ricans at the time. Before the arrival of PREPA, local communities were in control of their electrical generation, but this led to a huge disparity in availability across the island, with some communities having stable and adequate power supply, while others possessed none [1]. Centralizing the grid allowed many towns to become powered and brought new services and capabilities to many parts of the island [2]. However, with an increasing population on Puerto Rico, a concomitant strain was placed upon the aging and dilapidated grid. The island has multiple power plants dispersed across it, but the largest, publicly owned ones (500 MW capacity and greater) are mostly located along the coasts (Fig. 19.1). These larger plants are fueled by oil and natural gas, two resources which the island has none of and, therefore, must be imported. This creates two main problems when it comes to energy production and generation: (1) if oil/natural gas shipments are delayed or ceases entirely then generation decreases, and (2) if the island's poorly maintained and often remote distribution lines are

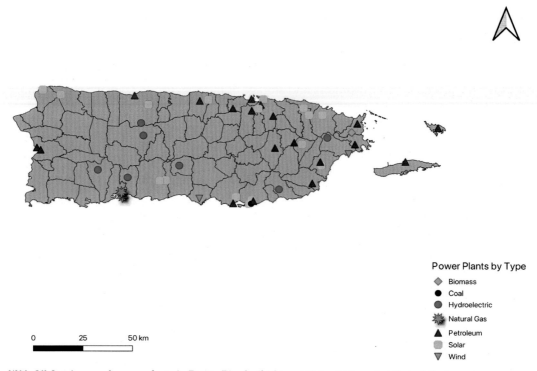

FIG. 19.1   A map of power plants in Puerto Rico by fuel type. It should be noted that while there are numerous solar plants, their overall generation is low compared to petroleum and natural gas. Map made by author with data from the US Energy Information Administration.

destroyed, then dissemination is hampered. Both of these problems have begun to manifest themselves more frequently in recent years due to a number of environmental, engineering, and economic issues that the island faces.

The environment of Puerto Rico is much more complex than most would think likely for a relatively small tropical island. Puerto Rico is 115 miles east to west and has a notably diverse and complex landscapes and biomes within its small size (Fig. 19.2). Such landscape complexity challenges energy dissemination. Geologically, the island has coastal plains, coastal escarpment, surrounding a mountainous interior. The climate of the island is very diverse, with the eastern side of the island being humid and covered in lush rainforests, while the southwestern side of the island is arid and hot. Trade winds dominate the island, flowing predominantly from east to west. During the summer months, the island experiences less cloud cover, but hurricanes can dominate during these months. Several other natural hazards are common, such as landslides and earthquakes. Landslides occur in the mountainous regions and can occur anytime of the year following a rain event. Earthquakes predominantly occur on the southwestern side of the island.

All of this natural complexity lends itself to the wide array of environmental challenges that make it difficult to build stable energy infrastructure. Power plants built on the eastern side of the island are susceptible to excessive damage from hurricanes. Whereas on the southwestern side, earthquakes can cause enough destruction to take plants offline, as was the case with the 2020 earthquakes [3]. While the risk of these impacts is statistically low, it makes maintaining a centralized grid challenging because a plant going offline means power shutoffs for large swaths of the island.

Engineering challenges are plentiful and mostly environmental in origin, but they also derive from the old and very complex grid transmission system. When the island's power system was centralized, many of the transmission lines that were installed ran through the most populous areas. Rural locations, on the other hand, might have but a single line extending from the closest hub. Hub stations are geographically concentrated and mostly center around the area where power is generated.

In the event, a natural disaster compromises this infrastructure, it can take days to reach these locations and make repairs since the road network on the island is often too damaged

FIG. 19.2 Topographical map of the archipelago of Puerto Rico. The eastern side of the island is wetter than the southwestern side of the island, which has more arid conditions. Map made by author using ArcGIS Online.

or obstructed [4]. In addition, critical components of the electrical grid such as lines, poles, transformers, and metering equipment are old, which leads to a high failure rate of equipment [4]. It is common for power outages to occur, even when there are no external pressures on the grid. This is a by-product of the outdated grid infrastructure that is currently in place [4]. Many of such engineering problems result from the economic challenges that both PREPA and Puerto Rico as a whole currently face.

While Puerto Rico is classified as a high-income economy by the World Bank [5], it has one of the largest financial deficits, and its GDP and GDP per capita are decreasing [6]. The reliance on imported goods and high annual spending are the predominant cause for Puerto Rico's economic deficit [7]. Furthermore, Puerto Rico's spending far exceeds its income, which increases its public debt almost every year and currently sits at 56.8 billion. PREPA is not helping, as the company spends more money to maintain operation than it makes in collecting payments for utilities. Data for PREPA's finances show that the company spends ~ 60% of its budget on fuel, ~ 18% on purchased power, and what is left is divided among maintenance (~ 5%) and other costs. Almost none of its budget is allocated to upgrades and modernization of its grid [8]. While a proposal was made in 2020 to expand the fossil fuel-based grid, it was rejected [9]. The company operates at a 4 billion deficit annually and makes up this deficit by pulling funds from the government. With little economic growth to stimulate change, the island's electrical utility will likely remain stagnant for the coming years, leaving Puerto Ricans with a very dilapidated and underperforming grid.

3 The rise of the microgrid

Despite many challenges Puerto Ricans face when it comes to their energy production, a new movement has emerged on the island stemming from the aftermath of Hurricanes Irma and Maria in 2017. A handful of cities across the island are turning to a microgrid solution much like what the island had in the years before the World War II era centralization. A microgrid is essentially a small-scale power production infrastructure. Instead of focusing on powering vast areas, a microgrid utilizes the best technology available to provide power to a specific region or zone. However, instead of running the electrical grid on purely fossil fuels, many of these locations seek to power the island by using renewable resources such as solar, wind, and biofuel [10,11]. Most microgrid solutions aim to increase grid reliability by utilizing resources that are affordable and readily available. For currently available, and viable, renewable technologies Puerto Rico has plenty of sun, coastal, and offshore space for wind, and potential for biofuel generation.

While solar has become the predominant focus of many communities, land availability and sporadic cloud cover can create difficulties. Solar panels can be placed on top of homes and in purchased land plots, but not all homes are conducive to solar installation due to poor rooftop infrastructure in some areas, this greatly exacerbates independent solar production potential on the island. In addition, solar sites placed too far away from a city or area run into the same issue as the island's centralized grid in that environmental challenges to grid logistics (such as transmission over complex terrain) can drastically increase the cost of power [12]. In addition, while the price of solar materials is decreasing, it is still too expensive for most cities to afford large enough infrastructure to provide stable power to their local municipality without external supplementation [13]. This leaves many cities with a limited microgrid potential.

The commercial development of wind power faces many trials on the island. For one, wind farms require large amounts of space so that turbines can be adequately distanced from each other. In addition, wind turbines need to be installed in a location that observes relatively laminar and consistent wind flow. While most coastal zones provide adequate wind, appropriate land for installation is sparse and there is a fear that offshore wind power installations would degrade tourism and hurricanes would destroy them. In the mountains, winds are too turbulent and would lead to faster degradation of turbine blades there is also difficulty in managing transmission line infrastructure to move power effectively [14,15].

In addition to these problems, the installation of wind farms can conflict with agriculture. The largest wind farm on the island was built on top of agricultural lands and removed a large amount of fertile soil from use for food growth [16]. Lastly, wind turbine installations are not allowed near or in cities, which make them incapable of being incorporated into a microgrid. Wind energy will face many challenges before it becomes a large portion of Puerto Rico's energy production in the future due to damage risk from tropical cyclones [17]. Adaptation into microgrid infrastructure will also pose a challenge as much of the island outside of the coast does not provide the necessary sustained winds to make them economical [18].

Biofuels have seen rising prominence on the island's microgrids but also pose challenges. Biofuels have an advantage of being a liquid fuel and can be easily stored and maintained by local municipalities. In addition, corn, switchgrass, and sugarcane can all be used to make biofuels and these resources grow easily on the island. However, there have been concerns related to biofuels such as carbon emissions [19], land plots needed for agricultural activity [20], and the investment capital required to build and maintain biofuel generators. The scalability of a biofuel plant also poses a twofold issue, as the fuel production will need to increase along with the size of the plant. This makes the option not as feasible as solar.

Many of the island's current microgrid solutions are extremely successful and quickly gaining popularity in the public eye. Casa Pueblo, located in Adjuntas, Puerto Rico is one of the best-known examples of microgrid adaptation for an entire city. The grassroots organization focuses on obtaining 100% energy independence by utilizing alternative energy solutions [13,21]. It was originally founded in 1980 as an environmental protection and cultural center with most of its early actions focused on protecting natural habitats of Puerto Rico against open-sky mining. After Hurricane Maria, Casa Pueblo was the only structure in the area that retained power, which allowed many locals to charge cell phones, radios, and other electronic equipment [19]. Due to the success they had with energy distribution in the post-hurricane months, the organization decided to make a commitment to provide stable energy without utilizing the grid. While the majority of Casa Pueblo's energy production comes from solar, battery, and biofuel, some buildings still need to be powered via the grid [13]. These sources are viewed as the cheapest and most efficient way to procure clean and reliable energy that will ultimately provide a grid with superior stability compared to the one they have now.

In addition, seeing as PREPA has done little to modernize the grid and provide clean energy alternatives to the inhabitants of Puerto Rico, Casa Pueblo believed it was necessary to expand energy independence to many of the island's inhabitants and has since gone on to work with several universities, non-profit organizations, and companies to bring clean and reliable power to the island. Other cities across Puerto Rico are now following the footsteps of Adjuntas and Casa Pueblo. Cities such as Humacao, Orocovis, Villalba, and Utuado have also developed some form of a microgrid to keep buildings that provide essential services operating [11].

4 Conclusions

While many of these microgrids are currently small and provide power to limited spatial areas, they could be expanded to serve as large-scale supplemental power supply or 1 day completely replace the need for grid-powered energy on the island. One of the major limitations as of now is cost, but even that will soon no longer serve as an impediment to independent and community managed grids on the island as costs come down. The success of microgrids on the island has become such an inspiration that even the island's centralized power manager PREPA is attempting to back a more renewable and regional-focused grid. In August 2020, the Energy Bureau backed a long-term plan that would increase renewable energy production on the island by 5.86 GW of energy (3.5 GW solar and 1.36 battery) [22]. This came alongside a plan to split the island into eight regional zones for distribution and grid management [23]. While this plan has been received by the majority of Puerto Ricans as a positive, many have seen that being in control of the electrical grid provides an opportunity for locals to pave their own electrical future. It is unknown if cities which have their own microgrids and continue to develop them will work with government agencies. We can be certain that much of Puerto Rico's future infrastructure will likely focus on renewables and emphasize local development as opposed to island-wide distribution.

Indeed, with increasing demand, diminishing trust in PREPA, and enduring confidence in local experts to provide stable microgrid solutions, Puerto Rico's energy future looks less morbid and more hopeful to many inhabitants of the island.

References

[1] Autoridad De Energia Electrica, Brushstrokes of Our History, AEE, 2020. https://aeepr.com/en-us/QuienesSomos/Pages/History.aspx.

[2] E. O'Neill-Carrillo, Una Nueva AEE: Energía Eléctrica para la Sociedad Puertorriqueña del Siglo XXI, 2010. http://aceer.uprm.edu/pdfs/Una_Nueva_AEE.pdf.

[3] P. McArdle, Puerto Rico's Electricity Generation Mix Changed Following Early 2020 Earthquakes, EIA, 2020. https://www.eia.gov/todayinenergy/detail.php?id=44216.

[4] M. Galluci, Rebuilding Puerto Rico's power grid: the inside story, IEEE Spectr. (2018). https://spectrum.ieee.org/energy/policy/rebuilding-puerto-ricos-power-grid-the-inside-story.

[5] World Bank, World Development Indicators: Puerto Rico, 2020. https://data.worldbank.org/country/PR.

[6] L. Merling, Puerto Rico's peculiar case: bankruptcy of an unincorporated territory, in: United Nations Conference on Trade and Development, September 2018. Center for Economic and Policy Research, 2018.

[7] S. Smith-Nonini, The debt/energy nexus behind Puerto Rico's long blackout: from fossil colonialism to new energy poverty, Lat. Am. Perspect. 47 (3) (2020) 64–86.

[8] Puerto Rico Fiscal Agency and Financial Advisory Authority, PREPA B2A Report FY2021—1Q FY2021, 2020. https://www.aafaf.pr.gov/financial-documents/prepa-quarterly-budget-to-actual-reports/.

[9] E. Merchant, In Blow to Natural Gas, Puerto Rico Regulators Affirm Solar-Centric Grid Overhaul, GTM, 2020. https://www.greentechmedia.com/articles/read/regulators-in-puerto-rico-approve-solar-and-storage-above-utilitys-resource-plan.

[10] E. O'Neill-Carrillo, I. Jordan, A. Irizarry-Rivera, R. Cintron, The long road to community microgrids: adapting to the necessary changes for renewable energy implementation, IEEE Electrif. Mag. 6 (4) (2018) 6–17.

[11] R.F. Jeffers, M.J. Baca, A. Wachtel, S. DeRosa, A. Staid, W.E. Fogleman, F.M. Currie, Analysis of microgrid locations benefitting community resilience for Puerto Rico (No. SAND-2018-11145), Sandia National Lab. (SNL-NM), Albuquerque, NM, United States, 2018.

[12] A. Kwasinski, F. Andrade, M.J. Castro-Sitiriche, E. O'Neill-Carrillo, Hurricane maria effects on Puerto Rico electric power infrastructure, IEEE Power Energy Technol. Syst. J. 6 (1) (2019) 85–94.

[13] N. Bickel, Energy Independence in Puerto Rico, Globalizations Michigan, 2020. https://global.umich.edu/newsroom/energy-independence-in-puerto-rico/.

[14] B. Dorminey, Renewable Energy World, 2012. https://www.energy.gov/sites/prod/files/2016/08/f33/UT_REPORT_2016-05-01.pdf.

[15] H. Kozmar, D. Allori, G. Bartoli, C. Borri, Complex terrain effects on wake characteristics of a parked wind turbine, Eng. Struct. 110 (2016) 363–374.

[16] D.S. Ramírez, R.R. Pérez, I.P. Roig, Terrenos Agrícolas y Energía Renovable: Caso de Estudio Pattern Energy Inc. en Santa Isabel, Rev. Administ. Públ. 46 (2015) 1–27.

[17] R.P. Worsnop, J.K. Lundquist, G.H. Bryan, R. Damiani, W. Musial, Gusts and shear within hurricane eyewalls can exceed offshore wind turbine design standards, Geophys. Res. Lett. 44 (12) (2017) 6413–6420.

[18] U.S. Energy Department, WindExchange, 2020. https://windexchange.energy.gov/states/pr.

[19] J.M. DeCicco, D.Y. Liu, J. Heo, R. Krishnan, A. Kurthen, L. Wang, Carbon balance effects of US biofuel production and use, Clim. Chang. 138 (3–4) (2016) 667–680.

[20] N. Huber, R. Hergert, B. Price, C. Zäch, A.M. Hersperger, M. Pütz, J. Bolliger, Renewable energy sources: conflicts and opportunities in a changing landscape, Reg. Environ. Chang. 17 (4) (2017) 1241–1255.

[21] A. Massol-Deyá, J.C. Stephens, J.L. Colón, Renewable energy for Puerto Rico, Science (2018). https://science.sciencemag.org/content/362/6410/7.summary.

[22] N. Klein, L. Feeney, The Battle for Paradise: Naomi Klein Reports From Puerto Rico, 2018. https://theintercept.com/2018/03/20/puerto-rico-hurricane-maria-recovery/.

[23] EIA, Puerto Rico Territory Energy Profile, 2020. https://www.eia.gov/state/print.php?sid=RQ.

Energy democracy movements in Japan

Setsuko Matsuzawa

Sociology and Anthropology, The College of Wooster, Wooster, OH, United States

1 Introduction

Japan's citizen-funded, renewable energy generation initially began in the early 2000s, led by consumer rights groups, who were opposed to nuclear energy from a food safety perspective. The Fukushima nuclear power plant crisis in 2011 caused Japan's general public to question their dependence on national energy policies and reimagine the ways in which they could produce, distribute, and consume energy. Since then, citizen-led renewable projects have expanded their support base beyond anti-nuclear food activists and have led to a full-fledged movement. The purpose of this chapter is to explore the development of Japan's energy democracy movement and its sociopolitical meaning.

Energy in Japan has long been dictated by the "nuclear village," a powerful network of the energy industry, the electric power industry, technocrats, and some academia. The nuclear village exercises what Niphi and Ramana, in this volume, call "narrative strategies" [1] to promote nuclear energy and undermine renewable energy. This long-standing political structure in Japan's energy sector has created "the democratic deficit of current policymaking" [2]. I argue that local organizing for citizen-led renewable projects across Japan produced not only new social relations between energy producers and consumers [3], but also a new power dynamic between the center and the periphery in energy politics. Unlike the rhetoric-based resistance described in Şorman and Turhan's chapter [4], citizens in Japan acted on renewable projects as a path for local autonomy from national energy policies.

First, I describe the emergence of citizen-led renewable projects. Second, I discuss renewable projects as a form of power in the post-Fukushima era. Third, I address Fukushima's commitment to rebuilding its economy based on energy democracy.

2 Renewables as social relations

Japan's first citizen-led wind turbine project came out of activism by the Seikatsu Club Consumers' Co-operative Union (SCCU), the world's largest cooperative organization, founded in the 1960s. After the Chernobyl nuclear accident in 1986, the SCCU ran a failed campaign against the planned construction of a third nuclear plant in Hokkaido, the largest and northernmost prefecture. In 1996, Seikatsu Club Hokkaido, one of the member cooperatives of the SCCU, turned its focus to renewable energy. The SCCU had learned from its food activism that dialogues between food producers and consumers lead to safer food. Thus, the SCCU hoped that cleaner energy could also be achieved by establishing visible connections between energy producers and consumers. Many Japanese had long treated energy as if it was automatically delivered from somewhere. External factors, such as the adoption of the 1997 Kyoto Protocol, as well as a 1995 amendment to Japan's Electric Act, also motivated the SCCU to focus on renewable energy.

In 1999, following a pilot program with co-op members, the SSCU established the Hokkaido Green Fund (HGF), a non-profit organization, with the help of the Institute for Sustainable Energy Policies (ISEP), a non-profit research organization. With publicity via a newspaper article, entitled "Citizens' Power to Build Wind Turbine," in a local daily, the HGF collected donations and investments from 217 citizens to fulfill 80% of project cost (141.5 million Japanese Yen) [5] for Japan's first citizen-funded wind turbine.

The project produced new social relations between energy producers and consumers. Local children named the wind power plants. Commemorative plaques with the names of investors and donors inscribed were placed at the plants. Energy became visible and participatory.

3 Renewables as people's power

Wide concerns over radiation and a sharp increase in greenhouse gas emissions in the post-Fukushima era prompted the Japanese government to introduce new programs (e.g., feed-in-tariff programs) to encourage renewable projects. Yet, for local people and evacuees in the inland City of Aizuwakamatsu, in Fukushima, renewable energy symbolized local autonomy and served as a way to resist national energy policies that exploited and damaged their localities with radiation. Within 4 months of the disaster, the people began organizing to discuss how to revive their localities. One meeting revealed that participants perceived renewable energy as a way to restart the "Freedom and People's Rights Movement," popular during Japan's Meiji Era (1868–1912) [6]. Its aim was to challenge the autocratic Meiji government by demanding a general election so that people at the periphery could influence policymaking at the center.

By this time, renewable energy experts and organizations, including the ISEP and the Hokkaido Green Fund, had been sharing their know-how and experiences with like-minded people across Japan. To encourage citizen participation in renewable energy, the Aizu Natural Energy Foundation and the Aizu Electric Power Co. (AiPOWER) were founded in 2013. The first megawatt solar power plant in the Aizu area was financed by local banks, the Green

Finance Organization Japan (or Green Fund), citizen funding, and the AiPOWER. In the following year, the Public Solar Fund of Aizu called for a total investment of 99.8 million Japanese Yen. Within 5months, the fund was invested by 125 citizens, 15.2% of whom were people from Fukushima, and the rest were from all over Japan [6].

The AiPOWER aims to create a sustainable society without nuclear power and with local energy independence. The AiPOWER's logo is a rainbow-colored fist, signifying people's power. The CEO's message reads "[W]e aim to prove to our future generations that the ability to create change is in the power of their own hands" [7]. People's power is also manifested in the AiPOWER's governance. Individuals hold 87% of the AiPOWER's stock nationally, and 61.6% of the AiPOWER's stock belongs to individuals in Fukushima [8].

4 Fukushima in global networks of community power

In addition to the non-government sector, the Fukushima prefectural government is committed to renewable energy. In 2014, the government announced its intention to achieve 100% renewable energy by 2040 under the two guiding principles: local production and consumption of energy; and democratization of energy via citizen-led renewable projects. Fukushima's goal sharply deviates from national energy policy, which currently focuses on hydrogen [9].

In November 2016, the city of Fukushima (the prefecture's capital city) hosted the first World Community Power Conference (WCPC) on the same day that the Paris Agreement came into force. The conference also marked the fifth year anniversary of the Fukushima nuclear power plant disaster and the 30th year anniversary of the Chernobyl nuclear accident. More than 600 people from over 30 countries participated in the conference and discussed the roles that community power has to play in a global shift toward 100% renewable energy. During the conference, the "Fukushima Community Power Declaration" was adopted. The Declaration reads, "Community power ensures that the local communities and their actors have democratic control of the renewable energy installations during the planning, installation, and operation period, and that the local communities and their actors get the majority of the economic and social benefits" [10].

The statement is especially meaningful to Fukushima because it had supplied one-third of electricity demand of the metropolitan area, prior to the nuclear disaster [11].

5 Conclusions

Nuclear disasters, such as the accidents at Chernobyl and at Fukushima, impacted Japan's energy democracy movement. Citizen-led renewable projects not only changed the social relations between energy producers and consumers, but also produced a new power dynamic in which citizens view renewables as a path toward local empowerment and autonomy. The Fukushima prefectural government's pledge to attain 100% renewable energy is clearly resistant to long-standing power, such as the nuclear village. Networks of local governments and citizens who pursue renewable energy nationally and globally will be the catalyst for democratizing energy in Japan.

References

[1] A. Niphi, M.V. Ramana, Talking points: narratives strategies to promote nuclear power in Turkey, in: J. Keahey, M. Pasqualetti, M. Nadesan (Eds.), Energy Democracies for Sustainable Futures, Elsevier, Amsterdam, 2022. In this issue.

[2] T. Feldhoff, Low carbon communities, energy policy and energy democracy in Japan, in: F. Caprotti, L. Yu (Eds.), Sustainable Cities in Asia, Routledge, Oxfordshire, Oxon, 2017, pp. 236–247.

[3] C. Hoffmann, Beyond the resource curse and pipeline conspiracies: energy as a social relation in the Middle East, Energy Res. Soc. Sci. 41 (2018) 39–47.

[4] A.H. Şorman, E. Turhan, The limits of authoritarian energy governance in the global south: energy, democracy and public contestation in Turkey, in: J. Keahey, M. Pasqualetti, M. Nadesan (Eds.), Energy Democracies for Sustainable Futures, Elsevier, Amsterdam, 2022. In this issue.

[5] C. Munetomo, The story of Japan's first citizen-funded wind power plant: the path to success, JFS Newsletter (165) (2016). https://www.japanfs.org/en/news/archives/news_id035575.html. (Accessed 1 December 2020).

[6] S. Furuya, Towards Local Autonomy—The Challenge of AiPOWER in Fukushima, Institute for Sustainable Energy Policies, 2015. https://www.energy-democracy.jp/751. (Accessed 1 December 2020).

[7] Aipower, Company Overview, 2019. https://aipower.co.jp/pdf/191011_aizuelectricpower-overview2019.pdf. (Accessed 16 February 2021).

[8] Aipower, Email Correspondence, February 14, 2021.

[9] G. Trencher, J. Van der Heijden, Contradictory but also complementary: national and local imaginaries in Japan and Fukushima around transitions to hydrogen and renewables, Energy Res. Soc. Sci. 49 (2019) 209–218.

[10] World Community Power Conference, Fukushima Community Power Declaration, 2016. http://www.wcpc2016.jp/wp-content/uploads/2016/11/Fukushima_Community_Power_Declaration_EN.pdf. (Accessed 10 March 2021).

[11] Fukushima Prefectural Government, Promotion of Renewable Energy, 2018. https://www.pref.fukushima.lg.jp/site/portal-english/en03-04.html. (Accessed 3 March 2021).

Participatory cartography as a means to facilitate democratic governance of offshore wind power in Brazil[*]

Thomaz Xavier[a], *Adryane Gorayeb*[a], *and Christian Brannstrom*[a,b]

[a]Department of Geography, Federal University of Ceará, Fortaleza, Brazil [b]Department of Geography, Texas A&M University, College Station, TX, United States

1 Introduction and background

Offshore wind farms are highly desirable for decarbonization initiatives because of relatively high wind velocity, lower turbulence, and strong reliability of wind over the sea as compared to land [1]. While onshore wind farms have an estimated energy density of $0.9\,W/m^2$ [2], estimates for offshore wind farms range between 3 and $5\,W/m^2$ [3–5]. Recently, Brazil's Energy Research Enterprise (Empresa de Pesquisa Energética; EPE) listed key challenges for offshore wind over its waters, which included "uncertainty regarding potential socio-economic conflicts that may arise between offshore wind and other activities" [6, p. 127]. However, other Brazilian experts have offered optimistic estimates for offshore wind without concern for host communities [3,7,8]. In January 2022, Brazil had 37 offshore projects in the licensing phase. Four wind farms in Ceará state are shown in Fig. 21.1:

[*] This chapter contains some material published in T. Xavier, A. Gorayeb, C. Brannstrom, "Energia eólica offshore e pesca artesanal: impactos e desafios na costa oeste do Ceará, Brasil," in D. Muehe, F. M. Lins de Barros, and L. Pinheiro, eds., *Geografia Marinha: Oceanos e Costas na Perspectiva de Geógrafos* (Rio de Janeiro: PGGM), pp. 608–630. Our research received financial support from Brazil's Coordenação de Aperfeiçoamento de Pessoal de Nível Superior (CAPES) and Conselho Nacional de Desenvolvimento Científico e Tecnológico (CNPq) through CAPES/PRINT 88887.312019/2018-00, PRONEM/FUNCAP/CNPq 88887.165948/2018-00 and Bolsa de Produtividade em Pesquisa (CNPq-PQ).

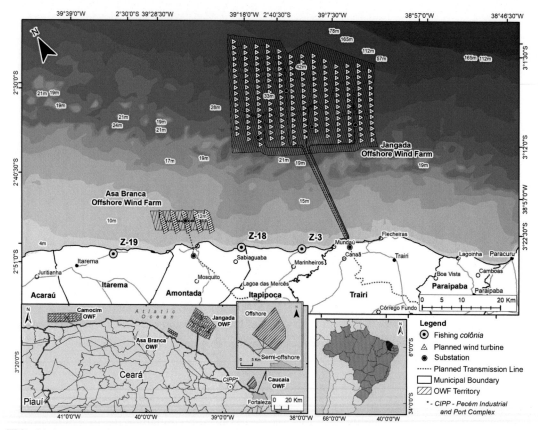

FIG. 21.1 Areas identified for the Asa Branca and Jangada offshore wind farms, Ceará state, Brazil. Inset map shows Camocim and Caucaia offshore wind farms.

- Caucaia (PEOC), a Brazilian-Italian project in the Caucaia municipality in Ceará for 59 turbines (598 MW capacity), which Brazil's federal environmental agency rejected in August 2020, after our fieldwork was completed (but a different group has proposed a new Caucaia project for 48 turbines and 576 MW installed capacity);
- Asa Branca I (CEMAB), supported by Brazilian investors, planned for Itarema and Amontada in Ceará (50 turbines; 400 MW capacity);
- Brazilian and Spanish investors proposed the Jangada (CEMJa) wind farm near Trairi and Itapipoca in Ceará, with 200 turbines (3 GW capacity);
- investors of the failed PEOC project filed plans for a 600-MW offshore wind farm in the Camocim municipality in the far west of Ceará state.

Protocols and instruments for ensuring the inclusion of host communities in democratic processes for offshore wind farm licensing are scarce. This scarcity is especially important in the Global South, where resource access rights, judicial independence, information asymmetries, and elite capture of regulatory institutions may favor renewable power investors over host communities and lead to injustices. Here we deploy participatory cartography and

SWOT-based workshops to identify resource uses among artisanal fishers in host communities near planned offshore wind farms in coastal Ceará state, Brazil. We evaluate the extent to which these techniques may enhance participatory or procedural justice, which focuses on fairness and justice of decision-making processes, while recognizing other forms of energy justice are outside the boundaries of this chapter [9–11]. As Heather Bedi discusses in Part II of this volume, energy transitions require attention to recognitional, procedural, and distributional justice. Our chapter adds to the growing field by showing how these forms of justice may be understood through participatory methods.

2 Materials and methods

We adopted participatory cartography methods [12,13] with broad parallels to participatory GIS developed in Anglo-American geography [14,15]. Participatory mapping helps identify and demarcate disputed territories between traditional communities and outside interests, given that traditional communities (such as artisanal fishers and farmers who use mainly traditional methods to obtain resources to sustain their livelihoods) occupy territories that have economic, social, cultural, and symbolic importance. The maps and associated political processes may enhance procedural justice, which concerns the participation of affected people in decision-making processes. Improved practices of participatory justice, in turn, may result in better outcomes relating to distributive justice, which focuses on the relative distribution of benefits and harms. Procedural and distributive justice are considered essential to improve democratic governance of resources claimed by traditional communities in coastal Brazil [16,17].

We conducted eight participatory workshops between November 2018 and March 2020 with fishers affiliated with fishing *colônias* in the Itapipoca (Z-3), Amontada (Z-18), and Itarema (Z-19) municipalities in Ceará state. These *colônias* (fishing associations or guilds, recognized in Brazilian federal law) include approximately 30 traditional communities that rely on small-scale fisheries [18–20]. For reference, Brazil's artisanal fishers are thought to be responsible for more than half of the country's total marine landings [21]. In the workshops, we deployed base maps and a SWOT (Strengths, Weaknesses, Opportunities, Threats) framework, which has been applied in many areas outside its origin as a strategic planning tool for organizations [22] in settings including artisanal fishers [23]. Participants included 68 fishers, all male, between 22 and 55 years of age. The dominance of men in the workshops reflects the rigid gender division of labor in artisanal fishing, which has strong cultural and legal foundations supporting men on fishing boats and women in shellfish collection and fish processing [21,24]. Participants were enrolled as follows: we contacted the leaders of the *colônias*; after they agreed to participate, we asked leaders to disseminate workshop information to associated fishers in their *colônias*; we offered meals during the workshop to encourage attendance for the 4–6 h period; we conducted extension workshops on GPS use in each *colônia*; we conducted an additional extension workshop on hygiene and healthy food with 20 children in Itarema.

Workshops were conducted with two to five researchers in the role of facilitators. First, we explained the workshop context, showing offshore wind farms planned in Brazil and implemented globally. Next, we facilitated discussion about local fishing context and then tasked participants with the creation of SWOT tables relating to local fishing conditions. Finally, we assisted in the production of sketch maps using participatory cartography techniques. Among all participants, five (7.4%) had prior knowledge of the offshore wind farms planned for coastal Ceará state. The remaining participants had no knowledge of offshore wind farms, so the facilitators treated the topic in the most neutral manner possible to avoid bias. To stimulate discussion on the local fishing context, we used these prompts:

- Which fish species are found in the area?
- Which tools are used to fish which types of fishes?
- Which types of boats are used in fishing?
- Is fishing dependent on wind or motors?
- What are the distance and directions used for fishing?

For the SWOT, we asked fishers to identify their strengths (desirable characteristics that distinguished their community), weaknesses (things that fishers needed to continue fishing), opportunities (things that motivated fishing), and threats (things that threatened fishing). Finally, we worked with fishers to create maps of their fishing environment using sketch mapping [13,25]. Throughout the workshops we audio recorded, with informed consent, participants and wrote field notes in journals. After fieldwork, the sketch maps were digitalized in QGIS software and qualitative data from the SWOT exercise were tabulated in spreadsheets. Audio recordings were transcribed and coded using the simple taxonomy in the SWOT scheme. We made simple grammatical corrections to transcripts and applied pseudonyms to participants. Our research protocols were approved by the Universidade Federal do Ceará (CAAE 06529217.1.0000.5054).

3 Findings and discussion

We discovered that fishers have little or no knowledge of offshore wind farm proposals, but fishers have precise knowledge about marine spaces desired by offshore wind farm investors. Small-scale fisheries overlap with offshore wind farms, especially in regard to fishing routes and fishing grounds.

Participatory mapping identified possible conflicts of ocean resources between planned offshore wind farms and small-scale fisheries (Fig. 21.2). We identified resource uses in the beach margin, including shrimping, non-boat fishing, and algae collection. Resources were also identified at the edge of the continental shelf for artisanal fishing 50–70 km offshore and professional and semi-professional fishing up to 800 nautical miles offshore. Our methods also obtained information about existing biodiversity in specific offshore fishing regions, such as locations of turtle nesting, dolphins, and whales. We noticed, for example, that the site of whale appearances is precisely where the Jangada (CEMJa) wind farm is planned. This suggests the need for offshore windfarm impacts to consider marine mammals, similar to recent suggestions [26,27].

Fishers indicated travel distances. For example, one fisher indicated that fishers stop heading offshore "when the fishing lines don't hit the bottom" (Fisher D, January 2020, Itapipoca); when this happens, fishers turn toward land because they are in small boats. At that moment, they reach the limit of the continental shelf. This makes it clear that the artisanal fishers use the entire continental shelf. Fishers also provided information on the sailing routes they take

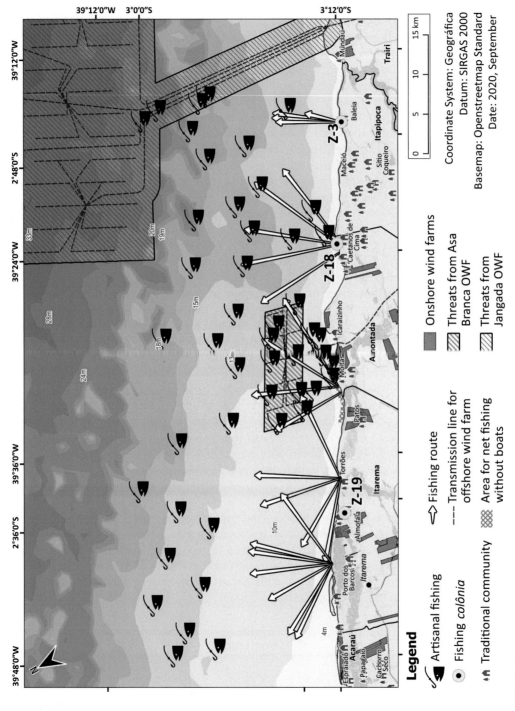

FIG. 21.2 Synthesized and aggregated results of participatory cartography workshops conducted in fishing *colônias* (associations or guilds) Z-3, Z-18, and Z-19.

for fishing (Fig. 21.2). Because they rely on sailing boats, the actual routes may be severely impacted by offshore wind farms. One fisher told us that their reliance on wind makes it impossible to easily adjust their routes (Fisher E, November 2018, Amontada). In other words, sailing around a wind farm could be an arduous task.

Fig. 21.2 also shows information about fixed fishing points in the sea that are traditionally known and used by artisanal fishers. During the participatory mapping process, fishers described locations in terms of substrate and/or rocks present on the seabed, the names of fishers, or the name of particularly abundant fish. Some of these sites are within the offshore wind farm area. This practice of naming fishing sites indicates the use of marine territory by fishers, ancestral relationships, and knowledge of ocean space. To find specific points, fishers use some rudimentary techniques, such as alignment with fixed points on land, and the use of geographical coordinates by satellite receivers, which are widely used among fishers. Fishers provided geographical coordinates to us, facilitating the mapping process. Projecting these data in QGIS revealed the overlap of offshore wind farms and fishing areas, especially the Asa Branca (CEMAB I) site, which is closer to land. This wind farm would alter the seabed, change fishing locations and possibly dislocate the fish shoals temporarily or permanently.

For example, fishers from *colônia* Z-18, especially in the Moitas community, identified that the Asa Branca (CEMAB I) wind farm is in a location that is heavily used by artisanal fishers. This shows that fishers were justified in their belief that offshore wind farms were a potential threat because of the enclosure of fishing grounds by the wind farm. Both offshore wind farms were seen as potential threats, similarly to how fishers also viewed the current terrestrial wind farms near their communities. In the coastal area of Itarema, Amontada, Itapipoca, and Trairi there are 38 onshore wind farms, many of which were built with minimal public consultation and flawed environmental analysis [28–30]. For this reason, when we questioned fishers about possible impacts from offshore wind farms, many respondents noted that there was already pressure from wind farms on territories used by communities for livelihoods. This suggests the potential for the accumulation of negative impacts of onshore and offshore wind farms. Some negative impacts mentioned by fishers regarding onshore wind farms included privatization and blocked access to land that had been subject to collective use; false promises of work or employment offered only during the construction phase; continuous fear of accidents owing to close proximity between turbines and residences.

The SWOT workshops offered insight into considering strategies and actions that could improve fishing, such as dock improvements, new boats or engines, educational programs to improve fisher knowledge or techniques, and better opportunities for selling their catch. SWOT and participatory cartography, as ways to document marine territoriality of fishing communities, also offer the potential to contribute to recognition and participatory justice. Not only do these methods offer visibility or recognition to fishers and the resources on which their livelihoods depend, but they also elevate fishers to the role of potential participants in decision-making about offshore wind farms.

4 Conclusions

How could participatory cartography and SWOT workshops inform democratic governances processes for offshore wind farm planning? Two preliminary conclusions may be offered. First, the maps, supported by discussions in SWOT-focused workshops, produced affirmations of territorial and biogeophysical information about the marine environment supporting livelihoods of traditional fishing villages. These maps could become part of formal public consultation procedures in line with concerns raised by Brazil's EPE [6] regarding possible conflicts between offshore wind farms and competing resource uses. Instead of being considered as an anachronistic group doomed to eventually disappear, artisanal fishers with maps should have a seat at the decision-making table among other stakeholders, who should consider more fully the possible impacts of offshore wind farms so that decarbonization initiatives are compatible with recognition, participatory, and distributive justice concerns of host communities.

Second, the participatory activities increased fisher knowledge regarding planned offshore wind farms, offering the potential to increase the quality of participation by host communities in discussions regarding offshore wind farms. Participatory mapping and SWOT-focused workshops may create conditions for increased participation of fishers in the licensing process, therefore increasing participatory justice and democratic decision-making. However, the terms of inclusion of host communities in public hearings may be determined by investors, regional elites, and state authorities to the exclusion of host communities. If host communities and their cartographic products are included in decision-making processes, there is no guarantee that distributive justice—compensation for harms caused by the offshore wind farm—will be addressed by investors or the state, or that benefits arrive in host communities owing to increased participation in planning and siting processes. As we have observed with regards to terrestrial wind farms, information asymmetries, dysfunctionality in the judicial system, and power of local elites may lead to undesirable relations between wind farms and host communities [28–30].

Participatory cartography and SWOT are well suited to the particular characteristics of offshore wind in Brazil. Unlike the terrestrial sphere, the offshore environment is a federal territory that must be ceded to wind farm investors [31], which in theory could break the grip of local elites on the licensing process. But in practice, we have observed similar patterns of elite control as compared to onshore wind farms. Local and regional elites seek influence and information from federal actors and agencies in offshore wind governance. The resources in the offshore tropical environment (e.g., fishes, turtles, and sediments) are mobile and not fixed in space, so they may appear to have no identifiable owner. Artisanal fishing activities organized around tropical marine resources may be easily characterized as archaic and destined to disappear, rather than legitimate livelihood strategies; moreover, artisanal fishing leaves no permanent or semi-permanent marker in the offshore environment. Participatory cartography helps to identify people and resources that are hidden from offshore wind farm planners and developers, while SWOT helps to give voice to the aspirations and concerns of those resource users.

Democratic governance of offshore wind farms would be improved by specific actions involving investors, state, and host communities. Investors should implement processes that fund rents or royalties to community associations, rather than to individuals, mitigate problems arising from the construction stage, use tax obligations to support local governments, offer technical training to host communities for improved employment opportunities, conduct transparent and science-based dialogue with host communities, facilitate permanent programs for education and other practices for host communities, and prepare strategies for marine restoration as compensation for environmental damages resulting from the offshore wind farm. State officials should guarantee the security of landholdings and resource uses (including access to marine fisheries) by traditional communities, serve as an honest broker between host communities and investors, enable participatory mechanisms to plan for offshore wind farms, facilitate participatory zoning processes, at municipal and state scales, to identify locations where offshore wind is not compatible with existing resource uses and users, and offer host communities discounts in electricity costs as a form of mitigation. Host communities also have a role to play in providing state authorities and investors with information and maps that illustrate the social and physical environment, collaborating with environmental and social analyses, and creating community councils that monitor and report on aspects surrounding the offshore wind farm.

References

[1] A. Possner, K. Caldeira, Geophysical potential for open ocean wind energy, Proc. Natl. Acad. Sci. 114 (2017) 11338–11343.

[2] L.M. Miller, D.W. Keith, Corrigendum: Observation-based solar and wind power capacity factors and power densities, Environ. Res. Lett. 14 (2019), 104008.

[3] L.F.A. Tavares, et al., Assessment of the offshore wind technical potential for the Brazilian Southeast and South regions, Energy 196 (2020), 117097.

[4] P.J.H. Volker, et al., Prospects for generating electricity by large onshore and offshore wind farms, Environ. Res. Lett. 12 (2017), 034022.

[5] J. Bosch, I. Staffell, A.D. Hawkes, Temporally explicit and spatially resolved global offshore wind energy potentials, Energy 163 (2018) 766–781.

[6] Empresa de Pesquisa Energética, ROADMAP Eólica Offshore Brasil: Perspectivas e caminhos para a energia eólica marítima, Rio de Janeiro, Brazil, 2020.

[7] M.S.S. Gomes, et al., Proposal of a methodology to use offshore wind energy on the southeast coast of Brazil, Energy 185 (2019) 327–336.

[8] D.K.S. Lima, et al., Estimating the offshore wind resources of the State of Ceará in Brazil, Renew. Energy 83 (2015) 203–221.

[9] U. Liebe, A. Bartczak, J. Meyerhoff, A turbine is not only a turbine: the role of social context and fairness characteristics for the local acceptance of wind power, Energy Policy 107 (2017) 300–308.

[10] B.K. Sovacool, et al., The whole systems energy injustice of four European low-carbon transitions, Glob. Environ. Change 58 (2019), 101958.

[11] K. Jenkins, B.K. Sovacool, D. McCauley, Humanizing sociotechnical transitions through energy justice: an ethical framework for global transformative change, Energy Policy 117 (2018) 66–74.

[12] H. Acselrad (Ed.), Cartografias sociais e território, Universidade Federal do Rio de Janeiro, Rio de Janeiro, 2008.

[13] A. Gorayeb, A.J.A. Meireles, E.V. da Silva, Cartografia Social e Cidadania: experiências de mapeamento participativo dos territórios de comunidades urbanas e rurais, Fortaleza, 2015.

[14] S. Ellwood, Thinking outside the box: engaging critical geographical information systems theory, practice and politics in human geography, Geogr. Compass 4 (2010) 45–60.

[15] S. Ellwood, Critical issues in participatory GIS: deconstructions, reconstructions, and new research directions, Trans. GIS 10 (2006) 693–708.

[16] A.C. Cortines, et al., Social cartography and the defense of the traditional Caiçara territory of Trindade (Paraty, RJ, Brazil), in: W.L. Filho, L.E. Freitas (Eds.), Climate Change Adaptation in Latin America: Managing Vulnerability, Fostering Resilience, Springer, 2017, pp. 445–456.

[17] C.N. Silva, et al., Modo de vida e territorialidades de pescadores da comunidade Cajueiro em Mosqueiro (Belém-Amazônia-Brasil), Revista NERA 20 (40) (2017) 246–272.

[18] A.C.E. Dias, A. Cinti, A.M. Parma, et al., Participatory monitoring of small-scale coastal fisheries in South America: use of fishers' knowledge and factors affecting participation, Rev. Fish Biol. Fish. 30 (2020) 313–333.

[19] S. Partelow, M. Glaser, S. Solano Arce, R.S.L. Barboza, A. Schlüter, Mangroves, fishers, and the struggle for adaptive comanagement: applying the social-ecological systems framework to a marine extractive reserve (RESEX) in Brazil, Ecol. Soc. 23 (3) (2018) 19.

[20] L.S. Queiroz, et al., The social and economic framework of artisanal fishing in the state of Ceará, Brazil, Geosaberes 11 (2020) 180–198.

[21] M. Vasconcellos, A.C. Diegues, D.C. Kalikoski, Coastal fisheries of Brazil, in: S. Salas, R. Chuenpagdee, A. Charles, J.C. Seijo (Eds.), Coastal Fisheries of Latin America and the Caribbean, FAO, Rome, 2011, pp. 73–116. FAO Fisheries and Aquaculture Technical Paper, No. 544.

[22] M.M. Helms, J. Nixon, Exploring SWOT analysis – where are we now? A review of academic research from the last decade, J. Strat. Manage. 3 (3) (2010) 215–251.

[23] M.D.C. Viegas, A.B. Moniz, P.T. Santos, Artisanal fishermen contribution for the integrated and sustainable coastal management – application of strategic SWOT analysis, Procedia – Social and Behavioral Sciences 120 (2014) 257–267.

[24] A.N. Santos, Fisheries as a way of life: gendered livelihoods, identities and perspectives of artisanal fisheries in eastern Brazil, Mar. Policy 62 (2015) 279–288.

[25] J. Corbett, et al., Overview: Mapping for Change – the emergence of a new practice, in: Partipatory Learning and Action 54: Mapping for Change: Practice, Technologies and Communication, International Institute for Environment and Development, 2006, pp. 13–19.

[26] L. Bergström, et al., Effects of offshore wind farms on marine wildlife—a generalized impact assessment, Environ. Res. Lett. 9 (3) (2014), 034012.

[27] L. Bray, et al., Expected effects of offshore wind farms on Mediterranean marine life, J. Mar. Sci. Eng. 4 (1) (2016) 18.

[28] J.C.H. Araújo, W.F. Souza, A.J.A. Meireles, C. Brannstrom, Sustainability challenges of wind-power deployment in coastal Ceará state, Brazil, Sustainability 12 (2020) 5562.

[29] C. Brannstrom, et al., Is Brazilian wind power development sustainable? Insights from a review of conflicts in Ceará state, Renew. Sust. Energy Rev. 67 (2017) 62–71.

[30] A. Gorayeb, et al., Wind power gone bad: critiquing wind power planning processes in northeastern Brazil, Energy Res. Soc. Sci. 40 (2018) 82–88.

[31] Brazil, Decreto No. 10.946, de 25 de janeiro de 2022, Diário Oficial Da União, 160 (17–b) (2022) 1–2.

Energy democracy cooperatives: Opportunities and challenges

Caroline G. Wright

Communication Studies, Arizona State University, Glendale, AZ, United States

1 Introduction

For most, work has become synonymous with being a part of a hierarchical organization where the labor is out of the workers' control and the compensation is decided by the higher-ups. This conception, however, excludes other forms of organization and a rich tradition of worker management. Employee stock ownership plans, communes, and worker cooperatives are three examples of alternative labor arrangements [1]. This chapter is primarily interested in worker cooperatives: how they function in the workplace and the wider socioeconomic system, with a particular focus on the energy sector.

Vieta et al. claim that worker cooperatives date back to the early 19th century, with the Rochdale Co-operative Manufacturing Society, a cotton mill located in England, being the first formal organization documented [2, p. 437]. Later in the century, cooperatives spread to the United States, driven by the activism of the Knights of Labor [2, p. 438]. Perhaps the most well-known example of this form of organization today is Mondragon, a "system of production cooperatives" located in Basque Country, Spain [3]. Founded in 1956, it "is a major exception to the time-honored belief that production firms organized on a cooperative basis of worker ownership are doomed to a short and precarious existence" [3]. To many progressive-minded people, this form of organization is preferred, but what defines worker cooperatives?

At a basic level, worker cooperatives are organizations that are to some extent owned and controlled by their workers [4]. On the opposite end of this spectrum is the corporation, where the ownership and management are in the hands of a small number of capitalists [4]. Artz and Kim explained that "there is not one universally accepted definition of a worker cooperative," but three defining characteristics bridge different organizations [5]. The first is that ownership shares are bought by each member, and the assets are shared among all members; the second is that the cooperative follows the "one person, one vote" principle, meaning that each individual is given equal voting power regardless of the size of their share; and finally, worker-owners provide the production input, while profits are allocated based on labor input of each member [5]. Artz and Kim differentiate cooperatives from typical capitalist firms in that they are created with the express purpose of providing employment to members, which they relate to the fact that cooperatives are typically set up during economic recessions to prevent high levels of unemployment [5]. A striking example of this phenomenon is the swell of cooperatives in Argentina since the early 2000s, corresponding to the country's great depression [6]. This trend ties into the claim of Webb and Cheney that the global economic crisis has created the need for new forms of organization, worker cooperatives being some of the most promising [7, p. 64].

Despite this, Artz and Kim revealed that capitalist firms greatly outnumber worker cooperatives in the United States [5]. Additionally, worker cooperatives typically employ less workers, the average being 11 to an organization [5]. It is easy to explain why capitalist firms are the most common form of organization in the United States and other places in the world, as capitalism is a dominant economic system in the global arena. This lack of worker or community control over industries is pronounced in the energy sector, as Abell found that energy cooperatives make up only 4% of the total number of worker cooperatives in the United States [8]. Further, the number of energy cooperatives in Europe has been declining in the past decade [9]. This trend is troubling because of the role the energy sector plays in industrial societies, where energy consumption has increased 10-fold or more since the Industrial Revolution [10]. Additionally, the energy that these societies rely on is primarily sourced from fossil fuels, which are a driving cause of climate change [10]. Due to this, as well as the decreasing confidence in nuclear energy after the Fukushima disaster [11], a transition to renewable sources of energy is becoming more and more of a necessity.

Unlike fossil fuels, renewable energy, such as sunlight, wind, or heat, relies on natural sources that replenish themselves and do not generate greenhouse gases [11]. Renewable energy sources are superior not only because of their ecological soundness but also because of their potential to foster democratic values. Tarhan points out that renewable energy can be locally sourced, and thus renewable energy systems can be owned, operated, and maintained by their consumers, in marked contrast to the corporate-controlled energy markets of the neoliberal period [12]. For all these reasons, the cooperative form of organization is fitting for the dissemination of renewable energy.

This chapter will argue that worker cooperatives have the potential to radically transform social relations, from individual workplaces to the society-wide organization of labor, and the implications this transformation would have on usage and dissemination of energy. Thus, it will first discuss the impact of the cooperative organization on workers and communities, and then detail what the available research reveals about successful cooperatives on the individual and economic levels. This chapter hopes to provide information that will guide the creation of future cooperatives with the explicit aim of democratizing society and energy systems.

2 The lived reality of worker cooperatives

Considering the description of worker cooperatives thus far, they would seem to be a preferable form of organization to capitalist firms. The cooperative values given by Webb and

Cheney paint a picture of a workplace based on voluntary and open membership, member economic participation, autonomy and interdependence, inter-organization cooperation, and community concern [7, p. 65]. While this list sounds appealing, an examination of the existing literature is necessary to discover whether cooperatives live up to these ideals in practice.

Many studies point to worker cooperatives providing material benefits to members compared to capitalist organizations. Burdin and Dean report that capitalist firms "exhibit a well-defined and negative relationship between wages and employment," while for worker cooperatives, "wages and employment move in the same direction" [13]. Garcia-Louzao reports that worker-owned enterprises provide more job security because even in times of financial hardship, members have more ability to control employment levels and thus find solutions other than firing workers [14]. There is also evidence that worker cooperatives while remaining competitive, have better pay and more benefits than capitalist organizations in the same sectors [15,16]. From these descriptions, it is clear that worker cooperatives lend themselves toward the welfare of workers, as opposed to capitalist firms which focus on maximizing profit.

Along with these upsides, worker cooperatives often offer more on-the-job training and skill acquisition for employees [5]. There is also a body of literature showing the experience of having a job in a worker cooperative is positive compared to capitalist firms. For example, Cornforth et al. point out that cooperative workers have more control over their work life, such as how many tasks they handle in a given day and how they go about their work [17, p. 56]. Meyers echoes this; since employees are in control of the allocation of work, they are able to take tasks they prefer, or at least equally share unpleasant work [16]. Additionally, the sense of shared ownership in a cooperative can increase the morale of employees, creating a positive work environment [5].

Hoffman discusses worker cooperatives in terms of emotions at work; employees reported that they felt freer to express themselves, and despite navigating interpersonal "feeling rules," they felt far less pressure to fake emotions than in a cooperative setting [18]. Elsewhere Hoffman has compared Welsh mine workers' descriptions of disputes under the capitalist and cooperative management. Reportedly, worker ownership drove the workers to accept difficult circumstances they would have complained about previously and created an "ethic of compromise" which decreased the number of disputes [19]. The many negatives of working in a corporate hierarchy, such as unfair bosses, pointless or overly difficult work, and powerlessness over one's circumstances, seem to be entirely sidestepped by worker cooperatives.

An economically minded individual's question might be whether this form of organization hurts productivity or profits. The research suggests that it does not. Rather, many claim that having a share in the company motivates employees to work harder and produce more [5]. Artz and Kim also report that due to having more open communication and less employee turnover, communication and competency are stronger in worker cooperatives, as workers are more experienced in their roles [5]. Since workers tend to stay at cooperatives longer and form bonds with one another, the synergy and performance of employees are greater. Worker participation in decision-making can also have a positive impact on productivity, as employees often know best what changes need to be made in their roles and decision-making power increases feelings of ownership over a firm [5]. Whether considering worker experience or financial success, worker cooperatives appear to be superior to capitalist organizations.

3 Cooperatives and communities

Aside from the benefits for workers, cooperatives can benefit wider communities. As mentioned in the introduction, cooperatives are often set up in times of economic crisis to provide jobs. The fact that these jobs are more secure, better paying, and more dignified can improve the lives of workers and their communities [20]. Logue and Yates argue that employee ownership "generates benefits for community life by creating concrete benefits, developing leaders and teaching the skills of participation and self-government" [21]. The cooperative form of organization is a potential method for safeguarding communities and transitioning toward a more democratic society. Cooperatives also protect communities by internalizing risks, rather than externalizing them as capitalist firms are known to do [20].

Cooperatives also affect community self-sufficiency and resilience. For instance, it has been noted that workers will accept less advantageous jobs if involved with a cooperative that benefits the local community, explaining why "cooperatives were historically the major providers of electricity and telephone service to sparsely populated rural areas in the U.S." [22]. Logue and Yates describe how worker ownership can create a snowball effect where the skills and experience gained in a cooperative workplace give workers the ability to spread those principles elsewhere, such as by setting up new cooperatives [21]. Zeuli and Radel discuss how outside of the United States, cooperatives are more likely to be created by communities to address multiple needs, leading them to call for cooperatives that ultimately eliminate the community's reliance on outside support [22]. Worker cooperatives, after all, are not the only kind of cooperative: there are also producer, consumer, and housing cooperatives [23]. One can envision a cooperative-based society that would provide greater security and freedom for all.

Energy is a sector where the benefits of cooperatives for communities are apparent. According to Herbes et al., "Community energy projects, especially renewable energy cooperatives (RECs), have become an increasingly important element of energy markets in many European countries" [24]. Energy cooperatives are typically user or consumer based and play an important role in energy generation, distribution, and provision [12]. One of the advantages of the cooperative model for energy is that partners are not liable for the system individually, making each member responsible for its maintenance [25]. Yildiz et al. argue that cooperatives tend to have values beyond increasing profit, such as ensuring the well-being of the community [25]. Research on energy cooperatives has found that democratic participation in energy distribution increased community social cohesion and ability to accomplish meaningful tasks, as well as bringing economic benefits such as energy sales, energy consumption, and new economic opportunities [12]. The promise of cooperatives for the energy sector makes it all the more urgent to provide resources that aid in the creation and maintenance of cooperatives.

4 Challenges cooperatives face

The research discussed thus far points to the democratizing power of worker cooperatives; however, those attempting to create such organizations will face many challenges. Work by Vieta et al. [2] describes the struggles involved in setting up a cooperative. Worker buyouts of failing companies come with a great deal of risk, as the workers are inheriting a business that crashed [2, p. 453]. Conversions of companies into cooperatives by an owner selling or giving it to the workers is both the rarest kind of cooperative and the kind most likely to succeed [2, p. 460]. These cooperatives do not have to deal with the struggles that most do; start-ups are the most common type of cooperative and require workers to put forth capital and find a place in a competitive market [2, p. 462].

Webb and Cheney [7, p. 74] give a helpful overview of common problems cooperatives run into after their creation, including weak or absent training programs, no consensus on what values are important, little research by members on how cooperatives work, insufficient rewards for members or clients, and a failure to organize in a way that differs from traditional capitalist methods. They identify the key difficulty as having to interact with a capitalist society while maintaining a cooperative nature [7, p. 74]. The struggle to remain cooperative in practice manifests in the decision-making process. The potential for a worker cooperative to be democratic in name only is a common pitfall [26]. Logue and Yates expand on this by explaining that "too often worker ownership is purely nominal," and cooperatives "may be too indirect and too stripped of the rights of ownership for employees to exercise any influence or control over the management and governance of the firm" [21]. Additionally, not all workers are interested in decision-making, as it comes with a burden of responsibility. Meetings and democratic processes take up additional time, leading some workers to prefer a more hierarchical form of organization [27].

Tied to this point is the fact that issues in wider society do not cease to exist simply because one adopts a cooperative structure. Research points to gender, racial, and other forms of inequality persisting in cooperatives [28]. Another example of this phenomenon is research showing that more white-collar workers than blue-collar workers are elected as representatives in cooperatives [29]. These inequalities are a stumbling block at the feet of any workplace that wishes to be truly democratic.

Another issue that cooperatives must contend with is degeneration. Pek defines degeneration as "the centralization of power within oligarchies comprised of workers not descriptively representative of the broader workforce, and worker apathy and reduced participation in democratic structures" [29]. The literature points to larger organizations suffering from more problems related to democratic processes, because the larger the firm, the less individuals feel a responsibility to participate [29]. The size of organizations also affects democracy because the sharing of information and the decision-making process becomes cumbersome [29]. Pek suggests that "apathy and low participation can reinforce each other," meaning the problem is likely to worsen over time [29].

Degeneration also impacts elections. In worker cooperatives, the same representatives will often be elected without contestation, and existing representatives will often select the new candidates [29]. Participation in elections is often nonexistent, low, or uncritical [29]. Pek also notes that when workers do vote, they may choose candidates based on "popularity assessments, workplace friendships, favoritism, and the desire to punish one's colleagues" [29]. From these findings, one can conclude that a form of organization that looks democratic on paper may fail to be in practice.

Further, there are many financial challenges that cooperatives face. Dow coined the term "horizon problem," which refers to the fact that cooperatives can suffer financially since workers cannot usually sell their share in the company, leading to underinvestment [30].

Cooperative workers also face more financial risk than regular employees. Ben-Ner et al. discuss how if a worker cooperative fails, the employees lose not only their salaries but also their savings, investments, and retirement plans all at once [31]. Despite cooperatives protecting workers from job termination, the financial precarity of the cooperative itself creates a form of job insecurity.

As mentioned previously, starting a cooperative takes a lot of capital, capital that workers often do not have. Thus, creating a cooperative may call for external sources of capital, but Artz and Kim [5] note that this may be difficult; for instance, investors may avoid giving loans to cooperatives, as they view them as high risk and may not know how to evaluate their chances of success. In the words of Elster, "Why would outside investors be attracted to a firm over which they have no control? For all they know, the cooperative might pay zero dividends year after year" [32]. Elster also argues that cooperatives may face outright discrimination from capitalist institutions which will lead them to suffer financially [32]. Artz and Kim echo this, stating that "tax policies and legislative statutes may favor other organizational forms" [5].

It is clear that cooperatives face a lack of institutional support for their ventures, which holds true for renewable energy cooperatives (RECs) as well [12]. RECs are often forced to rely on volunteer labor, capital donated by members or partnerships with the private sector to stay afloat [12]. Tarhan also outlines various barriers related to infrastructure that face RECs, such as limited access to the electricity grid and locations for the construction of facilities [12]. RECs must also contend with challenges to democracy, leading Tarhan to claim that "the nature of the process and outcome of community-owned energy projects seem to be a significant determinant of their social impact" [12]. To demonstrate this, Tarhan discussed an example of an unsuccessful wind cooperative in Wales. The cooperative was jointly owned by three farmers who made no attempt to involve the local community, which caused social division and decreased trust rather than bringing people together [12]. Whether one wishes to create a worker, producer, or consumer cooperative, all of the challenges discussed in this section must be kept in mind.

5 How to run a successful cooperative

There exist many examples of cooperatives that have overcome the challenges they faced to succeed. Whyte and Blasi provide insights from cooperatives in Yugoslavia, Israel, Mondragon, and the United States. Firstly, in some cases, job rotation proved to be an effective means of giving workers different experiences and counteracting the effect that division of labor may have had on the democratic ethos of the cooperative [33]. Whyte and Blasi argue that the creation of shared values is essential to the flourishing of a cooperative [33]. Another principle they include is that it is important for pay scale to be agreed upon by workers; depending on the workplace, that may mean employees with families receiving additional benefits, or those who take on dangerous work getting a higher wage [33]. Whyte and Blasi also promote the creation of a cross-organizational infrastructure to support the maintenance of cooperatives [33].

This finding is echoed by a study of cooperatives in Italy, Mondragon, and France. All of the successful organizations studied had either governmental or cooperative associations which provided support to start-ups [34]. The authors also note the importance of worker solidarity [34]. Finally, one worker was quoted as saying that "Strength breeds strength"—the longer a cooperative has been successful, the more it is respected, and its success will inspire future workers to organize on cooperative principles [34]. The existence of a wider cooperative movement greatly improves the chances of individual organizations; similarly for RECs, the more community members and institutions are involved in their planning, decision-making, and operation, the higher their chances of success [12].

To combat identity-based inequity in cooperative settings, Meyers and Vallas suggest "embracing a broadly multicultural conception of worker identity and embracing a diversity regime that actively encourage[s] learning across demographic lines" [28]. They insist that the issue of intergroup differences is not a minor one in creating a democratic culture [28]. Another paper addresses the issue of gender inequality. The authors argue that for cooperatives to combat sexism, there must be a conscious effort to do so on the part of the cooperative [1]. Additionally, they suggest that the cooperative must actively promote participation in decision-making for workers on all levels, as well as acknowledge and reward labor associated with women [1].

The problem of degeneration in worker cooperatives has been addressed in various ways. Some cooperatives have implemented mechanisms to keep an eye on representatives and recall them if needed, and others have broken into small groups to discuss what they want and picked representatives to carry out their will [29]. Pek argues for a move away from reliance on elections toward either direct democracy or sortition [29]. The practice of direct democracy, or each worker having a personal say on their workplace rather than relegating that task to a single individual, could eliminate the problem of degeneration but may contribute to slowing down the decision-making process. Sortition, on the other hand, would involve representatives being chosen by lottery rather than voting. Pek suggests that sortition would reduce the chance for power to become centralized in worker cooperatives, as representatives are chosen at random and are unlikely to be selected again [29]. Sortition could also address the problem of privileged representatives being more likely to win elections [29]. Though it has been argued that degeneration is an inevitability in large cooperatives, these solutions demonstrate that it can be avoided.

There also exists a body of research on how successful cooperatives handled financial issues. Artz and Kim point out that "problems of obtaining external financing are mitigated when large amounts of firm-specific capital is not required," such as in service and retail work, industries which make up the majority of American cooperatives [5]. Additionally, the issue of workers typically being unable to sell their shares can be overcome when the cooperative relies on assets that can be sold, such as motor vehicles [5]. These solutions do not apply to all organizations, but the formation of successful cooperatives in certain sectors can stimulate the creation of a cooperative movement.

Corcoran and Wilson discuss financial decisions available to all cooperatives. Cooperatives in Italy, Mondragon, and France set aside capital for workers' use; the capital pool can be loaned or granted to workers in times of crisis, preventing emergencies from tearing the cooperative apart [34]. These cooperatives also allowed non-members to invest in their organization and were supported by government subsidies and tax breaks [34]. Viardot found similar results in the sphere of energy cooperatives: the most successful organizations relied on outside initiatives to reduce the cost of entry into the field and employed communication strategies such as websites, seminars, lectures, educational tours, and exhibitions to increase community involvement [11]. These findings reinforce the fact that the success of cooperatives rests on outside support which must be built over time.

6 Cooperatives to change society

This chapter will conclude with a reflection on how cooperatives can transform society and what must be done to bring out their radical potential. Ranis discusses the possible impacts of worker cooperatives in post-depression Argentina. In his words, "As the workers proceed in the occupation and recuperation of their workplaces, they will be touching on fundamental questions concerning the direction of the neoliberal economy" [6]. However, without a strong national presence, Ranis argues that cooperatives will not be able to influence society, and if such a presence is built, capitalists may become threatened to the point that intense class warfare will erupt [6]. For such a conflict to lead to transformation in favor of democracy, the cooperative movement must be resilient and embedded in communities. To Ranis, cooperatives also can affect change on the cultural and ideological levels [6]. The hegemony of neoliberalism and its crushing effects on workers has created a discontent; a strong worker movement based on democratic principles should be able to gain massive support. Cooperatives provide "examples of worker autonomy" which "symbolize an alternative path to economic development that is predicated on worker solidarity and democracy in the workplace" [6]. A worker cooperative movement with strong leadership and clarity of purpose has the potential to lift the working class out of poverty and create labor conditions that are self-directed and dignified.

Krishna provides an in-depth analysis of how cooperatives can affect social change beyond individual workplaces; she discusses the domestic worker cooperatives UNITY and La Colectiva which mobilized for workers' rights through public outreach and lobbying [35]. The article identifies various challenges to such an endeavor, one being that since employees have more demands on them than in a capitalist institution, they are likely already working more than a full work week; having a further goal of social change simply adds on to this workload [35]. A potential way around this problem lies in worker incubators, organizations that exist to assist workers in creating a cooperative or direct them toward an existing one. One such organization, WAGES, takes on many of the difficult tasks involved in creating a cooperative so that workers can get their footing [35]. This point ties into the larger issue of needing infrastructure and inter-organization cooperation to create successful cooperatives.

Another challenge rests in the fact that many workers come to a cooperative for a stable job rather than to create systemic change [35]. Krishna suggests that this problem can be addressed through education on how social and economic issues affect workers personally in order to get them on board with the organization's mission [35]. Such education can be carried out at orientations or trainings; Krishna gives examples of La Colectiva's peer-led information sharing program and CFL's anti-oppression trainings for people of color [35]. Additionally, lawyers can speak to workers about their legal rights so they can better stand up for themselves [35]. Unified, educated workers are a powerful force, even in the face of hegemonic capitalism.

According to a study by Purtik, Zimmerling, and Welpe, cooperatives also have the potential to transition society toward eco-friendly and sustainable living [36]. The authors point out that cooperatives are conducive to this because they foster participation in social development, as well as comprehensive education [36]. In order to curtail climate change and ecological disaster, not only is a transition to renewable energy needed, but also a decrease in overall electricity consumption [12]. RECs offer a solution to both of these problems, as they facilitate the installation and maintenance of renewable energy infrastructure while also making the connection between energy generation and usage apparent. When communities are aware of the environmental, economic, and social implications of energy creation and consumption, they are more likely to conserve energy [12]. Rather than consumers, Tarhan argues that through RECs, community members become environmentally responsible prosumers [12]. Research has also shown that RECs can "increase public acceptance of the energy transition" [24]. Cooperatives can democratize both labor and energy, and, if pursued on a national level, can be a powerful tool in creating a more sustainable future.

7 Conclusions

Worker and energy cooperatives already exist, demonstrating that even within a hierarchical economic system, more democratic forms of organization can flourish. This fact is a promising one, one that ought to inspire workers to find ways to democratize their workplaces. This chapter sought to provide insight into the challenges that face cooperatives, potential solutions to these challenges, and the promise cooperatives hold for transforming labor and energy arrangements. The onus lies on workers and communities to find ways to transition toward a democratic society, whether through unionization, organizing on cooperative principles, or creating energy cooperatives.

References

[1] K. Sobering, J. Thomas, C.L. Williams, Gender in/equality in worker-owned businesses, Sociol. Compass 8 (11) (2014) 1242–1255.

[2] M. Vieta, J. Quarter, R. Spear, A. Moskovskaya, Participation in worker cooperatives, in: D.H. Smith, D. Horton, R.A. Stebbins, J. Grotz (Eds.), The Palgrave Handbook of Volunteering, Civic Participation, and Nonprofit Associations, Palgrave Macmillan, Hampshire, 2016, pp. 435–492.

[3] A.G. Johnson, W.F. Whyte, The Mondragon system of worker production cooperatives, Ind. Labor Relat. Rev. 31 (1) (1977) 18–30.

[4] J. Pencavel, L. Pistaferri, F. Schivardi, Wages, employment, and capital in capitalist and worker-owned firms, ILR Rev. 60 (1) (2006) 23–44.

[5] G.M. Artz, Y. Kim, Business ownership by workers: are worker cooperatives a viable option? Economics Working Papers, 2011.

[6] P. Ranis, Argentine worker cooperatives in civil society: a challenge to capital–labor relations, WorkingUSA 13 (1) (2010) 77–105.

[7] T. Webb, G. Cheney, Worker-owned-and-governed co-operatives and the wider co-operative movement, in: G. Cheney, V. Fournier, C. Land (Eds.), The Routledge Companion to Alternative Organization, Routledge, Abingdon, Oxon, 2014, pp. 64–88.

[8] H. Abell, Worker Cooperatives: Pathways to Scale, The Democracy Collaborative, Takoma Park, 2014.

[9] A. Wierling, V.J. Schwanitz, J.P. Zeiß, C. Bout, C. Candelise, W. Gilcrease, J.S. Gregg, Statistical evidence on the role of energy cooperatives for the energy transition in European countries, Sustainability 10 (9) (2018) 3331–3352.

[10] I. Capellán-Pérez, Á. Campos-Celador, J. Terés-Zubiaga, Renewable Energy Cooperatives as an instrument towards the energy transition in Spain, Energy Policy 12 (3) (2018) 215–229.

[11] E. Viardot, The role of cooperatives in overcoming the barriers to adoption of renewable energy, Energy Policy 63 (2013) 756–764.

[12] M. Tarhan, Renewable energy cooperatives: a review of demonstrated impacts and limitations, J. Entrepr. Organ. Divers. 4 (1) (2015) 104–120.

[13] G. Burdin, A. Dean, New evidence on wages and employment in worker cooperatives compared with capitalist firms, J. Comp. Econ. 37 (4) (2009) 517–533.

[14] J. Garcia-Louzao, Employment and Wages Over the Business Cycle in Worker-Owned Firms: Evidence From Spain, Bank of Lithuania, 2019.

[15] A. Hochner, C. Granrose, J. Goode, E. Simon, E. Appelbaum, M. Kalamazoo, et al., Job-Saving Strategies: Worker Buyouts and QWL, W. E. Upjohn Institute for Employment Research, 1988.

[16] J.S. Meyers, Workplace democracy comes of age: economic stability, growth, and workforce diversity, in: Worker Participation: Current Research and Future Trends, Semantic Scholar, 2005, pp. 205–237, https://doi.org/10.1016/S0277-2833(06)16008-2.

[17] C. Cornforth, A. Thomas, R. Spear, J. Lewis, Developing Successful Worker Co-operatives, SAGE Publications Ltd, 1988.

[18] E.A. Hoffmann, Emotions and emotional labor at worker-owned businesses: deep acting, surface acting, and genuine emotions, Sociol. Q. 57 (1) (2016) 152–173.

[19] E.A. Hoffmann, Confrontations and compromise: dispute resolution at a worker cooperative coal mine, Law Soc. Inq. 26 (3) (2001) 555–596.

[20] V. Pérotin, Worker cooperatives: good, sustainable jobs in the community, J. Entrepr. Organ. Divers. 2 (2) (2013) 34–47.

[21] J. Logue, J.S. Yates, Cooperatives, worker-owned enterprises, productivity and the International Labor Organization, Econ. Ind. Democr. 27 (4) (2006) 686–690.

[22] K.A. Zeuli, J. Radel, Cooperatives as a community development strategy: linking theory and practice, J. Reg. Anal. Policy 35 (2005) 1–12.

[23] A. McLeod, Types of cooperatives, in: Cooperative Starter Series, Northwest Cooperative Development Center, 2006. http://nwcdc.coop/wp-content/uploads/2012/09/CSS01-Types-of-Coops.pdf.

[24] C. Herbes, V. Brummer, J. Rognli, S. Blazejewski, N. Gericke, Responding to policy change: new business models for renewable energy cooperatives—barriers perceived by cooperatives' members, Energy Policy 109 (2017) 82–95.

[25] Ö. Yildiz, J. Rommel, S. Debor, L. Holstenkamp, F. Mey, J. Müller, et al., Renewable energy cooperatives as gatekeepers or facilitators? Recent developments in Germany and a multidisciplinary research agenda, Energy Res. Soc. Sci. 6 (2015) 59–73.

[26] I. Heras-Saizarbitoria, The ties that bind? Exploring the basic principles of worker-owned organizations in practice, Organization 21 (5) (2014) 645–665.

[27] C. Dickstein, The promise and problems of worker cooperatives, J. Plan. Lit. 6 (1) (1991) 16–33.

[28] J.S. Meyers, S.P. Vallas, Diversity regimes in worker cooperatives: workplace inequality under conditions of worker control, Sociol. Q. 57 (1) (2016) 98–128.

[29] S. Pek, Drawing out democracy: the role of sortition in preventing and overcoming organizational degeneration in worker-owned firms, J. Manage. Inq. 30 (2) (2019) 193–206, https://doi.org/10.1177/1056492619868030.

[30] G.K. Dow, Governing the Firm: Workers' Control in Theory and Practice, Cambridge University Press, 2003.

[31] A. Ben-Ner, W. Burns, G. Dow, L. Putterman, Employee ownership: an empirical exploration, in: The New Relationship: Human Capital in the American Corporation, Hamilton Digital Commons, 2000, p. 194. https://digitalcommons.hamilton.edu/chapters/128.

[32] J. Elster, From here to there; or, if cooperative ownership is so desirable, why are there so few cooperatives? Soc. Philos. Policy 6 (2) (1989) 93–111.

[33] W.F. Whyte, J.R. Blasi, Worker ownership, participation and control: toward a theoretical model, Policy. Sci. 14 (2) (1982) 137–163.

[34] H. Corcoran, D. Wilson, The Worker Co-operative Movements in Italy, Mondragon and France: Context, Success Factors and Lessons, Canadian Worker Cooperative Federation, Calgary, Alberta, 2010.

[35] G.J. Krishna, Worker cooperative creation as progressive lawyering: moving beyond the one-person, one-vote floor, Berkeley, J. Emp. Lab. L. 34 (2013).

[36] H. Purtik, E. Zimmerling, I.M. Welpe, Cooperatives as catalysts for sustainable neighborhoods–a qualitative analysis of the participatory development process toward a 2000-Watt Society, J. Clean. Prod. 134 (2016) 112–123.

Risks

Introduction to Part III: Energy risks

Majia Nadesan

Arizona State University, Glendale, AZ, United States

1 Introduction

In *The Next Catastrophe* (2007), sociologist Charles Perrow described how infrastructural risk is amplified by concentrations of energy (i.e., the concentration of hazardous activities and facilities), concentrations of populations, and concentrations of political and economic power, which centralize energy and decision-making, encouraging "over-reach" [1]. Concentrated power tends to self-replicate, seeking to expand its resources and influence as illustrated in carbon (coal, oil, and natural gas) and nuclear industries. Energy derived from these sources often demarcates the limits of democratic practices, as observed in both Michael Watts's *Accumulating Insecurity and Manufacturing Risk Along the Energy Frontier* [2] and Kate Brown's *Manual for Survival: An Environmental History of the Chernobyl Disaster* [3].

Carbon and nuclear energy landscapes are dominated by assemblages of organizations and processes whose centralization of decision-making perpetually erodes support for participative governance. The most undemocratic aspects of energy production are experienced by the most vulnerable members of energy supply chains, especially those involved in extracting and refining basic energy resources including petroleum, uranium, and rare earths. The marginalization of those most adversely impacted by contemporary energy assemblages enables huge externalizations of production costs and inures the public to disproportionate risk burdens and future impacts. Democratizing energy will require broad-scale institutional and cultural changes in energy ownership, energy production, and cultural attitudes toward energy security, access, and consumption.

2 Anachronistic assemblages

At the close of the 1960s the Seven Sisters' oil companies discussed in the Introduction to this collection still controlled about 85% of the global oil reserves. As described by *Al Jazeera*: "In their bid to dominate Africa, the Sisters installed a king in Libya, a dictator in Gabon, fought the nationalisation of oil resources in Algeria, and through corruption, war and assassinations, brought Nigeria to its knees" [4]. The logic of oil security dictated 20th century geopolitics, as illustrated by "Oil and the Decline of the West" published in *Foreign Affairs* in 1980 warning of looming shortages and decrying the monopolistic power of OPEC. In the contemporary era, the rise of state oil companies has arguably eclipsed the former power held by the Seven Sisters, but Western governments still maintain close relationships with their remaining national oil majors, despite risk-laden operations and severe externalities for eco-systems and human health.

Nuclear energy is similarly concentrated and even more hazardous. General Electric's CEO Jeff Immelt declared in June of 2012 that nuclear power is so expensive compared with other forms of energy that it is "really hard" to justify [5]. Nuclear power's green brand hinges on bracketing off uranium mining, refining, spent fuel storage, and plant decommissioning. Research published in the journal, *Atmospheric Chemistry and Physics*, calculated a severe nuclear accident every 10–20 years [6]. Yet another study predicted a 50% chance for another Fukushima-scale accident or larger in the next 50 years, a Chernobyl-scale event in the next 27 years, and a Three Mile Island scale event in the next 10 years [7]. Nuclear is catastrophically hazardous as the ongoing problems at the ruined Chernobyl and Fukushima reactors demonstrate.

Nuclear also requires subsidies, ranging from national government guarantees on lending to legislative caps on liability, such as the US Price–Anderson Nuclear Industries Indemnity Act. The Convention on Supplementary Compensation for Nuclear Damage (CSC) limits international liability for nuclear disasters by offering a uniform and limiting set of compensation standards for victims of nuclear disasters in impacted countries not the origin of the disaster [8]. The convention also exonerates manufacturers, placing liability exclusively on operators. The convention essentially limits only the liability, but not the incalculable risks, from nuclear accidents. The burden of managing radiation risks is shifted to individuals. The nuclear industry's revolving door relationships and the close connections between nuclear weapons and energy security together reinforce the nuclear complex and mitigate against realistic assessments of the myriad risks posed by the uranium supply chain, including the health and environmental risks of uranium mining, refining, fissioning, and spent waste management [9].

Uranium, petroleum, and now natural gas-derived energy perpetuate highly consolidated control over energy production, unequal access, and externalization of health and ecological costs. Recognition of these problems has prompted diverse groups and interests to advocate for energy democracy, although the meanings and mobilizations vary across contexts. What unites disparate ideas and organizational efforts are a shared negation of contemporary energy landscapes and extractive assemblages and a common affirmation of energy forms that prioritize human and ecological security over formulations of energy security dedicated to profiting energy speculators and extending status quo geopolitical relations.

Efforts to operationalize sustainability are fraught but Saha provides a key set of criteria for measuring organizational sustainability [10]:

- Stakeholder Engagement
- Approach and Commitment to Values
- Commitment to Sustainable Consumption and Sustainable Finance

- Leadership Approach toward Responsible and Sustainable Business
- Implementation Strategies toward Sustainable and Responsible Business across the Value-Chain
- Risk Management Strategies and Compliance Measures

This list is heuristic but is targeted organizationally, while energy justice movements are more broadly aimed at transforming energy landscapes. "Energy justice" movements tend to be embedded within particular energy landscapes, reinforcing the diversity of aims and means. Despite differences, an important unifying theme linking justice and sustainability is value-driven and expansive "participatory governance" across energy assemblages, up and down supply, distribution, and consumption chains.

Movements in energy justice and sustainability especially emphasize direct participation in decisions about local energy forms and governance. In this fashion, sustainability and environmental justice movements incorporate liberal and romantic ideals of democratic self-governance, with growing efforts to incorporate non-human ecological entities as citizens, such as rivers [11]. The juridical extension of rights to non-human entities illustrates how contemporary sustainability politics seek to securitize life.

As pointed out in the Introduction to this collection, democracy is especially relevant in the fulcrum of the Anthropocene, as argued by Frank Fischer in *Climate Crisis and the Democratic Prospect* [12]. Fischer identifies the limitations of expert-dominated techno-optimism and the prospects for ecological citizenship and environmental democracy in ecological cities and transition towns through participatory governance. Democratic environmentalism characterized by participatory governance captures the imagination, but what prospects and trajectories can deliver such imagined communities? So far, most experiments in energy democracy have involved local energy cooperatives, with success typically predicted by the extent of broader infrastructural support [13]. Financing such cooperatives can be a challenge [14].

Many utopic formulations of future sustainable energy landscapes prioritize cities and states in driving change adapted to the needs and interests of localities. Yet, national governments arguably play a more significant role in enabling and disabling future energy landscapes by supporting research, developing energy policies regarding financing infrastructural development and taxation, etc. Unfortunately, national governments' energy efforts have historically lacked democratic proceduralism, with policy often shaped by the interests and objectives of private corporations in both liberal democratic and non-liberal societies. Moreover, the climate crisis that increasingly dominates public consciousness greases receptivity to non-democratic energy solutions. Sustainable energy must not be a mere epiphenomenon of public relations branding, but rather must derive from transparent and inclusive risks assessments of entire supply chain impacts on ecologies and communities. Public funding for developing and evaluative alternative energy production should be extensive, transparent, and subject to democratic review processes. Human cooperative action dedicated to long-term survival is possible if barriers to change are identified, reformed, supplemented, and transcended.

Part III of this collection grapples with institutional inertia against the democratization of decision making, and the capture of renewables by established interests, further reinforcing rather than transforming unequal energy access and risk burdens. The first theme in Part III, "Assemblages" commences with broad critiques of unequal participation in global energy infrastructures and transitions offered by Breffní Lennon and Niall Dunphy in "Mind the gap:

Citizens, consumers and unequal participation in global energy transitions" and Veith Selk and Jörg Kemmerzell in "Worse than its reputation? Shortcomings of "energy democracy"." Contributors to this theme especially hone in on the entrenched authoritarian politics of nuclear energy, whose re-branding as "green" energy integral to natural security is challenged. Alevgül Sorman and Ethemcan Turhan address "The limits of authoritarian energy governance: Energy, democracy and public contestation in Turkey" while Avino Niphi and M. V. Ramana deconstruct nuclear branding with "Talking points: Narrative strategies to promote nuclear power in Turkey." Other contributors focus on the dynamics of energy dispossession as it is emerging in renewables, as illustrated by Lourdes Alonso-Serna's and Edgar Talledos-Sánchez's "Fossilizing renewable energy: The case of wind power in the Isthmus of Tehuantepec, Mexico." Bidtah N. Becker and Dana E. Powell point to problems surrounding energy sovereignty as fueling dispossession and suggest that change requires institutional and symbolic transformations aimed at self-determination and new forms of relationality in "Situating energy justice: Storytelling risk and resilience in the Navajo Nation."

Economic power and financial profits indisputably drive energy markets and energy futures but state actors have the power to support alternative energy futures through direct financing, subsidized lending, and tax incentives. Why do so many therefore remain enmired in archaic and undemocratic energy commitments? The second theme in Part III of this collection addresses the problem of security in sustaining non-democratic and polluting energy forms. Kacper Szulecki tackles the problem directly by posting the question: "Does security push democracy out of energy governance?" Barbara Wejnert and Cam Wejnert-Depue document the inextricable links across hazardous energy and state politics in "Hazard or survival: Politics of nuclear energy in Ukraine and Belorussia through the lens of energy democracy." Finally, Melisa Escosteguya and colleagues warn of the subordination of justice to the charged promises of "renewables" in "Will electro-mobility encourage injustices? The case of lithium production in the Argentine Puna."

Although concentrated economic and financial power and tired notions of state security play disproportionate roles in perpetuating undemocratic energy forms, there are also significant technological and cultural challenges to creating sustainable futures. Ekaterina Tarasova and Harald Rohracher point to the promises and contradictions in smart grids in empowering energy consumers in "Democratizing energy through smart grids? Discourses of empowerment vs practices of marginalization." Chad Walker and colleagues address the challenges of democratization when fundamental power differences exist between those individuals and organizations who produce and consume energy in "Contested scales of democratic decision-making and procedural justice in energy transitions." These authors point to concentrated control over energy as precluding authentic efforts at democratization. Although social activism can combat institutional inertia in some circumstances, building cultural support for sustainable and democratic transitions is tough. Peta Ashworth and Kathy Witt powerfully describe "Psychic numbing and the environment," asking if the described inurement to destruction is leading to unsustainable energy outcomes in Australia. Arthur Mason challenges the "Deluxe energy" enjoyed by the most powerfully situated in global societies. Finally, Ambika Opal and Jatin Nathwani grapple with intergenerational justice and our sense of disconnection from our children's children in "Global energy transition risks: Evaluating the intergenerational equity of energy transition costs." Democratic and sustainable energy futures hinge on resolving these challenges of centralized ownership and decision making and large-scale public apathy. Alexander Dunlap concludes that time is short and that action must be taken now.

References

[1] C. Perrow, The Next Catastrophe: Reducing Our Vulnerabilities to Natural, Industrial, and Terrorist Disasters, Princeton University Press, Princeton, NJ, 2007, p. 1.

[2] M. Watts, Accumulating insecurity and manufacturing risk along the energy frontier, in: S. Soederberg (Ed.), Risking Capitalism (Research in Political Economy), vol. 31, Emerald Group Publishing Limited, 2016.

[3] K. Brown, Manual for Survival: An Environmental History of the Chernobyl Disaster, WW Norton, New York, 2019.

[4] The secret of the Seven Sisters, Al Jazeera (2013). http://www.aljazeera.com/programmes/specialseries/2013/04/201344105231487582.html.

[5] P. Clark, Nuclear is 'hard to justify', GE says, Financial Times (2012) A1.

[6] J. Lelieveld, D. Kunkel, M.G. Lawrence, Global risk of radioactive fallout after major nuclear reactor accidents, Atmos. Chem. Phys. 12 (9) (2012) 4245, https://doi.org/10.5194/acp-12-4245-2012.

[7] S. Wheatley, B. Sovacool, D. Sornette, Of disasters and dragon kings: a statistical analysis of nuclear power incidents and accidents, Phys. Soc. (2015). arXiv:1504.02380 [physics.soc-ph].

[8] T. Nakagawa, Japan wants in on nuclear accident compensation pact, The Asahi Shimbun (2012). http://ajw.asahi.com/article/behind_news/politics/AJ201202030021. (Accessed 5 February 2012).

[9] M. Nadesan, Fukushima and the Privatization of Risk, Palgrave, London, 2013.

[10] I. Saha, A Case Study on BP's Sustainability and Ethical Performances – Reflecting on Its Leadership, 2015. https://www.linkedin.com/pulse/case-study-bps-sustainability-ethical-performances-reflecting-saha/.

[11] M. Safi, Ganges and Yamuna Rivers granted same legal rights as human beings, The Guardian (2017). https://www.theguardian.com/world/2017/mar/21/ganges-and-yamuna-rivers-granted-same-legal-rights-as-human-beings.

[12] F. Fischer, Climate Crisis and the Democratic Prospect, Oxford University Press, Oxford, UK, 2017.

[13] J.A.M. Hufen, J.F.M. Koppenjan, Local renewable energy cooperatives: revolution in disguise? Energy Sustain. Soc. 5 (2015) 18, https://doi.org/10.1186/s13705-015-0046-8.

[14] S. Hall, T.J. Foxon, R. Bolton, Financing the civic energy sector: how financial institutions affect ownership models in Germany and the United Kingdom, Energy Res. Soc. Sci. 12 (2015) 5–15, https://doi.org/10.1016/j.erss.2015.11.004.

Assemblages

23

Situating energy justice: Storytelling risk and resilience in the Navajo Nation

Dana E. Powell[a,b] and Bidtah Becker[c]

[a]Department of Anthropology, Appalachian State University, Boone, NC, United States [b]College of Indigenous Studies, National Dong Hwa University, Hualien, Taiwan [c]CalEPA, Sacramento, CA, United States

A pandemic carries many stories, only some of which involve the virus. (Redfield [1])

1 Introduction

In Navajo (Diné) territory, the current global pandemic carries stories of infrastructural precarity and climate change, entrenched in long-standing histories of extraction. The virus has been devastating by any public health measure: the Navajo Nation ranked third, on a per capita basis, after New York City and New Jersey, in April 2020, for rates of infection within the 27,000 square miles of its central and satellite reservations [2]. Rates of COVID-19 (referred to as *Dikos Ntsaaígíí-19* in the Navajo language) have soared with the Navajo Nation experiencing the loss of elders at a devastating rate.[a] At the same time, mid-century coal plants are being dismantled[b] and new solar projects are being installed, as everyday life on the reservation reorganizes itself in terms of curfews and shutdowns. Making sense of "democratizing energy" at this juncture requires, we argue, both an expansive definition of energy (beyond minerals, to include the vitality of life) and a critical eye to the ways in which exposure to the

---

[a] As of February 1, 2021, the Navajo Nation reported 28,388 cases and 1020 deaths, with 60% of cases in individuals who are 80 years or older. The loss of elders is devastating in any context; in the Navajo Nation, as in many Indigenous Nations, this loss is compounded by the loss of speakers of the Diné language as well as the loss of cultural knowledge. See Navajo Nation Department of Health: https://www.ndoh.navajo-nsn.gov/covid-19.

[b] See Jessica Kutz [3].

virus transports stories of other forms of historical exposure to the risks wrought by energy development and settler colonialism.[c] Discussions of energy equity in the future demand deeper analyses of injustice and the ways this is embodied, presently, in the virus itself.

The processes of settler colonialism shaping the Indigenous Southwest is complicated by both land dispossession and acquisition: since the Navajo Reservation was created in 1868, Diné people have steadily expanded the reservation's legal boundaries to become the largest trust land base in Indian Country, at 27,000 square miles. But while the virus in this population on this particular land base might be mitigated by a vaccine,[d] there is no swift inoculation to repair the longstanding environmental vulnerabilities on the Navajo Nation, which continue to shape contemporary life. At the level of the collective, many of these vulnerabilities were wrought by 20th-century uranium mining [6–8], oil, coal, and timber extraction [9–12], 21st-century drought [13,14], and hydraulic shale fracturing [15]. As a result, the precarity of energy systems has become a defining condition of life across the Reservation. Said otherwise, while science may deliver the answer to the virus by enhancing population immunity, technological innovation in extractive infrastructure has been uneven at best, and often toxic, in the Navajo Nation. Abetted by settler interests, energy science and technology have created the conditions for the Cold War era nuclear industry to leave a radioactive landscape in its wake, and the ongoing fossil fuel industry to reshape Diné homelands through oil and liquid natural gas wells, coal mines, and fracking. At the same time, these forces have generated a politics of neglect, such that up to 40% of Diné households have no running, potable water, and an estimated 30% of Diné households have no electricity [16].[e] This deficit is striking, given that less than 1% of homes in the United States lack some or all sanitation facilities. We take seriously what Jean Dennison calls the "colonial entanglements" [17] of land, governance, minerals, health, and the body politic. In short, we argue that the answer to the pandemic facing the Navajo Nation is not a technical or scientific solution, but rather, an entanglement of sociopolitical, decolonial measures.

In this chapter, we join minds as a Diné attorney and an Anglo (settler) anthropologist to reflect on the prospect of "democratizing energy" in a COVID-19 moment, that has profoundly impacted Diné life. Bidtah Becker's career has been dedicated to sound stewardship of Diné natural resources to improve reservation life and Dana Powell's has been dedicated to amplifying this kind of work. We have been in conversation over many years, together and in a wider network of scholars, practitioners, and knowledge keepers, regarding energy, sovereignty, and infrastructure, our differences making the exchange a convivial, reciprocal process of learning and co-thinking. Our collaboration articulates with wider scholarly debates in feminist and Indigenous political ecology [18–22] and Indigenous STS [23,24] that highlight

[c] Our theoretical approach here, in moving from minerals to material vitality and from democratizing energy to *decolonizing* energy, follows work in Black and Indigenous critical energy and environmental anthropology: see especially Myles Lennon [4] and Teresa Montoya [5].

[d] As of December 2020, the Pfizer BioNTeach vaccine was just beginning to reach health workers in the Navajo Nation. Some critics fear the second dose requirement of the vaccine may prove particularly difficult on the mostly rural reservation, where there is an existing mistrust of biomedicine and also significant challenges for many tribal members with transportation.

[e] NTUA estimates that the total cost of meeting the utility needs of all homes in the Navajo Nation reservation is at least $5.2 billion.

the complexities of environmental justice, tribal sovereignty, the coproduction of knowledge, and human-built infrastructure when pursuing energy and economic alternatives in American Indian territories. Given the particular political-legal, historical, and ontological relationships of Native peoples to territory, we know that "energy democracy" and its kindred aims (sustainability, development, just transitions, and so on) have certain limits and opportunities due to self-determination and federal primacy, which we say more about below.

We co-write to examine ways in which the SARS-CoV-2 virus has made an even more profound pandemic visible in the Navajo Nation. Anthropologist Peter Redfield's insight that "a pandemic carries many stories, only some of which involve the virus," emerges from his argument that "health care is never a singular proposition. Amid the most exceptional emergencies, people still have all manner of ordinary problems and complicating conditions—fractures, cancer, dementia, malnutrition" [1, p. 1]. The "ordinary problems" of the Navajo Nation are indeed structured by relations of coloniality and development—made increasingly visible by the "complicating conditions" of health, extraction, and climate. And to be sure, the Navajo Nation has been a contributor to climate change, with emissions from coal-fired power plants—midcentury infrastructures promising growth, offering to modernize the rural Nation and its transitioning economy. Diné leaders like former Navajo Nation Chairman Peter MacDonald pursued these projects well aware of their complicated risks and opportunities, believing they were doing their best to participate in the development designs of postwar growth [25].

Such "colonial entanglements" [17] complicate seemingly straightforward propositions to democratize energy, as this volume's editors invite us to consider. The more profound pandemic is the collision of infrastructural precarity with climate change (surely advanced by Navajo Nation's commitment to fossil fuels), a predicament that is cultural, political, ecological, and legal. As such, any meaningful immunity demands considerations of justice, equity, and continuance *for* Diné people *by* Diné people and is not distributable by securitized, refrigerated vials of Pfizer and Moderna's microbes.

The energy infrastructure stories the virus carries are indeed many—and far beyond what we can attend to here. We know some of these stories because we hear them from friends and family, read about them in the international media, or, in Becker's case, born of first-hand experience. For Powell, the stories are indeed more experientially distant; and yet they resonate, repeatedly, with stories she has heard from Diné people about energy systems vulnerabilities for more than two decades. We think together here, from our different locations and yet shared commitment to Diné energy justice and sovereignty, and an Indigenous feminist reading of the stories that point to profound challenges for democratizing energy—which we prefer to reframe, with critical energy anthropologist Myles Lennon, as "decolonizing energy" for energy justice—at a scalable impact [4].

---

[f] Bidtah observed how Chinese middleclass growth allowed Navajo vendors in the Lake Powell area to thrive. This is related to energy justice because of the water/energy nexus and the future financial success of those Navajo vendors is going to depend on resolving the energy/water nexus in that region. Pointedly, although only 2 miles from Lake Powell and across the street from the now demolished Navajo Generating Station, these Navajo vendors built extensive buildings to serve their customers with air-conditioned waiting rooms and cold ice cream in freezers but lack flushing toilets due to the lack of piped water. This was, of course, pre-pandemic. Time will tell when these businesses will be able to thrive again.

III. Risks

How do we pursue energy justice against the ongoing structures of settler colonialism and the increasing globalization of human lives and livelihood?[f] If energy justice anywhere is not just a matter of redistributing harms (e.g., distributive justice) but a matter of undoing forms of structural violence (such as anti-racism and decolonization), how does energy justice play out in places such as the Navajo Nation where extraction for energy purposes is both the backbone of the on-reservation private economy *and* a significant contributor to environmental harm? In the Navajo Nation, one of the most energy-rich Indigenous nations in North America, de-colonial energy justice cuts both ways: it is the ability for the tribe to claim and control its own energy resources (timber, coal, oil, wind, and solar) even when that means the development of fossils fuels, but it is also the reshaping of energetic modes of living to foster what Potawatomi philosopher Kyle Whyte has called the "collective continuance" of Indigenous peoples [22].

In Whyte's analysis, settler colonialism commits energy injustice by undermining the abilities of land-based collectives to self-determine their futures and build social resiliency. And collective continuance, no doubt, requires energy—energy as vitality and political life, as the ability to do work, and as the relations of exchange between and among beings that sustain life. Energy justice, in this manner, is not about (re)distribution for new modalities of consumption per se, but about the continuation of species, ways of life, and sovereign political bodies. We suggest energy justice comes to mean differently when situated historically and evaluated within the longstanding processes of settler colonialism and overlapping jurisdictions of multiple forms of state power, shaping and being shaped by the Navajo Nation. Indeed, there is not one state actor separate from the Navajo Nation. In some situations, the state actor *is* the Navajo Nation, as we accept that tribes are sovereigns.

As we write this essay during a recent surge of the COVID-19 pandemic in the United States, we also consider how the social geography of illness further implicates energy justice for Diné communities where water, food, and electricity—the triad of just energies for all communities—complicate social distancing and other measures for creating safe communities. The stakes of frequent handwashing to mitigate household transmission of the virus are particularly elevated, in a territory where water scarcity and potential toxicity go hand in hand [26]. In these circumstances, how do we apprehend the current public health crisis in critical relation to the longstanding socio-political crisis? How do we do so in a manner that mitigates the crisis to address individual injury and immunity and also the collective body politic and wider ecological environment? And how do we develop energy systems, in a manner consistent with a commitment to continuing Indigenous life and sovereignty?

2 An attorney's observations: Energy landscapes and lines of connection

Becker notes that it is estimated that 15,000 homes in the Navajo Nation lack electricity. While the energy created on the Nation for export is carbon emitting, the majority of energy delivered by NTUA to serve its customers is non-carbon emitting. The majority of NTUA power comes from the Colorado River Storage Project Act, a hydropower project. The next largest source of non-carbon emitting energy serving the Nation comes from the NTUA majority-owned Kayenta I and II Solar Projects, located on the Navajo Nation. This arrangement makes the majority of NTUA power non-carbon emitting, exceeding even the most progressive state standards for carbon emissions goals ([27]; see also www.cesa.org). Impressive as this claim may be, total electric consumption on the Nation is small, compared with any given state's total electric consumption. With the relatively low use of electricity in the Nation, some may argue that comparing NTUA's total electricity consumption to a state's renewable energy portfolio is not a fair comparison. Yet with NTUA's non-carbon emitting technologies, can we argue that the Navajo people at the household scale are not significantly contributing to climate change, even in light of Navajo-owned coal-fired power plants and oil and gas operations on the Navajo Nation. At the very least, the small use of electricity by Diné people to power their homes and businesses cast against the much higher use of electricity off the reservation powered by Navajo resources highlights that any attempt to address energy justice must be a socio-political, decolonial one.

Of particular interest, and perhaps as an example of decolonizing energy, NTUA is wholly owned by the Navajo people through the Navajo Nation. The Resources and Development Committee of the Navajo Nation Council exercises oversight authority over NTUA and a member of the Resources and Development Committee serves on the NTUA Management Board as a matter of Navajo law. As such, NTUA brings to its works the mandate that it must be doing everything it can to serve the Navajo people. NTUA takes some of the margins from the Kayenta Solar Projects and uses it to fund projects to electrify homes that are not electrified. It is important to note that while NTUA serves a significant portion of the Navajo Nation, several other utility providers provide electricity on the Navajo Nation: Jemez Mountain Electric Cooperative and Continental Divide Electric Cooperative in the east and Arizona Public Service Company in the west, to name a few. The infrastructural interface of these multiple private and public sector actors highlights the complexity of getting to "energy justice" in terms of distribution (of electrical power for consumption), remediation of energy-related environmental harms, and enacting tribal sovereignty, which underpins any meaningful notion of justice for Diné lives and livelihoods.

The energy landscape is, like most things in the Navajo Nation, complicated. Research has shown that in the first couple of months of the pandemic, there was a direct correlation between the rate of COVID-19 infections on American Indian Reservations and the lack of indoor plumbing (see Ref. [28]). In short, the water-energy nexus is further complicating energy justice.

Diné energy use includes the generation of government revenue, job creation, economic development, and support of Diné traditional and cultural practices. When Becker began this journey, she was working directly within the Navajo Nation government. In the beginning, the work focused on water: water in the courtroom and water in storage and clean drinking water projects. Very early on, Becker learned how water is a central resource in some forms of energy development and how energy is needed to clean water and move clean water to humans. Energy also helps move water to non-human environments to improve those environments. Water is necessary for all human life to exist; and while energy (as electricity) is not technically necessary for human life to exist, it has become necessary for human life in the modern world.

Over time and with changing jobs, her work focused on energy development on the Nation—both development through new models of solar development and protection of the existing energy activities on the Nation, largely coal-fired power generation with Navajo and Hopi coal. During her work with the Navajo Tribal Utility Authority, her focus is on bringing much-needed utilities to Navajo homes to combat the spread of the virus that causes COVID-19. During the drafting of this article, Becker had only been with NTUA for 18 months, and her duties changed dramatically due to the pandemic. As a result, NTUA has become more creative in finding ways to deliver utilities to unserved homes and has new infusions of substantial federal funding. The core work of electrifying homes, building new water lines to homes, ensuring success for the Navajo Gallup Water Supply Project in delivering new sources of surface water for domestic use, and supporting the expansion of broadband services continues but with a significantly different flavor and perhaps approach now that the larger systems understand the tangible impacts that occur from the lack of infrastructure.

Even though many Navajo homes lack electricity, as well as indoor plumbing, what is hardly missing from the Navajo landscape are power lines. They can be small power lines strung across miles and miles to serve individual homes or the large transmission lines crisscrossing the Nation, connecting the most rural Diné places with all corners of the southwestern United States, delivering electrical service to non-Navajo Nation, urban communities. To this day, when driving across the Navajo Nation, Becker always looks to see if a home with power lines has an outhouse, because in her experience, while people truly appreciate receiving electricity, it is clean water that is most desired. That being recognized, delivery of clean water depends on a reliable energy supply to take raw water, clean it, pipe it, and then safely dispose of it after it is used in the home. Pre-pandemic, she may have had to explain why the desire for clean water is so deep, but we are confident that the reader of this essay will be aware of the impact of the lack of clean water during a global pandemic; this is especially true in the Navajo Nation, given longstanding water contamination and increasing drought. As a personal side note, Becker also looks for abandoned homes with power lines and dreams about having the time to start a business rehabilitating these empty houses for young Diné people seeking housing for their families.[g]

Where there are no power lines, there may be some solar units installed at homes. Navajo Tribal Utility Authority has a solar unit program that comes with a monthly bill. That monthly bill ensures the solar unit operates for its "useful life." As Powell describes elsewhere, many well-intentioned efforts to install solar units never achieve their "useful life" and instead lay broken, dormant, and castaway, the material reminder of broken promises of development [11, pp. 118–112]. The current NTUA program is small as compared to the other utility services, less than several hundred customers and over time many of those customers moved onto the grid. The grid provides a secured 24/7 power supply at a much more powerful rate allowing homeowners to plug into the walls of their home a myriad of appliances. Most household solar units cannot provide the power that the grid provides. For example, NTUA provides families with a refrigerator appropriate for the solar unit power supply and strongly advises that medical equipment not be plugged into the walls of a home powered by a solar unit, given the precarity of the current.

3 An ethnographer's observations: The viral pandemic in a deeper historical context

Powell used to live just north of Wheatfields Lake, one of the largest bodies of water in the Navajo Nation, within what Diné people often call the "lungs" of the Navajo Nation: the

[g] The Navajo Housing Authority estimates that 34,000 individuals are in need of housing in the Navajo reservation [29]. (Earl Tulley, Navajo Housing Authority, personal communication; also see the Navajo Nation Housing Needs Assessment.)

forested Chuska Mountains. The tall ponderosa pines of the Chuskas and the off-grid, solar-powered home she lived in both eclipsed the view of the networks of transmission lines Bidtah describes above, carrying energy from Navajo coal off the reservation to consumers in southern Arizona and other urban hubs. These forests were once the wooded battlegrounds of the Navajo timber wars [12]; the surviving stumps offering reminders to future generations. As the shorelines of Wheatfields Lake receded over the years, due to persistent drought, Red Lake just to the south dried up entirely for a while, the pinon harvesting season shifted, and more and more feral horses foraged in the woods, causing the Navajo Nation to identify them as a central multispecies indicator of climate vulnerability [30].

One summer, many years after her time living near Wheatfields Lake, a sandstorm trapped her in the car—along with her mother and newborn son and they waited out the blinding dust in the parking lot of the Holiday Inn in Chinle, near Canyon de Chelly. Powell noticed these subtle climate changes over time and then heard them confirmed, in painful detail in June 2019, by sheepherders in Dilkon (Arizona) who described losing hundreds of livestock in one month alone to drought and sand dune encroachment. These were not "ordinary problems" (Redfield) but the alarming impacts of increasing desertification in the high desert, noted by climate scientists and land managers, alike [13,14]. The more "complicating conditions" included the network of transmission lines, unmitigated uranium tailings piles, coal ash pits, and heavy equipment that have made places like Wheatfields and Dilkon—though totally opposite biomes—intimately linked as sites where drought and energy extraction are lived experiences.

The landscape of energy production and use across the southwestern United States and the networks of power lines that Bidtah describes, tell these stories that the virus has carried. Transmission lines carry electricity—generated by Navajo coal—off of the reservation, to lighting, air conditioners, swimming pools, and appliances in the urban Southwest. Scholars have shown that the Sunbelt's metropolitan growth was built by the labor and resources of a colonial development design, yielding the social production of underdevelopment, which endures [10,25,31]. Many off-grid, rural families use kerosene for indoor lighting, haul laundry several hours to a border town laundromat, heat with hand-harvested wood, and keep plastic coolers outdoors to store perishable foods or prescription drugs. The everyday life of energy is rarely taken for granted in this terrain.

The CARES Act funding exposed yet another story: American Indian Nations had to apply to the United States government to receive federal funds, whereas US states did not. "The feds held $15 billion in our name," a Diné colleague friend told us, bitter that tribes were held to a different standard than states receiving similar aid. This inequitable process exposes the legal falsehood of the "Treatment as States" (TAS) status of Native Nations: TAS is a standard to determine if a Tribe can have treatment as a US state for primary jurisdiction in applying certain environmental laws. Tribes are treated "as states" only when this is in the interest of the federal government, which retains the ultimate power of primacy. In practice—as the CARES Act funding made clear—Native Nations are treated as "secondary" states with sovereignty that is "nested," as Audra Simpson argues [32], in the federal-states-tribes trifecta.[h] A Diné colleague lamented that the federal funding re-enacted the colonial logics of recognition, stating that, "the idea is, if we can get attention, we can get support from the state" (Tulley, personal

---

[h] Simpson's concept of "nested sovereignty" shows that when Indigenous political orders prevail in the present, they do so, seemingly paradoxically, "within and apart from settler governance" (2014, p. 11).

communication). On the other hand, the United States' legal relations with American Indian Tribes—despite its many failings—is often heralded as among the most sophisticated and just in the world. This contradiction between lived experiences of recognition and legal arrangements of the same gets at the relevance of auto-biographical and ethnographic analytics for illuminating some of the legal tensions, and legal fictions, of federal-tribal relations.

By most measures, the United States has failed miserably in this massive project of aid. Yet we take Simpson's caution seriously, lest we always "allow the state to determine what matters" [33], hitching our measures of success and failure, healing and suffering, to externally defined standards of population health. The virus has been devastating, to be sure, but it has also exposed or carried other stories that expose both the precarity *and* resilience in the Navajo Nation. For instance, in March 2020 when the mayor of Gallup (New Mexico), a border town to Navajo Nation, shut down the city, the head of the Department of Water Resources in the Navajo Nation asked the mayor where all the Diné people who get their water from Gallup ought to go? Gallup is the infrastructural hub for groceries, water, and fuel. The city's railroads move coal to other power plants, and its rolling hills, backed by sacred Mount Taylor, conceal some of the richest uranium deposits in the world—energy minerals which remain contested by citizens' movements [34].

4 Conclusions

In our separate work that has intersected over nearly twenty years, we have seen the need to situate understandings of "energy justice" within the complex "colonial entanglements" shaping Diné experience and the Navajo Nation. Since 2020, it is clear that the embodiment of injustice and environmental risk, well-established by decades of natural resource extraction, has been amplified and complicated by the global pandemic—posing a challenge to our analytics of "risk" and "resilience" as the Navajo Nation examines strategies for recovery from COVID-19 alongside adaptation and mitigation plans for climate change and energy transitions, as coal plants are decommissioned. We are haunted by a recent article, in which Diné storyteller Sunny Dooley suggests that Diné people have "the perfect human body for invasion" [35]. This is the case, Dooley narrates, not because of any kind of biological essentialism but due to the sociopolitical, geological, and technoscientific arrangements in the landscape that create infrastructural precarity in Diné everyday life. In Dooley's community, this plays out where naturally occurring arsenic and uranium—combined with the lack of resources to build a well, in the first place—such that Dooley and her relatives depend on hauling water from Gallup for their everyday use. "We have every social ill you can think of, and COVID has made these vulnerabilities more apparent. I look at it as a monster that is feasting on us—because we have built the perfect human for it to invade," says Dooley, indicating the sociopolitical roots of health disparities. Dooley continues: "COVID is revealing what happens when you displace a people from their roots" (Ibid). This kind of narrative situating upends any easy notion of "energy justice" that would rely, solely or primarily, on technoscientific reparations. The deeper significance of reconciliation that Dooley theorizes is about land and kinship justice: displacement from the roots.

The pandemic carries these stories of disconnection, intergenerational trauma, and infrastructural precarity. Yet it also carries stories of persistence and creative resistance, in people like Dooley (and scores of others we both know) who, in their work to dismantle economies of export and dependency, orient us toward understandings of *energy as vitality,* and *justice as relationality.* Work is underway across the Navajo Nation, led by primarily women-directed grassroots movements toward "just transition" and "restorative economies" (e.g., the work of Nicole Horseherder et al. [36]). Understanding displacement as foundational to colonialism, reconnecting relationships is central to mitigating displacement and establishing "continuance" [37]. The displacement that Dooley and so many others articulate is not remedied by inoculation, but by a host of collective, decolonial strategies, with energy at the center. This materialist interpretation of justice is evident in the transmission lines that Bidtah describes, the desiccated lakes and faltering livestock, and also in the knowledge keepers/makers and movements that push for just energies and energy justice, creating new kinds of stories that the pandemic might carry: stories of survival and resilience.

We have reflected upon energy, climate, and COVID-19 to place the present viral pandemic into deeper temporal horizons and argue that *energy justice*—more so than energy democracy—requires listening for the diverse stories the pandemic carries. These stories, in turn, reveal the more profound pandemic of the colonial condition as well as the vitalities and relations actively being organized in resistance to these forms of exploitation. This is the labor of environmental defenders, in a time of climate vulnerability, but also the labor of critical storytellers who critique the present within a deeper, often cyclical sense of time. With this, we aim—as attorney and ethnographer—to activate and amplify the stories that open up to Audra Simpson [38] calls "revenge": to expose what has created the conditions of the current moment and critique the logics that have restrained Indigenous life.

References

[1] P. Redfield, The danger of a single threat, Covid-19, Fieldsights (2020). May 22 https://culanth.org/fieldsights/the-danger-of-a-single-threat.

[2] J. Zelner, R. Trangucci, R. Naraharisetti, A. Cao, R. Malosh, K. Broen, N. Masters, P. Delamater, Racial disparities in coronavirus disease 2019 (COVID-19) mortality are driven by unequal infection risks, Clin. Infect. Dis. 72 (5) (2021) e88–e95, https://doi.org/10.1093/cid/ciaa1723.

[3] J. Kutz, The fight for an equitable energy economy for the Navajo Nation, High Country News (2021). https://www.hcn.org/issues/53.2/south-coal-the-fight-for-an-equitable-energy-economy-for-the-navajo-nation?utm_source=wcn1&utm_medium=email.

[4] M. Lennon, Decolonizing energy: black lives matter and technoscientific expertise amid solar transitions, Energy Res. Soc. Sci. (30) (2017) 18–27.

[5] T. Montoya, Yellow water: rupture and return one year after the Gold King Mine Spill, Anthropol. Now (9) (2017) 91–115.

[6] P. Eichstaedt, If You Poison Us: Uranium and Native Americans, Red Crane Books, Santa Fe, 1994.

[7] V.J. Kuletz, The Tainted Desert: Environmental Ruin in the American West, Routledge, New York, 1998.

[8] T.B. Voyles, Wastelanding: Legacies of Uranium Mining in Navajo Country, University of Minnesota Press, Minneapolis, 2015.

[9] K. Chamberlain, Under Sacred Ground: A History of Navajo Oil, 1922–1982, University of New Mexico Press, Albuquerque, 2000.

[10] A. Curley, *T'áá hwó ají t'éego* and the moral economy of Navajo coal workers, Ann. Am. Assoc. Geogr. 109 (1) (2019) 71–86, https://doi.org/10.1080/24694452.2018.1488576.

[11] D.E. Powell, Landscapes of Power: Politics of Energy in the Navajo Nation, Duke University Press, Durham, 2018.

[12] J.W. Sherry, Land, Wind, and Hard Words: A Story of Navajo Activism, University of New Mexico Press, Albuquerque, 2002.

[13] M.H. Redsteer, R.C. Bogle, J.M. Vogel, Monitoring and analysis of sand dune movement and growth on the Navajo Nation, Southwestern United States. US Geological Survey Fact Sheet, 2011, p. 3085.

[14] M.H. Redsteer, K. Bemis, K. Chief, M. Gautam, B.R. Middleton, R. Tsosie, Unique challenges facing Southwestern Tribes, in: G. Garfin, A. Jardine, R. Meredith, M. Black, S. LeRoy (Eds.), Assessment of Climate Change in the Southwest United States: A Report Prepared for the National Climate Assessment, Island Press, Washington, DC, 2013. A report by the Southwest Climate Alliance.

[15] S. Grant, Patchwork: Land, Law, and Extraction in the Greater Chaco (PhD dissertation), Department of Anthropology, University of Chicago, 2021.

[16] Navajo Nation, Tribal Utility Authority, Navajo Nation Utility infrastructure needs, White Paper, 2021.

[17] J. Dennison, Colonial Entanglements: Constitution a Twenty-First Century Osage Nation, University of North Carolina Press, Chapel Hill, 2012.

[18] C. Bauhardt, W. Harcourt (Eds.), Feminist Political Ecology and the Economics of Care, Routledge, London, 2019.

[19] D. Gilio-Whitaker, As Long as Grass Grows: The Indigenous Fight for Environmental Justice, From Colonization to Standing Rock, Beacon, New York, 2019.

[20] S. Hunt, Ontologies of Indigeneity: the politics of embodying a concept, Cult. Geogr. (2013), https://doi.org/10.1177/1474474013500226.

[21] B.R. Middleton, Jahát Jatítotòdom: toward an indigenous political ecology, in: R. Bryant (Ed.), The International Handbook of Political Ecology, Edward Elgar Publishing, Northampton, MA, 2015, pp. 561–576.

[22] K. Whyte, Settler colonialism, ecology, and environmental injustice, Environ. Soc. Adv. Res. (9) (2018) 125–144.

[23] N. Ishiyama, K. TallBear, Changing notions of environmental justice in the decision to host a nuclear storage facility on the Skull Valley Goshute Reservation, in: Paper Presented at Session 51: Equity and Environmental Justice, Waste Management 2001 Symposia, February 25–March 1, Tucson, AZ, 2001.

[24] K. TallBear, Native American DNA: Tribal Belonging and the False Promise of Genetic Science, University of Minnesota Press, Minneapolis, 2013.

[25] T.A. Needham, Power Lines: Phoenix and the Making of the Modern Southwest, Princeton University Press, Princeton, 2014.

[26] C. Tulley-Cordova, N. Tulley, B. Becker, K. Chief, Chronic wicked water problems in the Navajo Nation heightened by the COVID-19 pandemic, Water Resources IMPACT 23 (1) (2021) 16–18.

[27] W. Leon, The State of State Renewable Portfolio Standards, Clean Energy States Alliance, 2013. https://www.cesa.org/wp-content/uploads/State-of-State-RPSs-Report-Final-June-2013.pdf.

[28] D. Rodriguez-Lonebear, N.E. Barceló, R. Akee, S.R. Carroll, American Indian Reservations and COVID-19: correlates of early infection rates in the pandemic, J. Public Health Manage. Pract. 26 (4) (2020) 371–377.

[29] Navajo Nation, Navajo Housing Authority, Phase II Housing Needs Assessment and Demographic Analysis, 2011. https://www.navajohousingauthority.org/wp-content/uploads/2015/08/Navajo_Nation_Housing_Needs_Assessment_091311-1-PAGE-1-50.pdf.

[30] Navajo Nation Department of Fish and Wildlife, Climate Change Vulnerability Assessment for Priority Wildlife Species, Report prepared by the H. John Heinz III Center for Science, Economics, and the Environment, 2013.

[31] J. Redhouse, Removing the Overburden: The Continuing Long Walk, Unpublished ms. Redhouse/Wright Productions, John Redhouse papers, 1972–2013, Center for Southwest Research, University of New Mexico, Albuquerque, 1986.

[32] A. Simpson, Mohawk Interruptus: Political Life Across the Borders of Settler States, Duke University Press, Durham, 2014.

[33] A. Simpson, The sovereignty of critique, South Atl. Q. 119 (4) (2020) 685–699.

[34] T. DePree, The Life of the By-Product in the "Grants Uranium District" of Northwestern New Mexico (PhD dissertation), Department of Science and Technology Studies, Rensselaer Polytechnic Institute, 2019.

[35] S. Dooley, Coronavirus is attaching the Navajo 'because we have built the perfect human for it to invade', Scientific American (2020). July 8 https://www.scientificamerican.com/article/coronavirus-is-attacking-the-navajo-because-we-have-built-the-perfect-human-for-it-to-invade/.

[36] N. Horseherder, et al., 2021, Grist, https://grist.org/Array/what-would-a-just-transition-look-like-for-the-navajo-nation/.

[37] K. Whyte, Way beyond the lifeboat: an indigenous allegory of climate justice, SSRN Electron. J. (2017), https://doi.org/10.2139/ssrn.3003946.

[38] A. Simpson, Consent's Revenge, Cult. Anthropol. 31 (3) (2016) 326–333, https://doi.org/10.14506/ca31.3.02.

Will electro-mobility encourage injustices? The case of lithium production in the Argentine Puna

Melisa Escosteguy[a], Walter F. Diaz Paz[a], Martín A. Iribarnegaray[a], Araceli Clavijo[a], Carlos Ortega Insaurralde[a], Helen Stern[a], Cristian D. Venencia[a], Christian Brannstrom[b], Marc Hufty[c], and Lucas Seghezzo[a]

[a]Research Institute on Renewable Energy (INENCO), National Research Council of Argentina (CONICET), National University of Salta (UNSa), Salta, Argentina [b]Department of Geography, Texas A&M University, College Station, TX, United States [c]Graduate Institute of International and Development Studies, Geneva, Switzerland

1 Introduction

In an article published in *The Washington Post* in 2016 [1], the authors highlighted the contrast between impoverished residents in the Puna region of Northern Argentina and the "modern-day Silicon Valley treasure" that lies beneath their ancestral land. That treasure is lithium. The contrast could not be more dramatic. On the one hand, transnational corporations are rushing into the area to extract this metal, essential for lithium-ion batteries and extracted under the logic of the global capitalist accumulation. On the other, marginalized Indigenous communities are struggling to access basic services, maintain control of their territories, and preserve their livelihoods. The news article quoted a local leader complaining about mining companies not giving anything back to the communities, despite the fact that they "are taking millions of dollars from [their] lands" [1]. In response to those complaints, the companies' representatives insisted that they complied with all the rules and that they worked from the beginning "very, very closely with the local community." Regardless of who is right or who is wrong, it appears that the fairness of the distribution of social-environmental impacts across the lithium production network is, to say the least, contested.

Lithium is a critical element in the production of green technologies designed to miti-gate climate change through electro-mobility and power storage [2]. Lithium-ion batteries are widely used in electric vehicles (EVs), the dominant low emission technology in the car industry [3] and a key variable in projections linked to "peak demand" for oil [4]. Future lithium demand depends on the scale of EV deployment, which in turn is reliant on commit-ments by governments to decarbonization goals and manufacturers to low or zero emission automotive fleets. In this context, both the United States and the European Union included lithium among their lists of critical raw materials [5,6]. In a scenario with a high car depen-dency where global warming is limited to 2°C by 2100, the cumulated demand for lithium by 2050 would reach around 1.5 times the current level of known lithium reserves [7]. Although the massive deployment of electro-mobility would be technologically feasible, the lithium sector faces important economic, geopolitical, and environmental vulnerabilities [7,8].

Global lithium production is concentrated in Australia, China, and the salt flats of the Andean Plateau ("Puna") of Argentina, Bolivia, and Chile. The "lithium triangle" holds over half of the world's lithium reserves [9]. Argentina is the world's fourth largest lithium producer with two plants in operation and almost 1 million hectares under concession to lithium companies [10]. Mining operations in the Puna region overlap with lands occupied by Indigenous and peasant communities, whose main activities include pastoralism and small-scale subsistence agricul-ture. The extraction and processing of lithium can trigger social-environmental conflicts linked mainly to water usage, biodiversity loss, waste generation, and the hydrological balance of salt flats [11–16]. Furthermore, some authors have highlighted that, despite its role as a strategic resource for future technologies, lithium mining revives the economic models of the past since its revenues are unequally distributed [17]. Although there were some public initiatives to develop production linkages in Argentina, such as the fabrication of lithium-ion batteries or their components, they were not successful so far [6]. While the benefits of lithium mining go mainly to multinational firms, local communities have to deal with the environmental costs, face the impacts of climate change, and bear the uncertain consequences of extractivism in the context of the increased demand for transition minerals [17,18].

This chapter identifies justice concerns expressed in different types of documents relating to lithium production in the Argentine Puna, with special emphasis on communities near lithium extraction sites. Through the analysis of qualitative data, we show the prevalence of four different types of injustices at the local level and argue that these injustices reflect underlying power relations between powerful actors at different scales and these communi-ties. We conclude that a just transition to electro-mobility needs to begin with more inclusive decision-making processes at the sites of lithium extraction.

2 Materials and methods

We focused our analysis on the two projects currently in operation in Argentina: Salar de Olaroz and Fénix (Fig. 24.1). In our analysis of the social-environmental impacts of lithium production in Argentina, we applied an energy justice framework [19]. This framework con-tains four interconnected dimensions of modern justice theory [20–22]: (a) *distributive justice* identifies what is being distributed (costs and benefits, goods and ills), among whom (indi-vidual and collective actors), and how it is distributed; (b) *procedural justice* identifies who

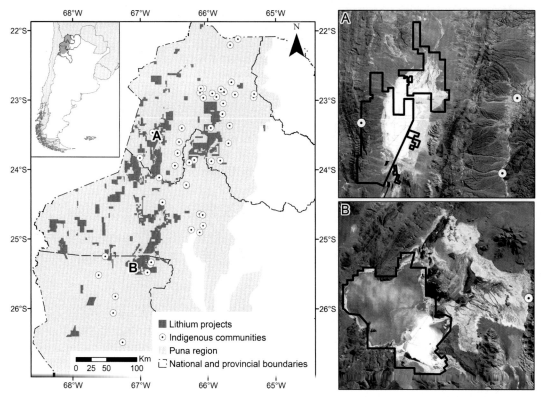

FIG. 24.1 Location of lithium projects and Indigenous communities in the Argentine Puna *(left)*. Location of the two projects selected for the case study: (A) Salar de Olaroz; (B) Fénix *(right)*.

plans and makes rules, laws, and decisions, who participate in decision-making, and the fairness of such processes; (c) *cosmopolitan justice* brings to the fore the well-being of individuals, taking a universal approach; (d) *recognition justice* identifies people whose vulnerability may be worsened by the processes of energy transition.

We used the four justice dimensions to analyze the distribution of negative impacts (considered here as injustices) associated with lithium production at the local scale, i.e., impacts affecting local livelihoods and the environment in the surroundings of the projects mentioned above. We relied on political ecology to understand power struggles at the interface between environmental and social issues. Several political ecology concerns informed the boundaries of our research topic, selection of respondents for interviews, and coding of qualitative data: unequal power relations (among states, firms, and communities), contested access to resources (for accumulation and livelihood reproduction), competing discursive framings among elites and community residents, and unequal access to information [23]. Local injustices were identified by coding 552 documents using MAXQDA Analytics Pro, a software package that allows the coding and analysis of qualitative data. We organized documents in 5 groups: (a) field reports (5 documents) included semi-structured interviews and conversations with local leaders held in four field trips (2019–20); (b) government reports

(12 documents); (c) corporate reports (101 documents) authored by the lithium companies located in the area; (d) newspaper articles (395 documents) including all the news reports that contain "lithium" published during the period April 2017–April 2019 by the 2 most read newspapers in Argentina and in the provinces where lithium extraction occurs; and (e) scientific and technical literature (39 documents) including published articles and theses related to lithium production in Argentina. Before the coding began, a codebook comprising a list of coding categories or *codes* was prepared. In qualitative analysis, a code is a construction (a word or a phrase) created by the researcher that helps to systematize the data and find patterns within it. Thus, the codebook was constructed based on the research question and literature on the topic. It was improved as the coding proceeded by adding or changing some categories according to the information provided in the documents.

3 Results and discussion

In this section, we briefly describe our case studies and the main results of the coding process following the energy justice framework. We also discuss our results in the context of the energy transition to electro-mobility.

3.1 Lithium production in the Argentine Puna

In the projects analyzed in this study (see Table 24.1), lithium from salt flats is recovered through evaporation techniques. Brine is pumped from the underground reservoirs into large open-air shallow evaporation ponds and, when the optimum concentration is reached, lithium carbonate is precipitated by the addition of soda ash [24]. Brine concentration evaporates on average $500\,m^3$ of water per ton of lithium carbonate, and around $50\,m^3$ of freshwater per ton are needed at different stages of the process [13].

Fénix lithium mine started its production in 1997 in the Salar del Hombre Muerto (located across the border between the provinces of Salta and Catamarca). The project was owned by the US firm FMC Lithium Corporation and operated through Minera del Altiplano S.A. (MdA), an Argentine subsidiary. In 2019, FMC created a spin-off company named Livent Corporation for its lithium operations. Livent is one of the world's five major lithium producers [25]. Extraction in the Salar de Olaroz (Province of Jujuy) started in 2015. It is operated by Sales de Jujuy S.A. (SdJ), a joint venture between the Australian company Orocobre Limited (66.5%), the Japanese Toyota Tsusho Corporation (25.0%), and Jujuy Energía y Minería Sociedad del Estado (JEMSE), a company of the Jujuy provincial government (8.5%) [10].

TABLE 24.1   Lithium projects currently in operation in Argentina.

| Name | Area (ha) | Capacity (ton LCE) | | | Main customers |
		Installed	Projected	Production	
Fénix	30,000	23,000	40,000	17,000	China, United States
				6000 ton LiCl	
Salar de Olaroz	63,000	17,500	42,500	12,605	China, United States, Japan, EU

LCE, lithium carbonate equivalent.

3.2 Justice dimensions of lithium production

As shown in Table 24.2, most of the injustices related to our case studies were identified in groups A, D, and E. In government and corporate reports, these issues were barely mentioned and, instead, they emphasized the positive impacts of lithium production. Observing different discourses from different stakeholders is a clear indication of the existence of conflicting social, political, and economic interests around lithium.

Results also reveal that distributive justice concerns were predominant in the coding, which is probably linked with the context of structural poverty in which communities live. In terms of distributive justice, we detected three main concerns. First, frequent claims related to (un)employment issues such as (a) local people are only hired for the stage of construction as low-skill workers; (b) the employer is not the lithium company itself but a sub-contractor; (c) the workforce employed in the plants is largely composed by workers from outside the region, and (d) members of communities not adjacent to extraction sites are hardly ever hired. Second, communities, scholars, and part of the civil society are highly concerned about environmental impacts, mainly water consumption and availability of surface water. Community residents also believe that the construction of wells to pump brine can cause groundwater pollution and that trucks and vehicles circulating on unpaved roads are responsible for air pollution. Third, we found that local people regularly complained about the lack of basic infrastructure and limited access to basic services. Some mining royalties, which amount to up to 3% of the price of the mineral extracted in bulk, are redistributed to the municipalities where lithium extraction occurs, but not all communities receive an equal share. Residents believe that environmental impacts from lithium production activities in the area are disproportionately higher than the few benefits they receive in exchange.

TABLE 24.2   Codes and types of documents used in this study.

Codes	Coded segments per type of document						Frequency
	A	B	C	D	E	Total	(%)
Views on lithium production	20	30	71	235	92	448	20.5
Positive impacts	80	85	144	180	27	516	23.6
Perceptions about the salt flats	7	6	24	13	44	94	4.3
Governance	71	56	206	201	90	624	28.5
Negative impacts (injustices)	60	2	11	225	208	506	23.1
Distributive justice	22	2	1	30	47		
Procedural justice	5	1	0	21	9		
Cosmopolitan justice	3	0	0	14	7		
Recognition justice	3	0	1	7	7		
Total	238	179	456	854	461	2188	100

A, field reports; B, government reports; C, corporate reports; D, newspaper articles; E, scientific and technical literature.

Regarding procedural justice, the principle of free, prior, and informed consent is not complied with. According to the National Constitution and international treaties signed by Argentina, such as the International Labor Organization (ILO) Indigenous and Tribal Peoples Convention No. 169 adopted in 1989, Indigenous communities have the right to be consulted by the state before an extractive process starts in their territories. The documents analyzed revealed that no consultation was carried out before MdA's Fénix started extracting lithium in the Salar del Hombre Muerto, while the consultation for the Salar de Olaroz was carried out by SdJ itself, which fails to respect the principle of neutrality in the consultation process. Some coded segments stated clearly that this situation generated some conflicts. People living near Fénix protested at the beginning of 2020 because they had not been consulted about the construction of an aqueduct that would provide water to the project and allegedly affect the flow of a nearby river. Similarly, in Salar de Olaroz, members of different communities organized themselves in a group called La Apacheta to ask for a full consultation to be carried out.

These results are in line with the coded segments related to cosmopolitan justice. We found several references to an event that involved families living in the surroundings of Fénix. MdA tore down some wire fences that delimited some families' lands, arguing that they needed more space for vehicles to easily access the lithium processing plant. The concerned families tried to stop this intrusion into their lands and as a response received an eviction order. Some residents were even arrested by the provincial police. Similar situations were reported in other salt flats in the Puna region. Indigenous communities of Salinas Grandes and Laguna Guayatayoc, for example, sued the Argentine State for the violation of their rights before the Inter-American Commission on Human Rights. Different potential injustices related to working conditions were also identified. Due to the COVID-19 pandemic and the temporary halt in the construction of some works scheduled for 2020, many workers were fired or furloughed without compensation. Some workers claimed that despite the extreme working conditions at salt flats, there are no permanent doctors in mining camps and they have no permanent health insurance, while hygiene and safety measures are not fully respected. Other workers pointed out that in many cases they were not paid for working overtime.

In relation to recognition justice, we detected several communities with no secure land tenure even though they have lived there for centuries. Most land in the area, which is legally owned by provincial states, has already been given in concession to mining companies in a process that ignored or underestimated the presence and livelihoods of local communities.

3.3 Toward a just transition to electro-mobility

Our results clearly confirm the existence of a number of injustices that can be associated with the extraction and production of lithium products on communities at the local scale in the Argentine Puna, in line with cosmopolitan injustices identified in other electro-mobility transitions [22,26]. They certainly compromise the fairness of the transition to electro-mobility based on lithium-ion technologies. Our results add to the key sustainability challenges identified for the industries that will supply the metals and minerals for low-carbon technologies [2].

Although our analysis focuses on the negative impacts, it is worth mentioning that positive impacts appear often in the coding. Lithium seems to be surrounded by ambivalence: while some local people recognize the injustices and resist them, others, particularly in Jujuy, are openly in favor of and have many positive expectations about lithium production.

In any case, we believe that a just transition to electro-mobility will only be achieved if extraction activities are accompanied by specific and strong actions from the outset, such as but not limited to (a) Identification of direct and indirect livelihood impacts with full participation of host communities; (b) Critical analysis of corporate social responsibility actions and independent identification and assessment of impacts and mitigation measures; (c) Analysis of the full range of positive and negative social perspectives toward lithium held by different stakeholders, with a clear unpacking of all underlying reasons, including a gender perspective; (d) Strengthening information networks among Puna community leaders to share best practices and avoid co-option by state authorities and firm representatives; (e) Improving participatory decision-making processes, recognizing community land rights, and respecting their right to free, prior, and informed consent; and (f) Improving working conditions, education and access to health care for all Puna communities directly or indirectly affected by lithium projects.

4 Conclusions

Local communities in the Argentina Puna region, often marginalized and vulnerable, who are highly dependent on the water and grazing land provided by their territories, have not been sufficiently included in decision-making processes related to lithium mining, in violation of recognition and procedural justice principles. A number of distributive justice concerns detected in this study are worth being examined more thoroughly, such as (a) What materials are extracted through which practices and technologies, leading to which types of negative environmental and livelihood impacts?; (b) How are costs and benefits distributed among Puna communities and other stakeholders?; and (c) Which mitigation measures are in place for addressing the negative impacts of lithium extraction? Recognizing neighboring communities as rightful actors in the much-needed transition to electro-mobility is the only way to make this transition more just and more sustainable locally and globally.

Acknowledgments

We acknowledge funding by the Swiss National Science Foundation through project LITHIUM (The global political ecology of lithium commodity chain). Funding by the National Council of Scientific and Technical Research of Argentina (Consejo Nacional de Investigaciones Científicas y Técnicas—CONICET) is also acknowledged.

References

[1] T.C. Frankel, P. Whoriskey, Tossed aside in the 'white gold' rush, Wash. Post (December 19) (2016).
[2] B.K. Sovacool, S.H. Ali, M. Bazilian, et al., Sustainable minerals and metals for a low carbon future, Science 367 (6473) (2020) 30–33.
[3] D.G. Victor, F.W. Geels, S. Sharpe, Accelerating the Low Carbon Transition: The Case for Stronger, More Targeted and Coordinated International Action, The Brookings Institution, 2019.
[4] A.M. Jaffe, The role of the US in the geopolitics of climate policy and stranded oil reserves, Nat. Energy 10 (2016) 16158.
[5] P.I. Vásquez, The lithium triangle: the case for post-pandemic optimism, in: Wilson International Center Working Paper, 2020.

[6] M. Obaya, A. López, P. Pascuini, Curb your enthusiasm. Challenges to the development of lithium-based link-ages in Argentina, Res. Policy 70 (2021), 101912.

[7] E. Hache, G.S. Seck, M. Simoen, et al., Critical raw materials and transportation sector electrification: a detailed bottom-up analysis in world transport, Appl. Energy 240 (2019) 6–25.

[8] H. Ambrose, A. Kendall, Understanding the future of lithium: part 1, resource model, J. Ind. Ecol. 24 (1) (2020) 80–89.

[9] J. Sterba, A. Krzemień, P.R. Fernández, et al., Lithium mining: accelerating the transition to sustainable energy, Res. Policy 62 (2019) 416–426.

[10] U.S. Geological Survey, Argentina Lithium Map, 2018.

[11] M. Argento, F. Puente, Entre el boom del litio y la defensa de la vida. Salares, agua, territorios y comunidades en la región atacameña, in: B. Fornillo (Ed.), Litio en Sudamérica, CLACSO, 2019, pp. 173–220 (In Spanish).

[12] S. Babidge, F. Kalazich, M. Prieto, et al., "That's the problem with that lake; it changes sides": mapping ex-traction and ecological exhaustion in the Atacama, J. Polit. Econ. 26 (1) (2019) 738–760.

[13] C.F. Baspineiro, J. Franco, V. Flexer, Potential water recovery during lithium mining from high salinity brines, Sci. Total Environ. 720 (2020), 137523.

[14] V. Flexer, C.F. Baspineiro, C.I. Galli, Lithium recovery from brines: a vital raw material for green energies with a potential environmental impact in its mining and processing, Sci. Total Environ. 639 (2018) 1188–1204.

[15] M.A. Marazuela, E. Vázquez-Suñé, C. Ayora, et al., The effect of brine pumping on the natural hydrodynamics of the Salar de Atacama: the damping capacity of salt flats, Sci. Total Environ. 654 (2019) 1118–1131.

[16] W. Liu, D.B. Agusdinata, S.W. Myint, Spatiotemporal patterns of lithium mining and environmental degrada-tion in the Atacama Salt Flat, Chile, Int. J. Appl. Earth Obs. Geoinf. 80 (2019) 45–156.

[17] F.M. Dorn, F. Ruiz Peyré, Lithium as a strategic resource: geopolitics, industrialization, and mining in Argentina, J. Lat. Am. Geogr. 19 (4) (2020) 68–90.

[18] R. Morales Balcazar, Crisis y minería del litio en el Salar de Atacama. La necesidad de una mirada desde la jus-ticia climática, in: Observatorio Plurinacional de Salares Andinos (Ed.), Salares Andinos: Ecología de Saberes por la Protección de Nuestros Salares y Humedales, Fundación Tantí, 2021, pp. 69–82 (In Spanish).

[19] B.K. Sovacool, J. Kester, L. Noel, et al., Energy injustice and Nordic electric mobility: inequality, elitism, and externalities in the electrification of Vehicle-to-Grid (V2G) transport, Ecol. Econ. 157 (2019) 205–217.

[20] D. McCauley, V. Ramasar, R.J. Heffron, et al., Energy justice in the transition to low carbon energy systems: exploring key themes in interdisciplinary research, Appl. Energy 233 (2019) 916–921.

[21] B.K. Sovacool, The political ecology and justice of energy, in: T. Van de Graaf, B.K. Sovacool, A. Gosh, et al. (Eds.), The Palgrave Handbook of the International Political Economy of Energy, Palgrave Macmillan, 2016, pp. 529–558.

[22] B.K. Sovacool, M. Martiskainen, A. Hook, et al., Decarbonisation and its discontents: a critical energy justice perspective on four low-carbon transitions, Climate Change 155 (2019) 591–619.

[23] T. Perreault, G. Bridge, J. McCarthy (Eds.), The Routledge Handbook of Political Ecology, Routledge, 2015.

[24] C.H. Díaz Nieto, N.A. Palacios, K. Verbeeck, et al., Membrane electrolysis for the removal of Mg^{2+} and Ca^{2+} from lithium rich brines, Water Res. 154 (2019) 117–124.

[25] A.R. Quinteros-Condoretty, L. Albaredac, B. Barbiellinia, et al., A socio-technical transition of sustainable lith-ium industry in Latin America, Procedia Manuf. 51 (2020) 1737–1747.

[26] B. Jerez, I. Garcés, R. Torres, Lithium extractivism and water injustices in the Salar de Atacama, Chile: the colo-nial shadow of green electromobility, Polit. Geogr. 87 (2021), 102382.

The limits of authoritarian energy governance: Energy, democracy and public contestation in Turkey

Alevgül H. Şorman[a,b] *and Ethemcan Turhan*[c]

[a]Basque Centre for Climate Change (BC3), Leioa, Spain [b]IKERBASQUE, Basque Foundation for Science, Bilbao, Spain [c]Department of Spatial Planning and Environment, University of Groningen, Groningen, the Netherlands

1 Introduction

The shock of the COVID-19 pandemic appears to have hit global energy trajectories in ways that disrupt the much-anticipated decarbonization pathways [1]. Although complying with climate pledges might be somewhat delayed [2], it has also become apparent that a different interplay of energy and society assemblies is indeed possible, under extreme, unexpected conditions. Moreover, the ex-post COVID wave shock is now coupled with the slashing of gas imports from Russia in light of Russia's invasion of Ukraine, serving as a wakeup call for breaking away from dependence on authoritarian regimes. Projections of European demand on Russian gas imports are anticipating cuts by a half [3] or even by two thirds [4]. However, despite the light of recent events, the bounce back of emissions is already under way. After dropping 5.4% in 2020 due to the pandemic, global emissions are expected to increase 4.9% 2021 [5] in the absence of rapid structural changes in global economic, transport, or energy systems [6]. While renewables might soon reach their highest levels in terms of output and share, it has also been acknowledged that a transformation in the power sector alone will only result in a third of the commitments to achieving net-zero emission targets [7]. Thus, it is imperative that governance schemes either lead or defer these winds of change. The steady rise of authoritarian populisms and social polarization coupled with ambiguous energy futures imply different political, social, and ecological challenges and lock-ins regarding energy transitions. Such transitions are experienced differently across both the Global North and

Global South [8,9] while also being multifaceted across how different axes of justice, politics, and power play out [10] now witnessed across many sectors across the global social fabric in line with increasing oil and gas prices [11].

Against such a backdrop, while the building global momentum is calling for decarbonizing energy systems as in the cases of the Green New Deal [12] (for a partisan polarization over this deal see Ref. [13]), the European Green Deal [14], and more radical transformation options [15], Turkey's energy futures hang on a tightrope. As a country once hailed as an economic miracle, Turkey today witnesses a failing economic model based on credit-expansion-driven domestic demand, shattered democratic checks-and-balances as well as booms and busts of construction, extractivism, and energy rush based on clientelist relations [16]. Old fashioned energy production schemes, such as combined cycle natural gas and coal, continue to form the central constituents powering the economy, depending heavily on energy imports skyrocketing from 52% to 69% between 1990 and 2019 [17]. Fossil fuel-based production accounted for almost half (48.3%) of the country's installed capacity in 2020; with 32.3% share of hydropower, leaving renewables capacity at a mere 19.4% [17]. These figures become even more problematic when one considers Russian invasion of Ukraine in 2022 and Turkey's heavy dependence on Russian oil, natural gas and coal making up for 35.1% of its all fossil fuel imports [17]. Fuel combustion from energy industries had the biggest share in Turkey's cumulative GHG emissions in 2018 (around 38% of total emissions with over 157 million tons of CO_2), while the transport sector accounted for around 20%, followed by manufacturing industries and construction at around 14% of total CO_2 emissions [18]. Such infrastructures and respective emissions are a mere reflection of the fossil lock-in to the colossal project of authoritarian developmentalism via carbon-intensive industrialization and modernization based on aggressive neoliberal policies [19].

In the political sphere, the Justice and Development Party (AKP) led by president Erdoğan has been governing the country for two decades with increasing tones of authoritarian, clientelist, and populist policies. Following the constitutional referendum in 2017 which led to executive aggrandizement by concentrating political power in Erdoğan's hands, already dubious energy and economic growth targets started to appear bleak. Erdoğan's autocratic governance has also given way to authoritarian impulses within environmental policies, expanded securitization of fossil fuel resources, mirrored through the offshore gas rush in the eastern Mediterranean [20], along with a national promise of extracting a recently discovered natural gas reserve in the Black Sea [21]. Securitization has primarily placed both fossil fuel and renewable energy choices beyond public debate, thereby enabling decision-making on premises of impulse, urgency, anxiety, and a willingness to sacrifice while also locking in, legitimizing and amplifying Turkey's overall energetic metabolism. As such, current policies have not only led to the subjugation of the country's unique natural ecosystems with a hunger for more energy acquisition but also continue to threaten and repress the dissenting voices including those of women, rural communities, ethnic minorities, and youth. Often, such social forces at play are suppressed with the trinity of authoritarian developmentalism, chauvinist populism, and coercive state power [22].

In this chapter, we bring in several knowledge claims from critical biophysical economics and political ecology to situate the undemocratic nature of Turkey's energy predicament. The combination of these disciplines put emphases on biophysical limits [23], on asymmetrical power relations [24], on the notion that energy is a social relation [25], and eventually leading to the conclusion that the condition of the ecological system is inextricably linked to the status of the social system [26]. For doing so, we inquire the socio-metabolic intensification in Turkey across the different sectors making up the economy, with critical attention to how the uneven praxis of power in the last two decades in Turkey has led to deep transformation of socio-natures [27]. We argue that the domination of authoritarian, developmentalist politics over socio-natures have led to a two-way trend. We reckon that the top-down, and growth-oriented neoliberal developmentalist model in Turkey has not only deepened under Erdoğan but has achieved its current status due to the normalization of coercion against environmental dissent, emboldened clientelism, a disregard for already faltering legal mechanisms, and abrupt regulatory arrangements to favor private capital [28]. This in turn has also fueled what Arsel et al. refer to as the "environmentalism of the malcontent" [29], a socio-ecological consciousness revival in the society due to rampant environmental destruction and unmet welfare expectations.

2 The multiple manifestations of authoritarian energy governance

Historically, the topic of energy has been omnipresent in the transformation of economic regime in Turkey since the 1970s. First from state-led import-substitution industrialization to a technocratic neoliberal regime and thereafter to its current reincarnation as cronyism-driven, market-friendly autocratic governance under Erdoğan, energy has played a crucial role at each step of the transformation of political and economic power in the country. The most notable shift occurred in the aftermath of the 2001 crisis, in which an International Monetary Fund—World Bank-led structural adjustment program reconfigured the regime of accumulation around an export-led growth strategy manifested as privatization of public assets, elimination of agricultural subsidies and promotion of subcontracting public services [30]. As a result of these colossal changes, domestic and foreign capital skyrocketed private energy investments owing to deregulation while energy-related decisions were "insulated from public participation, resulting in intense sociospatial and socio-economic inequalities and conflicts" [31, p. 1]. As Erensü [32, p. 156] also reminds us, this transformation "almost tripled the country's electricity generation capacity, created a lucrative market for the capital owners but also created deep social and ecological conflicts as energy infrastructures aggressively grappled land and water in the countryside" in the past two decades.

2.1 Socio-metabolic intensification

When looking into Turkey's societal metabolism, the scrutiny of how continued flows of energy and materials have been sustaining the biophysical backbone of the economic system [33,34], we can see that the country's energetic metabolism has persistently been intensifying. Within the time span of almost two decades since 1990, Turkey's final energy use increased by over 2.6 times in total size. While qualitative transformation in household energy use has not shown much increase, the commercial and public services sphere has shown tremendous growth. An increase of about 15-folds in energy throughput in commerce and public services went hand in hand with the market liberalization and the country's integration into the global energy markets. Aiming at being a regional "energy hub," the Turkish state cleared the ground for the market formation and intensive corporatization in service of its

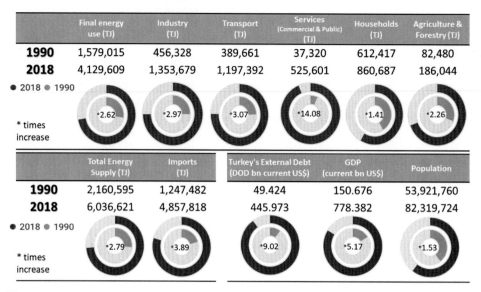

	Final energy use (TJ)	Industry (TJ)	Transport (TJ)	Services (Commercial & Public) (TJ)	Households (TJ)	Agriculture & Forestry (TJ)
1990	1,579,015	456,328	389,661	37,320	612,417	82,480
2018	4,129,609	1,353,679	1,197,392	525,601	860,687	186,044
* times increase	•2.62	•2.97	•3.07	•14.08	•1.41	•2.26

	Total Energy Supply (TJ)	Imports (TJ)	Turkey's External Debt (DOD bn current US$)	GDP (current bn US$)	Population
1990	2,160,595	1,247,482	49.424	150.676	53,921,760
2018	6,036,621	4,857,818	445.973	778.382	82,319,724
* times increase	•2.79	•3.89	•9.02	•5.17	•1.53

FIG. 25.1   Turkey's energetic metabolism by subsectors, supply and imports (in TJ) coupled with socio-economic indicators. Energy data from Ref. [35]; Socio-economic data from Refs. [36, 37].

energy security-oriented calculus of grandiose geopolitical aspirations [31]. In terms of over-all weight, however, the industrial and transport sectors have driven the overall energetic metabolism both tripling in size. Fig. 25.1 shows the shift in the energetic metabolism of the country between 1990 and 2018 in comparison to other socio-economic indicators.

Such an intensification can be seen as a promoter of economic growth, yet for a nuanced analytical assessment, it is vital to investigate where these resources are coming from, how they are produced, and by whom these products are extracted [38,39]. Scrutinized under a political-economic lens, it can be argued that such a metabolic expansion has been linked to extensive marketization and an inflating construction-mining bubble [40] coupled with mounting foreign debt obligations [41] increasing almost 10 times in size.

2.2 Expanding energy frontiers

In the name of keeping up with this escalation, domestic energy resources have been treated as commodities par excellence to be appropriated at all costs. While energetic me-tabolism intensified per subsectors, the means of energy provision (the total primary energy supply) has quadrupled. This quadrupling manifested itself in terms of expanding spatial energy frontiers domestically, as well as transnationally; as shown in the case of accelerating fossil fuel imports from places as far as Colombia [42], Nigeria, Algeria, and Qatar [43].

The tentacles of energy projects in Turkey have extended from the urban to the rural in the form of ecological sacrifice zones built to deliver energy services to the metropolitan hubs, externalizing urban energy demand and material footprints to rural areas [19]. The country's Vision 2023 strategy document [44] foresaw almost a doubling of the total installed power ca-pacity to 110 GW via thermal and power plants, small-scale hydropower projects, and renew-able energy projects together with two nuclear power plants, one already under construction by the Russian Rosatom in Akkuyu [45]. (For a detailed discussion on narrative strategies on nuclear power in Turkey see Niphi and Ramana [46] of this collection.) Striving to overcome the country's energy import dependence, the strategy of the Ministry of Energy and Natural Resources set its eye on expanding the installed capacity powered by domestic and renewable sources from 59% to 65% between 2019 and 2023, including 50% increase in domestic coal and 100% increase in solar energy capacity [47]. This rush over energy resources [48] in service of economic and geopolitical goals has therefore subjugated the environment and the people as secondary resulting in numerous socio-ecological conflicts as mapped out in the environmental justice atlas [19,49] (see Section 3).

2.3 Transforming landscapes

Manifestation of power, intensive resource extraction, and increasing energy throughput have been put forward in the form of megaprojects most evident in urban cities. A new massive airport in Istanbul aiming to be the world's biggest airport despite proven oversized and sunken with the collapse of air traffic due to COVID-19, a new bridge with its connecting highway infrastructure perturbing forest land, and a ludicrous project in planning, a Suez Canal-type waterway parallel to the Bosphorus, the *Kanal İstanbul* is expected to result in a colossal amount of land excavation and ecological destruction while creating windfall profits for landowners in the surrounding areas [50]. Some commentators claim that the underlying reason behind these megaprojects is to construct a new city in the north of Istanbul [51], quite an energy-intensive ambition.

Given their scale and the jaw-dropping resources spent on megaprojects, such ventures often bear the risk of not living out expectations (i.e., Turkey claiming to have lost $208 million in revenues due to technical problems between the period 2007 and 2009 from the Baku-Tblisi-Ceyhan oil pipeline) while also possibly reinforcing corruption schemes and leading to further erosion of democracy [52]. Riding on the credit expansion tide in the aftermath of the 2008 global economic meltdown, Erdoğan's regime further accelerated its energy rush by pushing and pursuing the all-of-the-above strategies. As Fırat [53, p. 89] underlines, "relying mainly on borrowed finance (domestic or foreign), imported pipes and other critical materials, and cheap domestic labor, the construction of certain large-scale energy and transport infrastructures, such as pipelines, does not look sustainable from the perspective of sound energy policy nor from an economic perspective."

3 The emergence of new energy politics via socio-ecological conflicts and resistance

Often, where socio-ecological conflicts emerge, there also surfaces an embedded need to explore the consequences of struggles over authority [54], scrutinizing the types of new subjectivities and priorities put in place [55–57]. Despite the tumultuous history of environmental dissent in terms of its victories [49], in Turkey, we observe a wave of new energy politics out of contemporary socio-ecological conflicts. This wave comes as a response to Erdoğan's authoritarian energy agenda making the winners and losers of the actually existing energy governance explicit. At its core, such new energy politics position themselves at distance from established channels of power, thereby questioning the legitimacy of contemporary energy politics [57,58].

One clear manifestation of this reckoning is the rise of new political identities born out of environmental movements against reckless energy investments. Over the years, Turkey has witnessed a rise in mobilizations against the disproportionate social, environmental, and health impacts of electricity generation projects [59,60]. Starting in opposition to small-scale hydropower and nuclear, then expanding and spreading over to coal-fired energy investments and most recently emerging as a solid resistance to geographically intensive geothermal and onshore wind projects, Turkey's environmental movements waged popular struggles against the undemocratic, top-down, and technocratic nature of Erdoğan's aggressive and coercive energy agenda. While one can observe continuity of the state-led developmentalisms insistence on large-scale energy infrastructures (from the much-contested Southeastern Anatolia Project to Yatağan Power Plant from the 1980s), we also note that the ground conditions have become much more inhospitable to any type of opposition in the past two decades.

Many of these singular and isolated local environmental movements are in fact deeply inter-related in their dissent against the exploitative nature of the country's energy policies. Despite their spatial differences, the confluence of these movements against mining and energy investments have given way to new political subjectivities and thereby destabilized the inherent developmentalist assumptions across the left-right spectrum in Turkish politics. These fragmented pockets of opposition also succeeded to a large extent to expose the actors of dispossession as illustrated in the Networks of Dispossession project [61]. Meanwhile, stories of these movements have been documented in detail by rooted NGO campaigns (e.g., The real cost of coal in Muğla by Climate Action Network [62] or Greenpeace capturing coal stories [63]) and amplified by local extensions of global campaigns (e.g., Break Free from Fossil Fuels [64] and 350.org Türkiye [65]). Such struggles, in essence, have created ad hoc pockets of resistance thereby constituting a larger systematic struggle of movements against the repressive energy governance [66].

All in all, environmental opposition in Turkey offers a means of politicizing energy planning beyond the state's geopolitical calculus and thereby helping reframe energy as a social relation [26,67] while extending spaces and meanings around an ecologist identity [31]. Although mostly reactive, these movements also have brought forth debates on democratic, place-based, and collective [68] ownership of small-scale energy production via citizen involvement: a matter so far kept at bay by the Erdoğan regime's preferential treatment of large-scale infrastructures built by its cronies.

4 Conclusions

Turkey's energy landscape has been dominated by an import-based, growth-at-all-costs strategy in the past two decades despite repeated claims of self-sufficiency in high-level politics [69]. Consequently, a "critically insufficient" climate policy has become the collateral damage of this all-out energy expansion. This critical insufficiency has also been clearly manifested through gradual phases of socio-metabolic intensification, expansion of energy frontiers, and transforming socio-natures, leading to irreparable harm. Yet, emerging environmental movements draw the lines of insubordination against these instances of extraction and commodification, while also questioning the uneven distribution of its benefits.

From this perspective, we suggest that energy investments of all sorts are not so much inevitable economic necessities but rather deliberate political choices grounded in authoritarianism and technocracy in the Turkish experience. Following Hoffmann's [25] call to frame energy as a social relation, we contend that spatial and material dimensions of energy investments co-evolve within historically specific socio-natural relations situated in a broader political-economic context. Therefore, rather than framing energy as a commodity, one that is seen as an input to accumulation regime in the hands of the government's cronies, we argue that framing energy as social relation might lead to a renewal of socio-ecological relations in Turkey. A recent study suggesting that transformation of open-pit coal mines into solar powerhouses in the country could provide energy to 6.9 million households annually is a case in this point insofar as these solar investments are controlled and managed democratically by, for and with the communities in these regions [70]. This observation requires us to go beyond the simplistic narratives such as meeting the rising demand via more investment, security of supply and bigger and better technocratic assemblages. Instead, we argue that new energy politics in the country call for a thorough re-thinking of the fruits of metabolic expansion asking the essential by and for whom, why, where, and how questions. These questions moreover require maximum attention and immediate answers amidst the geopolitical chaos of imports to and from Russia.

In sum, we claim that distinct social values [71] are emerging from environmental movements in the face of Turkey's authoritarian energy governance. These social values blend with new energy politics in embracing a transformative vision and channeling the malcontent to political change. However, we should also acknowledge the limits of hope these movements may create beyond the rhetoric. Ultimately, environmental movements producing new subjectivities and connecting local grievances against an authoritarian energy agenda can only yield transformative results as long as they succeed in bringing multi-dimensional, intersectional, and socially acceptable solutions to their woes.

Acknowledgments

Ethemcan Turhan gratefully acknowledges the project "*Harnessing the heat below our feet: Promises, pitfalls and spatialization of geothermal energy as a decarbonization strategy*" funded by FORMAS (Swedish Research Council, Project no: 2020-00825).

References

[1] C. Kuzemko, M. Bradshaw, G. Bridge, A. Goldthau, J. Jewell, I. Overland, D. Scholten, T. Van de Graaf, K. Westphal, Covid-19 and the politics of sustainable energy transitions, Energy Res. Soc. Sci. 68 (2020), 101685.

[2] C. Farand, Governments still due to submit tougher climate plans in 2020, despite Cop26 delay, Climate Home News (2nd April) (2020). https://www.climatechangenews.com/2020/04/02/governments-still-due-submit-tougher-climate-plans-2020-despite-cop26-delay/. (Accessed 01/06/2020).

[3] IEA, A 10-Point Plan to Reduce the European Union's Reliance on Russian Natural Gas, 2022, Available from: https://www.iea.org/reports/a-10-point-plan-to-reduce-the-european-unions-reliance-on-russian-natural-gas. (Accessed 22 March 2022).

[4] EC Press Release, REPowerEU: Joint European Action for More Affordable, Secure and Sustainable Energy, 2022, Available from: https://ec.europa.eu/commission/presscorner/detail/en/ip_22_1511. (Accessed 22 March 2022).

[5] The Global Carbon Project, 2022. Available from: https://www.globalcarbonproject.org/. (Accessed 22 March 2022).

[6] C. Le Quéré, R.B. Jackson, M.W. Jones, A.J. Smith, S. Abernethy, R.M. Andrew, A.J. De-Gol, D.R. Willis, Y. Shan, J.G. Canadell, P. Friedlingstein, Temporary reduction in daily global CO2 emissions during the COVID-19 forced confinement, Nat. Clim. Chang. (2020) 1–7.

[7] IEA, Energy Technology Perspectives 2020, IEA, Paris, 2020. https://www.iea.org/reports/energy-technology-perspectives-2020. (Accessed 27/09/2020).

[8] J. Marquardt, L.L. Delina, Reimagining energy futures: contributions from community sustainable energy transitions in Thailand and the Philippines, Energy Res. Soc. Sci. 49 (2019) 91–102, https://doi.org/10.1016/j.erss.2018.10.028.

[9] L.L. Delina, B.K. Sovacool, Of temporality and plurality: an epistemic and governance agenda for accelerating just transitions for energy access and sustainable development, Curr. Opin. Environ. Sustain. 34 (2018) 1–6.

[10] A. Jerneck, L. Olsson, B. Ness, S. Anderberg, M. Baier, E. Clark, T. Hickler, A. Hornborg, A. Kronsell, E. Lövbrand, J. Persson, Structuring sustainability science, Sustain. Sci. 6 (1) (2011) 69–82.

[11] D. Detomasi, Why Gasoline Prices have Soared to Record Highs, 2022, Available from: https://theconversation.com/why-gasoline-prices-have-soared-to-record-highs-178707. (Accessed 22 March 2022).

[12] H. Res, 109. Recognizing the Duty of the Federal Government to Create a Green New Deal (116th United States Congress, 2019). https://www.congress.gov/bill/116th-congress/house-resolution/109/text.

[13] A. Gustafson, S.A. Rosenthal, M.T. Ballew, M.H. Goldberg, P. Bergquist, J.E. Kotcher, E.W. Maibach, A. Leiserowitz, The development of partisan polarization over the Green New Deal, Nat. Clim. Chang. 9 (12) (2019) 940–944.

[14] European Commission, The European Green Deal, Communication from the Commission to the European Parliament, the European Council, the Council, the European Economic and Social Committee and the Committee of the Regions, Brussels, 2019, p. 24. https://ec.europa.eu/info/strategy/priorities-2019-2024/european-green-deal_en. (Accessed 27/09/2020).

[15] P. Tschakert, P.J. Das, N.S. Pradhan, M. Machado, A. Lamadrid, M. Buragohain, M.A. Hazarika, Micropolitics in collective learning spaces for adaptive decision making, Glob. Environ. Chang. 40 (2016) 182–194.

[16] T. Subasat, The political economy of Turkey's economic miracles and crisis, in: Turkey's Political Economy in the 21st Century, Palgrave Macmillan, Cham, 2020, pp. 31–62.

[17] MMO (Makine Mühendisleri Odası), Türkiye'nin Enerji Görünümü, 2021. https://www.mmo.org.tr/sites/default/files/gonderi_dosya_ekleri/TegNisan2021Sunumu_0.pdf. (Accessed 21 March 2022).

[18] TurkStat, Turkey's Greenhouse Gas Emissions by Sectors (CO2 equivalent), 1990–2018, 2020.

[19] C.İ. Aydin, Identifying ecological distribution conflicts around the inter-regional flow of energy in Turkey: a mapping exercise, Front. Energy Res. 7 (2019) 33.

[20] P. Wintour, How a rush for Mediterranean gas threatens to push Greece and Turkey into war, The Guardian (11th September) (2020). https://www.theguardian.com/world/2020/sep/11/mediterranean-gas-greece-turkey-dispute-nato. (Accessed 27/09/2020).

[21] BBC, Turkey's Erdogan hails huge natural gas, BBC Business (21st August) (2020). https://www.bbc.com/news/business-53866217. (Accessed 27/09/2020).

[22] F. Adaman, B. Akbulut, Erdoğan's three-pillared neoliberalism: authoritarianism, populism and developmentalism, Geoforum (2020), https://doi.org/10.1016/j.geoforum.2019.12.013.

[23] C.J. Cleveland, R. Costanza, C.A. Hall, R. Kaufmann, Energy and the US economy: a biophysical perspective, Science 225 (4665) (1984) 890–897.

[24] A. Stirling, Transforming power: social science and the politics of energy choices, Energy Res. Soc. Sci. 1 (2014) 83–95.

[25] C. Hoffmann, Beyond the resource curse and pipeline conspiracies: energy as a social relation in the Middle East, Energy Res. Soc. Sci. 41 (2018) 39–47.

[26] T.J. Bassett, A.W. Peimer, Political ecological perspectives on socioecological relations, Nat. Sci. Soc. 23 (2015) 157–165. https://www.cairn.info/revue-natures-sciences-societes-2015-2-page-157.htm#. (Accessed 06/07/2020).

[27] O. İnal, E. Turhan (Eds.), Transforming Socio-Natures in Turkey: Landscapes, State and Environmental Movements, Routledge, 2019.

[28] O. Bayulgen, Byzantine energy politics: the complex tale of low carbon energy in Turkey, in: R. Mills, L. Sim (Eds.), Low Carbon Energy in the Middle East and North Africa, Palgrave Macmillan, Cham, 2021, pp. 155–183.

[29] M. Arsel, B. Akbulut, F. Adaman, Environmentalism of the malcontent: anatomy of an anti-coal power plant struggle in Turkey, J. Peasant Stud. 42 (2) (2015) 371–395.

[30] S. Erensü, Powering neoliberalization: energy and politics in the making of a new Turkey, Energy Res. Soc. Sci. 41 (2018) 148–157.

[31] B. Özkaynak, E. Turhan, C.İ. Aydın, in: G.M. Tezcür (Ed.), The politics of energy in Turkey: running engines on geopolitical, discursive, and coercive power, The Oxford Handbook of Turkish Politics Oxford University Press, 2020, https://doi.org/10.1093/oxfordhb/9780190064891.013.29.

[32] S. Erensu, Y.M. Madra, Neoliberal politics in Turkey, in: G.M. Tezcür (Ed.), The Oxford Handbook of Turkish Politics, 2020, https://doi.org/10.1093/oxfordhb/9780190064891.013.17.

[33] M. Giampietro, K. Mayumi, A.H. Sorman, The Metabolic Pattern of Societies: Where Economists Fall Short, vol. 15, Routledge, 2011.

[34] A.H. Sorman, Societal Metabolism. Degrowth: A Vocabulary for a New Era, Routledge, London, 2014.

[35] Eurostat, Simplified Energy Balances [nrg_bal_s] Turkey Available from: eurostat.ec.europa.eu (Extracted on: 12.08.2020).

[36] World Bank, World Development Indicators database, GDP (current US$) ID: NY.GDP.MKTP.CD; Population ID: SP.POP.TOTL Available from: https://data.worldbank.org/country/turkey; (Extracted on: 13.04.2021).

[37] World Bank, World Development Indicators database, External debt stocks, total (DOD, current US$) External Debt Statistics database Available from: http://datatopics.worldbank.org/debt/qeds (Extracted on: 13.04.2021).

[38] A.H. Sorman, E. Turhan, M. Rosas-Casals, Democratizing energy, energizing democracy: central dimensions surfacing in the debate, Front. Energy Res. (2020), https://doi.org/10.3389/fenrg.2020.499888.

[39] A. Scheidel, A. Schaffartzik, A socio-metabolic perspective on environmental justice and degrowth movements, Ecol. Econ. 161 (2019) 330–333.

[40] F. Adaman, B. Akbulut, Y. Madra, Ş. Pamuk, Hitting the wall: Erdoğan's construction-based, finance-led growth regime, Middle East Lond. 10 (3) (2014) 7–8.

[41] J. Taskinsoy, A Hiccup in Turkey's Prolonged Credit Fueled Economic Transition: A Comparative Analysis of Before and After the August Rout, 2019. SSRN 3431079.

[42] A. Cardoso, E. Turhan, Examining new geographies of coal: dissenting energyscapes in Colombia and Turkey, Appl. Energy 224 (2018) 398–408.

[43] O.G. Austvik, G. Rzayeva, Turkey in the geopolitics of energy, Energy Policy 107 (2017) 539–547.

[44] Turkey's Strategic Vision Project Available from: http://www.tsv2023.org/index.php/en/ (Accessed 30/09/20).

[45] C.İ. Aydın, Nuclear energy debate in Turkey: stakeholders, policy alternatives, and governance issues, Energy Policy 136 (2020), 111041.

[46] A. Niphi, M.V. Ramana, Talking Points: Narrative Strategies to Promote Nuclear Power in Turkey. Democratizing Energy: Imaginaries, Transitions, Risks, Elsevier, 2022.

[47] MENR (Ministry of Energy and Natural Resources), 2019–2023 Strategic Plan, 2019. https://sp.enerji.gov.tr/ETKB_2019_2023_Stratejik_Plani.pdf. (in Turkish) (Accessed 30/09/20).

[48] R.L. Bryant, Political ecology: an emerging research agenda in third-world studies, Polit. Geogr. 11 (1) (1992) 12–36.

[49] Environmental Justice Atlas: Available from: https://ejatlas.org/ (Accessed 01/06/20).

[50] E. Dogan, A. Stupar, The limits of growth: a case study of three mega-projects in Istanbul, Cities 60 (2017) 281–288.

[51] M. Sönmez, Mega projelerin ekonomi politiği, Mimarlık. Ist 58 (2017) 32–36 (in Turkish).

[52] B.K. Sovacool, C.J. Cooper, The Governance of Energy Megaprojects: Politics, Hubris and Energy Security, Edward Elgar Publishing, 2013.

[53] B. Firat, The most eastern of the west, the most western of the east: energy transport infrastructures and regional politics of the periphery in Turkey, Econ. Anthropol. 3 (1) (2016) 81–93.

[54] J.C. Ribot, Representation, citizenship and the public domain in democratic decentralization, Development 50 (1) (2007) 43–49.

[55] A.J. Nightingale, Power and politics in climate change adaptation efforts: struggles over authority and recognition in the context of political instability, Geoforum 84 (2017) 11–20.

[56] A.J. Nightingale, H.R. Ojha, Rethinking power and authority: symbolic violence and subjectivity in Nepal's Terai forests, Dev. Chang. 44 (1) (2013) 29–51.

[57] I. Ruostetsaari, Governance and political consumerism in Finnish energy policy-making, Energy Policy 37 (1) (2009) 102–110.

[58] R. Hague, M. Harrop, M., Comparative Government and Politics, vol. 6, Palgrave Macmillan, Nueva York, 2004.

[59] B. Özkaynak, C.İ. Aydın, P. Ertör-Akyazı, I. Ertör, The Gezi Park resistance from an environmental justice and social metabolism perspective, Capital. Nat. Social. 26 (1) (2015) 99–114.

[60] F. Adaman, B. Akbulut, M. Arsel (Eds.), Neoliberal Turkey and its Discontents: Economic Policy and the Environment under Erdogan, Bloomsbury Publishing, 2017.

[61] Z. Ustun, Networks of dispossession II, Graph Comm. J. (2014). Available from: https://blog.graphcommons. com/networks-of-dispossession-version-ii/. (Accessed 30/09/2020).

[62] Climate Action Network (CAN), The real cost of coal in Muğla, 2019, Available from: http://www.caneurope. org/content/uploads/2019/07/september-Final_CostsofCoal_Pdf.pdf. (Accessed 13.04.2021).

[63] Komur Hikayeleri Greenpeace, Coal Stories Available from: https://komurhikayeleri.org/ (Accessed 13.04.2021).

[64] Fosil Yakitlardan Kurtul, Break Free from Fossil Fuels, 2016, Available from: https://fosilyakitlardankurtul. org/. (Accessed 13.04.2021).

[65] 350.org Türkiye Available from: https://350.org/sweet-dreams-are-made-of-this-turkeys-coal-to-disagree/ (Accessed 13.04.2021).

[66] P.E. Dönmez, Against austerity and repression: historical and contemporary manifestations of progressive politicisation in Turkey, Environ. Plan. C Politics Space 39 (3) (2021) 512–535.

[67] A. Stirling, "Opening up" and "closing down" power, participation, and pluralism in the social appraisal of technology, Sci. Technol. Hum. Values 33 (2) (2008) 262–294.

[68] R.C. Sayan, Exploring place-based approaches and energy justice: ecology, social movements, and hydropower in Turkey, Energy Res. Soc. Sci. 57 (2019), 101234.

[69] Ministry of Development, Tenth Development Plan 2014–2018. Republic of Turkey, 2013. https://sbb.gov.tr/ wp-content/uploads/2018/11/The_Tenth_Development_Plan_2014-2018.pdf. (Accessed 30/09/2020).

[70] SOLAR3GW, Kömür Sahalarının Güneş Potansiyeli, 2022, Available from: https://ekosfer.org/wp-content/ uploads/2022/03/komur-sahalarinin-gunes-potansiyeli.pdf. (Accessed 23 March 2022).

[71] E. Leff, Power-knowledge relations in the field of political ecology, Ambient. Soc. 20 (3) (2017) 225–256.

26

Hazard or survival: Politics of nuclear energy in Ukraine and Belorussia through the lens of energy democracy

Barbara Wejnert[a] and Cam Wejnert-Depue[b,c]

[a]Department of Environment and Sustainability, University at Buffalo, Buffalo, NY, United States [b]Environmental Sciences and Policy, Krieger School of Arts and Sciences, John Hopkins University, Baltimore, MD, United States [c]Post-Masters Research Associate, Pacific Northwest National Laboratory and University of Maryland's Joint Global Change Research Initiative, Washington, DC, United States

1 Introduction

Nuclear energy is the second-best (after onshore wind energy) source of clean energy. It reduces greenhouse gas emissions by avoiding the release of billions of tons of carbon dioxide (contrary to coal or gas), making countries less carbon-emitting. Operating nuclear power plants do not produce direct carbon dioxide emissions. Nuclear energy has an emissions footprint of 4g of CO_2 equivalent (gCO_2e/kWh) over the lifetime compared to coal which emits 1.142kg of CO_2 per kWh, and gas 0.572kg of CO_2 per kWh of generated electricity [1]. It outperforms biomass (230g/kWh), solar energy (48g/kWh), and hydropower (24g/kWh) [2]. Furthermore, nuclear power has none of the intermittency problems that wind and solar energies have (either too much production or none); therefore, it does not increase consumer energy prices. Unsurprisingly, nuclear power is currently the largest electricity production source in several European countries, including 80% of total electricity production in France, 51.3% in Belgium, 53.7% in Slovakia, and 44% in Ukraine.

Regardless of the indisputable benefits of nuclear energy, the substantial risks of such energy production appear to overwhelm the benefits. The high costs of building and operating nuclear power plants make nuclear energy less economical than renewable energy sources, especially given the sharp decline of renewable energy costs [3, p. 42096]. Moreover, there are still unsolved problems of storing radioactive waste that pose a high risk of radioactive proliferation and endanger public health, wildlife, contamination of air, water, and soil. The high costs of building storage facilities for nuclear waste and finding a location for such facilities constitute a problem. Also, the existing older reactors require costly and substantial modification or need to be pulled out of energy production after reaching their expected operation lifetime. They also produce more waste. However, the most concerning are the severe long-term effects of radiation on population health and the environment at the time of nuclear accidents. The intricate balance of benefits and hazards of nuclear energy illustrated in Table 26.1 poses a political dilemma of pursuing development.

This chapter explores countries' nuclear energy politics and the opposition to it by scientists, environmentalists, and the public while focusing on Ukraine and Belorussia. This region was exposed to dire consequences of the nuclear power plant explosion in Chernobyl in 1986. In particular, *first*, the chapter explains the catastrophic public health and environmental consequences of Chernobyl's nuclear disaster. *Second*, it critically examines transitions to Energy Democracy across Europe in the aftermath of the Chernobyl disaster. *Third*, it investigates Ukraine's and Belorussia's energy politics and the public opposition to it. *Fourth*, it claims that

TABLE 26.1 Comparison of benefits and hazards of nuclear energy production when focusing on nuclear power plant explosion in Chernobyl, in 1986.

Benefits of nuclear energy production[a]	Hazards of nuclear energy production[b] (Chernobyl example)
Second best source of clean energy (after onshore wind)	Permanent relocation of 200,000 residents; 800,000 people-liquidators relocated to clean up the side of explosion
Do not produce direct carbon dioxide emission nor air pollution while operating	High mortality rate: (1) Red Cross reported hundreds of thousand deaths due to radiation; (2) mortality rate among liquidators increased from 3.5 to 17.5 per 1000 over 15 years, (3) mortality rate in radiation-contaminated areas increased from 16 to 26 per 1000 over 15 years; and (4) five million people were exposed to radiation in Europe
Halts carbon dioxide emission, help reducing greenhouse gases emissions	Decline of population's health: (1) high rate of mortality due to diabetes and thyroid cancer, (2) severe effects on reproductive health—miscarriages, premature births, stillbirths, delivery-related complications, and infertility, and (3) high number of births of retarded or physically disable children
Only 4 g of carbon footprint (4 g of CO_2) per kWh energy	Environmental degradation: (1) contaminated 57,915 mi^2; (2) uninhabitable 1544 mi^2 explosion zone; and (3) air pollution equivalent to 350 Hiroshima bombs
Outperforms biomass, solar and hydropower	Wildfires; soil of burned forest remains radioactive for over few decades
No intermittency problems as solar or wind energy have (too little or too much production)	Produced radioactive waste remains dangerous to human health for thousands of years[c]
New generation of nuclear power plants are much safer than in 1986	

[a] *Source: US Energy Information Agency, Nuclear power and the environment. Report of January 15, 2020. https://www.eia.gov/ energyexplained/nuclear/nuclear-power-and-the-environment.php#:~:text=Nuclear%20power%20reactors%20do%20not,or%20carbon%20 dioxide%20while%20operating.&text=Nuclear%20power%20plants%20also%20have,amounts%20of%20energy%20to%20manufacture.*
[b] *Hazardous outcome of Chernobyl disaster.*
[c] *Also applies to general hazardous outcomes of nuclear energy production.*

both in Ukraine and Belorussia, the top-down governmental approach to energy politics and discounted concerns of the environmentalists, scientists, and the public still reliving the horror and consequences of the Chernobyl disaster kindled a shift to Energy Democracy. *Finally*, the chapter concludes with a discussion on the nature of Energy Democracy and its role in helping people to advocate changes in energy politics and transition from Energy Dominance to Energy Democracy as defined by Szulecki [4] and Stephens et al. [5], and Burke and Stephens [6].

2 Nuclear disaster in Chernobyl

The devastating explosions in Chernobyl nuclear power plants intensified public fear and cautioned scientists about nuclear energy production. It was the most severe nuclear disaster in human history. "The exposed, burning reactor spewed radioactive isotopes into the atmosphere. The fallout reached across Europe as far as Scandinavia and Great Britain—but the worst-hit regions were Ukraine, Belarus, and Russia," as described by one of the witnesses [7, p. 195]. It took 2 weeks to contain the fire from the main reactor. The disaster had particularly severe consequences on public health and the environment.

2.1 Public health consequences

According to expert accounts, the disaster claimed hundreds of thousands of lives [7], destroyed the health and livelihood of a few million people, and contaminated and destroyed the natural habitat of nearly 60,000 mi^2 of land. After the collapse of the Soviet Union, the National Research Centre for Radiation Medicine (NRCRM) in Kyiv, Ukraine reported that among residents living in regions near Chernobyl, as compared to those living in the non-contaminated area, the mortality rate exceeded birth rates by 20% [8]. The high number of casualties included cleanup teams known as "liquidators." There were nearly 800,000 firefighters, engineers, military troops, police, miners, cleaners, medical personnel sent immediately into the disaster area to control the fire and core meltdown and prevent radioactive material from spreading further into the environment. The "liquidators" shoveled the radioactive debris into man-made ditches with gas masks as the only protective gear. According to the post-disaster assessment, 180,000 "liquidators" died due to health problems related to exposure to radiation [9,10].

From 1988 to 2012, mortality rates among liquidators increased from 3.5 to 17.5 deaths per 1000 people, and by 2008 the disability rate among liquidators reached nearly 90%. Over 43,000 liquidators suffered from cancers by 2008 (with the most common being leukemia and thyroid cancer) and many from diabetes and other terminal illnesses [11]. As of January 2018, 1.8 million people in Ukraine and Belorussia, including 377,589 children, were regarded as Chernobyl victims. In radiation-contaminated areas considered safe for residency, the pre-accident average mortality rate of 16 persons per 1000 almost doubled, increasing to 26 per 1000 by 2018 [12].

Women's health was also severely impacted. Devastating consequences included miscarriages, premature delivery, stillbirths, congenital disabilities, delivery-related complications, and infertility. As far as Norway, the miscarriage fraction of all pregnancies increased by 16.3% from November 1986 to January 1987 [13]. In Ukraine, Belorussia, and eastern Poland, hundreds of thousands of women ended their pregnancies after Chernobyl, some due to fear and radiophobia that spread within weeks following a disaster, some following doctors' advice about potential congenital disabilities. Decades later, children were born severely physically handicapped or mentally retarded [14,15].

The Soviet government augmented the catastrophic outcomes of the explosion by covering up the accident. It informed the international community only after high radiation levels reached Sweden 2 days following the explosion [12,16]. Moscow also released wildly misleading information of only 31 deaths related to the Chernobyl accident [17, p. 260]. In political authorities' opinion, the cover-up that included jail sentences for Chernobyl directors was justified "to prevent a panic" [18].

Soon after the accident, however, the Soviet government resettled nearly 200,000 Chernobyl residents. The effect of Chernobyl on population health is still visible today. Nearly five million people still live in contaminated areas of Ukraine, Belarus, and Russia. In Belorussia, the number of leukemia and thyroid cancer patients is steadily rising, and maternal health complications persist [19,20].

2.2 Environmental impact

Environmental destruction was the second grave consequence of the disaster. According to some experts, it will take another 20,000 years before the area immediately surrounding the Chernobyl plant—called the exclusion zone from which nearly 200,000 people were evacuated—will again be fit for human habitation. Today, nearly 57,915 mi^2 of land in Belarus, Russia, and Ukraine remain contaminated, and the 4000 km^2 (1544 mi^2) exclusion zone lingers uninhabited. Approximately 150,000 km^2 of land between Belarus, Ukraine, and Russia (an area larger than the state of New York) were contaminated so severely that 8 million people suffered severe land use restrictions or relocation. The radioactive fallout, carried by winds, scattered over much of the Northern Hemisphere. Pollution of plants and grasslands in Britain led to strict restrictions on lamb sale and other sheep products for several years [20,21].

The Chernobyl disaster caused "irreversible damage to the environment that will last for thousands of years," explained Greenpeace in their 2016 study of the accident. "Never in human history has such a large quantity of long-lived radioisotopes been released into the environment by a single event" [21]. Nuclear scientist, Platonov, estimated the strength of pollution from the Chernobyl explosion as equivalent to 350 Hiroshima Bombs, releasing 3000 micro-roentgen per hour, exposing residents in nearby towns and villages and the local environment to radioactive contamination [7, p. 138]. Ihor Gramotkin, director of the Chernobyl power plant, added that it would take another 20,000 years before the area immediately surrounding the plant will again be fit for human habitation [22]. Released by the accident, long-lived radionuclides still circulate, and the Chernobyl radioactivity continues. By 2021, the "exclusion zone" surrounding the Chernobyl nuclear power plant stays heavily contaminated with cesium-137, strontium-90, americium-241, plutonium-238, and plutonium-239. Decades after the accident, 2.2 million people or one in five Belarussians live in contaminated land (including 700,000 children) [22].

During the accident, the surrounding forest did not burn, but "the pines and evergreens around the reactor turned red, then orange," explained Sergei Sobolev, deputy head of the Executive Committee of the Shield of Chernobyl Association. These orange trees were cut down in 150 ha around the reactors and placed in ditches. Eight hundred burial sites filled with orange forest and various radioactive debris from the reactor were erected around Chernobyl [7, p. 138]. To stop the persistent nuclear contamination, it took 206 days to build the first sarcophagus of 400,000 cubic meters of concrete and 7300 tons of metal framework. In 2007 was completed the two-decades-long international cooperation on the construction of the second sarcophagus led by the International Atomic Energy Agency. The second sarcophagus is larger than Wembley Stadium and taller than the Statue of Liberty. It should seal in the entire disaster site for 100 years [23], but the intense radiation around the nuclear disaster zone persists. According to some scientists, three decades after the nuclear disaster, towns and villages within the exclusion zone, e.g., the city of Pripyat, Ukraine or the village of Babchin, Belarus, are still 3000 years away from resettlement [22], despite the observed return of wildlife to the area [24,25].

Furthermore, yearly recurring wildfires within the exclusion zone magnify the nuclear pollution. The most severe wildfires of 2020 consumed 57,000 ha (22%) of the Chernobyl exclusion zone. According to Greenpeace, in 2021, one of the centers of the wildfires was located only 4 km (approximately 2.5 mi) from the second Chernobyl sarcophagus. The smog of burning wildfire reached Norway and the Ukrainian capital Kyiv, carrying contaminated air with cesium-137 and other radionuclides that increase cancer risk [26].

3 Democracy energy in post-Chernobyl Europe

Across Europe, the Chernobyl disaster sparked the Energy Democracy movement addressing climate change, social equity, and citizens' inclusive engagement in the energy politics that overrides the ultimate goal of profit maximization [4,5]. Pressed by concerned citizens, governments of individual countries responded to public demands and altered the adopted domestic energy policy. The majority of European countries having nuclear power stations recommended inspection of the installations, particularly of those of older types, notably in Germany, Switzerland, and France. Switzerland held a referendum in 1990 where citizens supported a 10-year moratorium on building new Nuclear Power Plants (NPPs), but the four existing plants continued in operation [27]. In 1997, Green Party and the Environmental Movement compelled the French government to halt the building of an advanced Superphénix "breeder reactor." Opponents of the atomic power industry and environmentalists lobbied to refocus energy on green energy sources in France and other European countries.

Poland, located in Chernobyl's vicinity, held a referendum in 1986, where 86% of Poles voted against nuclear power and demanded to freeze construction of its first NPP in Żarnowiec. By 1990, Poland became one of the strongly critical of nuclear power countries [28]. The newly democratic Ukrainian government passed a moratorium on the construction of NPPs in 1990. The moratorium included completing two units of the Khmelnytskyi nuclear power plants that were 75% and 28% completed [29]. After the disaster, responding to citizens' demands, democratizing Belarus suspended the construction of its only nuclear power plant in Astravyets. Anti-nuclear protests and public anti-nuclear demands led to the suspension of the growing worldwide constructions of nuclear plants. Subsequently, the global nuclear energy production dropped from 17.5% in 1996 to circa 10% in 2019 [28].

Discussion concerning nuclear installations' safety flared up again after Japan's earthquake and tsunami that led to a nuclear reactor crisis in Fukushima in April 2011. The strong momentum of the global environmental movement pushed many governments into anti-nuclear sentiment. Some countries halted the construction of nuclear power plants. German Chancellor Angela Merkel's *Energiewende* policy aimed to increase renewable energy capacity while phasing out nuclear energy. In 2017, pressed by popular demands, the Swiss government rejected building new NPPs in the foreseeable future. Italy and Finland have also considered a more cautious approach to nuclear energy [30,31]. Since 2011, in Lithuania, 88% of the public show no support for nuclear energy production [32].

Simultaneously, the growth of energy needs, high oil prices, demands for carbon reduction, and the price of CO_2 emission permits instituted to combat global climate change swayed several countries to reconsider reliance on nuclear energy. Dynamically developing India, China, Turkey, and the United Arab Emirates declared the renewal of nuclear energy programs in the 2000s. The Visegrad Group countries, the Czech Republic, Hungary, Poland, and Slovakia, heavily dependent on their abundant coal resources (e.g., coal accounts for 70% of Polish energy) and having limited renewable energy sources, also reevaluated dependence on nuclear energy to meet the European Union's (EU) clean energy mandate [33–36].

4 Energy dominance and energy democracy in Ukraine and Belorussia

Despite the trauma of the nuclear catastrophe and still lasting disastrous outcomes to public health and the environment, the governments of democratic Ukraine and non-democratic Belorussia—the countries most affected by the Chernobyl disaster, turned the hazardous fuel into the backbone of their energy portfolio in the 2000s [37]. In both countries, these decisions met with vivid protests of scientists, environmentalists, and the general public opposing the governments' decisions and signaling the birth of Energy Democracy.

4.1 Ukraine's nuclear energy expansion

During the energy crisis of 1993, the Ukraine government lifted its moratorium to construct nuclear power plants. The Ukrainian public demanded a secured flow of energy. In 2019, President Volodymyr Zelensky won the presidency by promising to deliver secure and inexpensive energy to Ukrainian homes [38]. To meet this promise, he endorsed the rebirth of the Ukrainian nuclear energy program claiming that nuclear power plays a significant role in the EU climate and energy policy of the European Green Deal [39].

Zelensky's government presented several arguments to justify the reopening of the nuclear energy program; *first*, according to the President, Ukraine's energy demanded independence from Russia's energy supply, especially after the Russian annexation of Crimea [40]. *Second*, nuclear power was supposed to revive the economy and be "the future of Ukraine" that could place Ukraine among the leading nuclear energy producers, both in Europe and globally. *Third*, Ukraine's nuclear energy could play a pivotal role in strengthening ties with the European Union. The expansion was planned to be the "Power Bridge Ukraine—European Union" that would start supplying electricity to the EU network as early as 2019 with project completion by 2025 [41,42]. *Finally*, Frank Timmermans, the European Commission Executive Vice-President for the European Green Deal initiative, supported the Ukrainian nuclear energy production. He perceived "the incredible opportunities of the new economy and to create new jobs for the people who are still working in the mining sector." Also, the Director-General of the International Atomic Energy Agency (IAEA), Rafael Mariano Grossi, an enthusiast of nuclear energy, supported the project as an innovative solution to decarbonization [43].

4.2 Opposition in Ukraine

Zelensky's energy politics alarmed Ukrainian scientists, environmentalists, and constituencies concerned with the environmental and public health hazard of nuclear energy production. The opposition to Zelensky's energy politics grew slowly [40,42]. As scientists demonstrated, by 2020, 12 out of 15 of Ukraine's active nuclear reactors from the Soviet Era producing nearly half of Ukraine's energy are supposed to retire in 2020. Scientists claim that aging reactors were prone to accidents and malfunctions, operated only on half of their capacity, and produced relatively more nuclear waste [44,45].

The concern of Ukrainian nuclear experts and environmentalists increased when two of these reactors had been given lifetime extensions without any safety upgrades in 2013. This decision led to the United Nations Espoo (EIA) rule that Ukraine had breached the convention rules of assessing the major projects of significant adverse impact on the environment across countries [46]. By 2020, the lifetimes of an additional six reactors had been expended without necessary upgrades or risks assessment. The neighboring countries, particularly Lithuania and Germany, raised the alarm about a fleet of elderly Soviet-era reactors reaching retirement age.

The National Ecological Centre of Ukraine claimed that Ukraine became a nuclear time bomb. Subsequently, in March 2021, the National Ecological Centre asked the European Investment Bank not to support nuclear development in Ukraine. The ecologists claimed that the Ukrainian government plans to keep all reactors running for at least 10 years beyond their expiry date. In 2020, the government created an independent State Nuclear Regulatory Committee of Ukraine to inspect and evaluate the safety conditions of existing power plants. Nonetheless, representatives of the environmental nongovernmental organizations and scientists were skeptical about the inspectorate's independence [47].

Also dangerous to public and environmental safety was radioactive waste amassing in only one existing nuclear waste storage facility. The waste was initially sent to storage facilities in Russia, costing Ukraine nearly 200,000 million dollars yearly. However, in 2018, following Ukraine's entry into the EU Deep and Comprehensive Free Trade Agreement (DCFTA), Russia began returning nuclear waste to Ukraine. By 2021, Ukraine's radioactive waste was the second largest in Europe.

Moreover, scientists and environmentalists opposed President Zelensky's plan to restart the construction of the NPP Khmelnytskyi. The unfinished units of the power plant were planned to be constructed by reinforcing the old concrete structures of construction that were suspended in 1990 [48,49]. Moreover, scientists raised concerns about a potential new catastrophe due to retrofitting Soviet Era reactors to new nuclear fuel produced by American Westinghouse [50]. In 2021, the Industrial Union of Nuclear Energy Industry (Профспілка атомної промисловості та енергетики) issued a letter to the Ministry of Energy expressing concerns about the government decision to increase the capacity of nuclear production of aging reactors. Although the President argued that such modification met safety standards reviewed by experts of the Ukrainian government, the reliability of the assessment was questioned [51].

Subsequently, to silence domestic dissent and stifle opposition to the expansion of nuclear energy, President Zelensky dismissed the opposition's concerns on safety grounds as baseless. "We understand that if professionals are doing the construction, if the state is working on the safety of nuclear power plants, then there is no threat either to the environment or the climate," explained Zelensky [44]. The construction of the units was expected to commence in 2021–22.

4.3 Belorussia's nuclear energy expansion

After a short pause in constructing its first nuclear power plant, Astravyets, Belorussia also revitalized its nuclear energy program in the 2000s. The Russia-Belarus energy dispute fueled the drive for a revival of the construction of Astravyets. In 2007, Russia hiked the payment for its oil sold to Belorussia, and in return, Belorussia demanded transit fees from Russia for oil transit to Europe across the Belorussian territory. The dispute was settled with the decision to restart Astravyets construction using Russian technology and its nuclear fuel.

After 2007, President Lukashenko decided to nuclearize Belorussia to prepare for future energy needs and secure energy independence. The use of Russian nuclear technology, however, proved to have accidental failures. For example, in November 2020, electricity production in the inaugurated pilot industrial operation of Power Unit 1 of Astravyets was halted due to an voltage transformers explosion [52,53]. The completion of the construction was nevertheless planned for 2022.

Being an authoritarian leader, President Lukashenko vividly opposed the engagement of the public or scientists in his decision-making regarding energy politics. Unable to silence the growing opposition, the President wanted to prove that nuclear energy is safe and started settlements in the Chernobyl region on less contaminated land. His government reclaimed uninhabitable territories and opened parts to the exclusion zone for short-term tourist visits. However, these actions did not convince environmentalists or broader public or scientists about the safety and benefits of nuclear energy. The opposition to the country's nuclearization grew.

4.4 Opposition in Belorussia

Scientists and environmentalists were profoundly concerned and firmly opposed the construction of the Astravyets nuclear plant, noticing that other renewable energy sources are safer and cheaper. They also distrusted Russian nuclear technology and demanded the country's denuclearization. Many Belarusians continue to experience the consequences of Chernobyl, including a steady increase of cancer patients, congenital disabilities, limited availability of uncontaminated food products, and residency in contaminated land. By 2020, 10% of the Belorussian population (1 million) lived near the exclusion zone. Concurrently, the independent opposition newspaper "Will of Nation" published experts' opinions that areas near the exclusion zone will never be wholly decontaminated [54].

To oppose the government's plans, scientists and anti-nuclear activists formed an opposition group, the National Front, demanding public involvement in the government's decision-making regarding nuclear energy [55]. Each year, on the anniversary of the Chernobyl accident, the National Front organized protest marches of activists and scientists, traveling between Chernobyl and Minsk carrying banners that say "No to Construction of Nuclear Power Plant." The marches were called the "Chernobyl Route." Moreover, since 1990, on the central square of the capital Minsk, the opposition organized protests and memorial meetings marking the anniversary of the Chernobyl disaster. In 2011, faced with growing criticism, President Lukashenko banned the organization of the city meeting [55]. Nonetheless, the government was unable to silence the growing opposition.

5 Conclusions: Energy democracy as an opportunity for Ukraine and Belorussia

According to the hierarchical power structure, both Ukraine and Belorussia guided their energy policies, with little concern for public health and environment safety, and distribution of benefits to a selected few, e.g., oligarchs and regime supporters, i.e., the principle of Energy Dominance (paternalistic energy) [56]. However, by the 2000s, the top-down energy politics that discounted public involvement in nuclear energy decision-making changed, and an opportunity for anti-nuclear opposition to demand equity and opportunity for public engagement opened. The opposition included scientists and environmentalists who vividly remembered the Soviet race for nuclear energy, which led to the Chernobyl disaster. They worry about nuclear power safety, the enormous costs of nuclear power plant constructions, and accumulated nuclear waste. When the World Nuclear Industry Status Report indicated that renewals like solar and wind energy are much cheaper energy sources, the opposition resisted nuclear and fossil-fuel-dominant energy to advance renewable energy transitions [57,58].

Specifically, the opposition demanded that energy politics accounted for the principle of no harm to the environment, climate change, community control, social equity, contribution to the health and well-being of all people, and public participation in energy decision-making. These demands signaled the introduction of guidance of energy politics according to Energy Democracy. The new guidance addressed not only energy safety and security, but also demanded community participation and citizens' inclusive engagement that overrides the ultimate goal of profit maximization that benefited the selected few. In this sense, in the future, the nuclear energy policy of Ukraine's and Belorussia's governments could foster citizens' demands for a shift from Energy Dominance to Energy Democracy. Energy Democracy became an emergent social movement that could potentially reclaim and democratically restructure regimes' energy approach in both countries, stabilizing and securing energy production while strengthening democracy in its process.

References

[1] World Meteorological Organization, Landmark United in Science Report Informs Climate Action Summit, 22 September 2019. https://public.wmo.int/en/media/press-release/landmark-united-science-report-informs-climate-action-summit. (Accessed 3 March 2021).
[2] US Energy Information Administration, Nuclear Power and the Environment, 15 January 2020. https://www.eia.gov/energyexplained/nuclear/nuclear-power-and-the-environment.php. (Accessed 3 March 2021).
[3] M.V. Ramana, Small modular and advanced nuclear reactors: a reality check, IEEE Access 9 (2021) 42090–42099.
[4] K. Szulecki, Conceptualizing energy democracy, Environ. Polit. 27 (2018) 21–41, https://doi.org/10.1080/0964 4016.2017.1387294.

[5]   J. Stephens, M. Burke, B. Gibian, E. Jordi, R. Watts, Operationalizing energy democracy: challenges and opportunities in Vermont's renewable energy transformation, Front. Commun. (3 October 2018), https://doi.org/10.3389/fcomm.2018.00043.

[6]   M. Burke, J. Stephens, Energy democracy: goals and policy instruments for sociotechnical transitions, Energy Res. Soc. Sci. 33 (2017) 35–48.

[7]   S. Alexievich, Voices From Chernobyl, Picador, New York, 2006.

[8]   D. Bazyka, V. Sushko, A. Chumak, V. Buzunov, V. Talko, L. Yanovych, National Research Center for radiation medicine of the National Academy of Medical Sciences of Ukraine—research activities and scientific advance in 2016, Probl. Radiac. Med. Radiobiol. (December 22) (2017) 15–22.

[9]   R. Gray, Chernobyl covered up by the secretive Soviet Union at the time, the true number of deaths and illnesses caused by the nuclear accident is only now becoming clear, BBC Future (25 July 2019).

[10] P. Celej, Chernobyl's Cathastrophy. 180 thousands liquidators died (Katastrofa w Czarnobylu. Zmarło 180 tys. osób likwidujących szkody), Gazeta Wyborcza (16 April 2015) (in Polish).

[11] Over 1.7 million Ukrainian residents are handicap due to accident in Chronobyls nuclear power station (Ponad 1,7 mln mieszkańców Ukrainy ma obecnie status poszkodowanych w związku z katastrofą w elektrowni jądrowej w Czarnobylu), Dziennik Gazeta Prawna (6 April 2021) (in Polish).

[12] R. Gray, Chernobyl covered up by a secretive Soviet Union at the time, the true number of deaths and illnesses caused by the nuclear accident are only now becoming clear, BBC Future (25 July 2019).

[13] M. Ulstein, T.S. Jensen, L.M. Irgens, R.T. Lie, E. Sivertsen, F.E. Skjeldestad, Pregnancy outcomes in some Norwegian counties before and after the Chernobyl accident, Tidsskr. Nor. Laegeforen. 110 (3) (1990) 359–362. https://pubmed.ncbi.nlm.nih.gov/2309180/.

[14] B. Ferreira, Why hundreds of thousands of women ended their pregnancies after Chernobyl, Vice (3 June 2019).

[15] J. Cwikel, R. Sergienko, G. Gutvirtz, R. Abramovitz, D. Slusky, M. Quastel, E. Sheiner, Reproductive effects of exposure to low-dose ionizing radiation: a long-term follow-up of immigrant women exposed to the Chernobyl accident, J. Clin. Med. 6 (2020) 1786.

[16] K. Brown, Manual for Survival. A Chernobyl Guide to the Future, Allen Lane, New York, 2019.

[17] E. Sawa-Czajka, Development of the atomic power industry in Poland: public debate on environmental conservation versus economic needs, Res. Polit. Sociol. 21 (2011) 253–271.

[18] T. Shanker, Chernobyl officials are sentenced to labor camp, Chi. Trib. (30 July 1987).

[19] Green Facts, Facts on Health and Environment, Chernobyl Nuclear Accident, 16 February 2021, Available from: https://www.greenfacts.org/en/chernobyl/l-2/2-health-effects-chernobyl.htm.

[20] Unit of Radiation and Cancer International Agency for Research on Cancer, Reconstruction of doses for Chernobyl liquidators, Final Performance Report, Lyon, France, 2003.

[21] V. Fedosenko, Chernobyl will be inhabitable for at least 3,000 years, say nuclear experts, Routers News (24 April 2016).

[22] L.N. Maskalchuk, Soil contamination in Belarus, 25 years later, Nucl. Eng. Int. Mag. (2 February 2012).

[23] C. Borys, Chernobyl's new sarcophagus took two decades to make. Bigger than Wembley Stadium and taller than the Statue of Liberty, it will seal in the entire disaster side for 100 years, BBC Future (3 January 2017).

[24] S. Thompson, Trees and other kinds of vegetation proven to be remarkably resilient to the intense radiation around the nuclear disaster zone, BBC Future (1 July 2019).

[25] B. Thompson, Is wildlife thriving in Chernobyl's radioactive landscape? The Christian Monitor Daily (21 April 2016).

[26] R. Alimov, Chernobyl still burns, 23 April 2020. Green Peace Blog. Available from: https://www.greenpeace.org/international/story/30198/chernobyl-still-burns-forest-fires-ukraine-nuclear-radiation/.

[27] A. Prokip, Anniversary of Chernobyl finds European nuclear power at a crossroad, Blog Post, Kennan Institute, 24 April 2020.

[28] M. Schneider, A. Froggatt, The world nuclear industry status report 2020, France. Tech. Rep., Mycle Schneider Consulting, Paris, September 2020. Available from: https://www.worldnuclearreport.org.

[29] World Nuclear News, Construction work resumes on Khmelnitsky units, World Nuclear News (30 November 2020).

[30] S. Furfori, Europe's Nuclear Energy Debate, EURACTIV.sk, 29 January 2020.

[31] P. Szalai, Foratom chief. balance of power is shifting against nuclear in the EU, Int. Issues Slovak Foreign Policy Aff. XXVII (3–4) (2018) 46–50.

[32] Kresy, Litwini przeciwko elektrowni atamowej, Kresy (12 October 2012) (in Polish).

[33] J. Żylińska, Kontrakt na atom za niecałe dwa lata, Dziennik Gazeta Prawna (6 April 2021) (in Polish).
[34] Energetyka, Report, March 2011, Available from: http://energetyka.wnp,pl. (in Polish).
[35] M. Markiewicz, Not on my courtyard, Politics 13 (2011) 20–22 (in Polish).
[36] L. Turski, We are correcting mistaken reports, Gazeta Wyborcza (6 April 2011) (in Polish).
[37] D. Zaks, Ukraine clings to nuclear power despite Chernobyl trauma, Phys. Org. (30 April 2017).
[38] A. Prokip, Tottering gas market reforms in Ukraine: paternalism and political games, Blog post, Wilson Center, 11 February 2021.
[39] World Nuclear News, Ministers gear up for COP26, World Nuclear News (21 April 2021).
[40] Ukraine is getting ready to build atomic power stations units quickly and without Russia, Economicheskaya Pravda (22 October 2014) (in Ukrainian).
[41] World Nuclear News, Khmelnitsky expansion part of European 'renaissance' says Energoatom chief, Word Nuclear News (8 January 2021).
[42] World Nuclear News, Ukraine must expand nuclear energy, says president, interview with president Zelensky, World Nuclear News (5 October 2020).
[43] R.M. Grossi, Grossi: nuclear already plays key role in decarbonization, World Nuclear News (31 March 2021).
[44] T. Kasperski, Nuclear power in Ukraine: crisis or path to energy independence? Bull. At. Sci. (1 July 2015), https://doi.org/10.1177/0096340215590793.
[45] World Nuclear News, Ukraine prepares to reduce output during pandemic, World Nuclear News (30 April 2020) (in Polish).
[46] National Ecological Center in Ukraine, Don't turn Ukraine into a nuclear energy source for the European Union!, 29 March, 2011. Blog post Bankwatch.org.
[47] Eco.org, The nightmare of nuclear catastrophe circulate in Ukraine, Eco.Org News (21 April 2016) (in Polish).
[48] M. Markiewicz, Engineers from nuclear building site, Politics 34 (2009) 81–83 (in Polish).
[49] Khmelnitsky nuclear power plant, Power Technology Newsletter (10 April 2021). Available from: https://www.power-technology.com/projects/khmelnytskyi-nuclear-power-plant/.
[50] A. Kublik, Ukraine is purchasing American nuclear fuel, Gazeta Wyborcza (13 April 2014) (in Polish).
[51] P. Kost, Reforms, chaos, retreat-Ukraine's energy, Energetyka 24 (2021) 1 (in Polish).
[52] I. Zylinska, M. Cedro, The fear of Belorussian Chernobyl, Dziennik Gazeta Prawna (15 March 2021) (in Polish).
[53] Radio Free Europe/RL's Belarus Service, Belarus nuclear plant taken offline after 'Protection System Activated', Radio Free Europe News (16 January 2021).
[54] Belorussin opposition demands participation in debates about nuclear power stations (Bialoruska opozycja chce debaty nt. Budowy elektrowni atomowej), Bankier Portal Finansowy (24 April 2008).
[55] S. Dolzhenko, Regime and opposition celebrate Chernobyl's anniversary, Wiadomości Onet (26 April 2011) (in Polish).
[56] How did the Ukrainian oligarchy keep going after Euromaidan? VoxUkraine (22 February 2021) (in Polish).
[57] O. Aliieva, Doubtful prospects for nuclear energy in Ukraine, Henrich Böll Stiftung, 22 October 2020. Blog post.
[58] O. Zaika, Ukraine's nuclear impasses, Henrich Böll Stiftung, 12 October 2020. Blog post.

Talking points: Narrative strategies to promote nuclear power in Turkey

Avino Niphi[a] and M.V. Ramana[b]

[a]Indian Institute of Technology, Madras, Chennai, Tamil Nadu, India [b]University of British Columbia, Vancouver, BC, Canada

1 Introduction

Located on the southern coast of Turkey, Akkuyu is the site of a massive construction project. Starting in April 2018, the Russian state-owned company Rosatom has been building a series of nuclear reactors. The site has a long history. In the second of six attempts to introduce nuclear power to Turkey, Akkuyu was selected in 1976 to host a nuclear power plant [1,2].

Turkey's nuclear ambitions go back to at least 1955, when the country became the first to conclude an agreement for cooperation with the United States under the Atoms for Peace program [3]. Earlier attempts to start nuclear power generation in Turkey were unsuccessful, until the Justice and Development Party (AKP) consolidated political power. However, a small research reactor called TR-1 was built by a US company called American Machine and Foundry between 1959 and 1962 [4]. This reactor has been shut down but Turkey continues to operate two research reactors that started functioning in 1979 and 1982.

Despite this long-standing interest, there are good reasons for Turkey to not proceed with nuclear construction. The Akkuyu site is located in a seismically active region [5] and could be vulnerable to Tsunamis [6], raising the possibility of a severe nuclear accident. For decades, there has been strong opposition to nuclear reactor construction at Akkuyu and elsewhere [4,7]. Multiple surveys have shown very high percentages of citizens being opposed to nuclear power [8–12]. In contrast, there are high levels of support for solar and wind energy (somewhat analogous to the case in Japan as discussed by Setsuko Matsuzawa [13]).

There have also been problems with the construction of the plant [14], which has led to concerns in neighboring countries [15]. In 2015, the Chamber of Turkish Engineers and Architects (TMMOB) sued the Environment Ministry when the latter approved the environmental report on Akkuyu [16], and it has shown that electricity from the plant will be much more expensive than alternatives [17].

Energy Democracies for Sustainable Futures
https://doi.org/10.1016/B978-0-12-822796-1.00027-9

Given this opposition and the multiple problems associated with nuclear power, how does the current government publicly justify continuing with its nuclear program? We outline the discursive elements at play in the Turkish government's promotion of a pronuclear agenda. The narrative strategies utilized involve overestimating Turkey's energy demand and positing nuclear power as the main solution; downplaying the risks associated with nuclear energy; discrediting renewables and belittling their capacity; highlighting nuclear power as a marker of Turkey's strength and prestige; and promising jobs and opportunities. While many of these are demonstrably wrong, such narratives have implications for material investments in the infrastructure and energy policy which pose environmental and financial risks.

2 Discursive elements

There is a long history of discursive strategies being used to justify the inclusion and prioritization of nuclear energy in Turkey's energy mix [18,19]. The use of such strategies to promote nuclear power is by no means restricted to Turkey [20–23], and they are effective in gaining support from elites in different countries for nuclear power. We outline below some of the key themes utilized in Turkey.

2.1 Scarcity and growth

Turkey's efforts to build nuclear power plants, going back to the 1960s, have involved a staple argument: Turkey's growing energy needs cannot be met without nuclear power. A study from that period carried out by the Nuclear Energy Institute of the Technical University of Istanbul, on behalf of the Electric Power Resources and Survey Administration, concluded that "after 1982 national energy resources would be insufficient for the supply of rapidly increasing electricity demand, and the first nuclear power plant should start operating in 1977" [24]. By the 1970s, Turkey featured in market surveys for nuclear power plants carried out by the International Atomic Energy Agency, and one such survey projected Turkey having between 1200 and 3200 MW of nuclear capacity by 1989 [25].

These projections for nuclear power did not materialize, nor did Turkey run out of energy. Nevertheless, the same trope continues to figure prominently in arguments for nuclear power. In 2000, for example, the Turkish Electricity Generation-Transmission Corporation projected that "Total electrical energy consumption...is estimated to reach 200 billion kWh in 2005, 290 billion kWh in 2010, and 547 billion kWh in 2020...Nuclear energy is therefore necessary for the diversification of Turkey's electric power resources" [26].

Writing in 2020, it is evident that nuclear energy was unnecessary and the prediction for energy demand utilized to argue for nuclear power was grossly inaccurate: Turkey's electricity consumption in 2019 was 272 billion kWh [27], half of what was expected in 2000. In other words, the projection used to justify the acquisition of nuclear power was wrong by 100%. That these inflated projections of energy demand and nuclear power capacities don't ever seem to materialize seems to be unimportant to its advocates, who continue to invoke the notion of scarcity to justify the acquisition of nuclear power.

The pattern of arguing for nuclear power on the basis of exaggerated projections of future energy demand continues. In April 2018, while speaking at the groundbreaking ceremony of the first power unit of the Akkuyu Nuclear Power Plant, President Recep Tayyip Erdoğan stated:

> The Turkish economy has been growing by 5.8 percent on average from 2003 to 2017. In 2017, our economy grew by 7.4 percent. By 2023, when we celebrate the 100th anniversary of our country as a republic, we have set ourselves the task of becoming one of the top 10 largest economies in the world. This means we will need more energy – oil and natural gas, as well as renewable energy sources. Over the past 15 years, we have carried out previously unthinkable major projects…This nuclear power plant is of great importance for our future in this regard. *Official Website of the President of Russia [28]*

Note how energy is posited as a significant anchor for this vision of rapid growth and development, to the exclusion of many other elements.

The argument offered by nuclear advocates is simple—nuclear power is an absolute necessity if Turkey is not to run out of energy—even though, as we have shown, these demand projections typically do not hold in reality.

2.2 Modernity and recognition

Officials often portray nuclear power as a marker of modernity or development, allowing others to see Turkey as a great power. Conversely, lack of nuclear power is seen as indicating an inferior status. Thus, in 2015, at one of the launch functions for the Akkuyu plant, Turkish Energy Minister Taner Yıldız proclaimed, "Development cannot take place in a country without nuclear energy" [29]. And in 2007, a AKP parliament member Mustafa Ozturk stated:

> Nuclear power plants reflect the strength, the level of development, and the prestige of a country. We have been late for 40 years in shifting to nuclear technology, thus, we have to be successful in bringing this high-tech to our country. *Balkan–Sahin [18]*

Nuclear power has also played into President Erdoğan's attempts at positioning himself as the leader of a great state. While talking about Turkey, he consistently draws upon Turkey's Ottoman legacy and uses phrases such as "thousands of years old state experience," "magnificent civilizational history," and "a-thousand-year-long dominance in its region" [30]. Nuclear power has been roped into this portrayal. In recent years, this has in large part been through including nuclear plants into Turkey's Vision 2023. Vision 2023 is a set of goals laid out for the centennial of the formation of the Turkish Republic. The leading goal is to make Turkey into one of the top 10 economies in the world.

The linkage between nuclear power and the occasion in Turkey's history was made clear by President Erdoğan while speaking at the groundbreaking ceremony of the first power unit of the Akkuyu Nuclear Power Plant: "In 2023, we will commission the first reactor at this plant, and Turkey will thus join the countries that use atomic energy. In 2023, we will mark the 100th anniversary of our republic with the successful completion of this project" [28]. Note also the invocation of "countries that use atomic energy" as a special category and the implicit celebration of Turkey joining that group.

As with the narrative around scarcity and growth, proponents of Vision 2023 posit a requirement for a large increase in the amount of energy used by Turkey (e.g., Ref. [31]) and position nuclear power as a major element in this growth. Because nuclear power projects are, by their nature, designed to produce large amounts of electricity, they fit well within this picture.

This "mega" nature of nuclear power also ties into the Vision 2023 narrative. The Investment Office of the Presidency of the Republic of Turkey states that "mega projects" are what are going to "propel the country towards 2023 targets" and the pride of place on that list of mega projects goes to the Akkuyu nuclear plant [32]. Such "mega projects" have "provided an effective avenue for building patronage networks… [and] served to fulfill Erdogan's promises of a new and powerful Turkey reminiscent of the golden days of the Ottoman imperial era" [33].

Nuclear projects have been financed using different arrangements. The Akkuyu Nuclear Power Plant uses a build-own-operate (BOO) model where Russia's state-owned Rosatom finances the construction but has an agreement to sell 50% of the power produced at a guaranteed price, with the rest sold on the electricity market. The proposed Sinop Nuclear Power Plant in northern Turkey, on the other hand, was to be funded through a mixture of capital from France and Japan and Turkey's state-run power producer Elektrik Üretim AŞ. Such financing mechanisms were partly due to the country's economy being restructured along neoliberal lines, allowing private investments to flow into public undertakings, including energy related mega projects, and extracting profits.

During the early years of Akkuyu's construction, there was the hope that the plant would be ready well ahead of time as would be the ones at the Sinop and at the İğneada sites. Thus, in 2015, the government projected, as part of Vision 2023, "Turkey plans to have three operational nuclear power plants by 2023" [32].

As of 2020, it is unlikely that even the first unit of Akkuyu will be generating electricity by 2023 [34]. An agreement between the governments of Turkey and Japan to construct a nuclear plant at Sinop has fallen through, largely for financial reasons, as have efforts to find a builder for the İğneada site. However, in what can be best described as a farce, the environmental impact assessment process for the Sinop site is moving forward *without* a reference reactor [35].

2.3 Environmental desirability

Official discourse about nuclear energy in Turkey also justifies it on environmental grounds, especially by invoking the climate crisis. This idea is illustrated by President Erdoğan's remarks at the groundbreaking ceremony of the Akkuyu nuclear plant: "Nuclear power plants do not emit carbon dioxide; they produce clean and environmentally safe energy. This nuclear power plant will…play a big role in combating climate change" [28]. Opponents of nuclear power in Turkey and elsewhere reject this framing [19,36]. Nevertheless, framing nuclear energy as an alternative to fossil fuels has been an important discursive strategy [26,37].

Alongside promoting nuclear energy, officials often belittle renewable energy sources. For instance, Party officials often cite how Turkey's great economic goals until 2023 cannot be met with renewables alone [18,38]. Energy Minister Taner Yildiz argued: "They ask why there is no investment in solar energy or wind energy? This is because when the wind does not blow or the sun does not shine you cannot produce energy. This is what we call energy diversification" [39].

These statements frame intermittency in renewable fuel supply as undesirable and pass off renewables as not adequate to Turkey's increased energy demands. Despite a fairly rapid increase in the capacity of wind and solar energy sources in Turkey [40], government policies have often favored and publicly legitimized large, established, and centralized technologies [41]. Once again, this is not unique to Turkey, and there is a long history of renewables being dismissed as unreliable [42,43].

2.4 Job creation

Advocates for nuclear power also emphasize the project's potential for job creation. For example, in 2017, Energy and Natural Resources Ministry Undersecretary Fatih Dönmez told the Turkish pro-government Sabah newspaper that "About 10,000 people will be employed while the Akkuyu NPP's construction is most intensive, and about 3500 jobs will be provided during operation. The majority will consist of Turkish citizens" [44]. These jobs are framed as desirable, by highlighting the technical training involved. President Erdoğan articulated this during the groundbreaking ceremony for the Akkuyu project: "This experience will undoubtedly lead to a greater number of employees with innovative skill sets. Importantly, hundreds of our students are currently studying at Russian universities. They will return home upon completion of their studies and start working here" [28]. In June 2018, during a social media broadcast, President Erdoğan announced plans to send students to France, Japan, and China as well [45].

For their part, Rosatom officials paint a picture of Turkey becoming involved in the global nuclear export business by promising "Thousands of professionals will be involved in the Akkuyu Nuclear Power Plant project in Turkey and Turkish companies will gain relevant experience to participate in tenders for the construction of nuclear power plants in different countries" [46].

2.5 Downplaying risk

Alongside these positive framings of nuclear power has been the official downplaying of risks attached to nuclear power. The risk that is of greatest concern to groups in Turkey is that of severe accidents [36], such as the ones in Chernobyl in 1986 or Fukushima in 2011. The likelihood of such accidents is increased by the construction of the Akkuyu nuclear power plant in close proximity to an area prone to earthquakes [26,47]. The Akkuyu project has also drawn criticism from its neighbors, particularly Cyprus and Greece, and from the European Parliament [48]. The transboundary risks of nuclear power have been widely acknowledged since the Chernobyl accident.

Turkey's government has downplayed these risks, often by using inapplicable examples. For instance, dismissing public concerns following Fukushima, President Erdoğan said, "In that case, let's not bring gas canisters to our homes, let's not install natural gas, let's not stream crude oil through our country" [49]. Likewise, in 2017, the Minister for Energy and Natural Resources tried to trivialize the risks to health by claiming that the "amount of radiation one can absorb from an x-ray machine exceeds by far the amount of radiation that one will receive by living nearby a plant for one year" [50].

Such tendencies are not unique to Turkey, and there is a long history of dismissing concerns about the safety of nuclear plants or the risks of radiation [51–53].

3 Hidden factors

A number of other factors might also play into the decision to build nuclear power in Turkey.

3.1 Domestic capital

Although nuclear power production has been a long-standing goal of successive Turkish governments, what is different between these early attempts and today is the virtual domination of the Justice and Development Party (AKP) in domestic politics and the advent of neoliberalism. The AKP "committed itself to the liberalization of the energy sector and diversification of energy resources, including the use of nuclear power" [18], and this strategy has led to a massive accumulation of capital within the energy sector. Alongside finance and manufacturing, energy has consistently emerged as the highest foreign direct investment attracting sector in Turkey [54].

The high cost of the Akkuyu and other nuclear plants (preliminary estimates of more than $20 billion) means that if even a fraction of this money were to flow to Turkish companies, it could result in great profits. Some companies have signed up contracts with Rosatom to provide services [28,55]. It is not a wonder therefore that big capital groups such as the Turkish Industry and Business Association (TUSIAD) and the Independent Industrialists and Businessmen's Association (MUSIAD) have strongly promoted nuclear energy [18,56]. There is also the possibility of corruption [57].

3.2 A possible weapons program?

Outside of Turkey, a common narrative has been to link the country's interest in nuclear power with a desire to acquire nuclear weapons (see, e.g., Refs. [58–60]). This view is supported, at least in part, by various official statements. In September 2019, President Erdoğan stated: "Some countries have missiles with nuclear warheads, not one or two. But (they tell us) we can't have them. This, I cannot accept… There is no developed nation in the world that doesn't have them" [61]. Likewise, at a speech in the same year at the United Nations General Assembly, he said "Nuclear [military] power should be forbidden for all or should be permissible for all" [62].

Such oblique statements notwithstanding, there is no evidence that Turkey is involved in acquiring nuclear weapons. However, the linkage between nuclear weapons and nuclear energy is a bonus, a positive argument, for groups supporting nuclear energy. There are many instances of this linkage being used to lobby for state support of nuclear energy programs in countries like the United States [63,64], and the United Kingdom [65]. While unlikely to be publicly articulated, the potential for the nuclear energy infrastructure to be useful to acquiring nuclear weapons might well be an argument used within domestic policy circles.

4 Conclusions

The philosopher Randall Marlin defined propaganda as "the organized attempt through communication to affect belief or action or inculcate attitudes in a large audience in ways that circumvent or suppress an individual's adequately informed, rational, reflective judgment" [66].

The narrative strategies underlying Turkey's nuclear program illustrate such an attempt to affect public attitudes about nuclear technology.

An informed observer, on the other hand, would see that these narrative elements, for the most part, do not really have any basis in historical or contemporary fact. Projections for energy demand growth, and especially the rate of nuclear power growth, are unlikely to materialize, just as in the past. Many "modern countries" have no nuclear power plants or are phasing them out. Renewable energy sources such as wind and solar are much cheaper than nuclear power [34,67], and an important 2018 study from Sabancı University showed that a much larger capacity of renewables can be incorporated into the Turkish electricity grid with modest investments and planning [68]. Academic studies have also shown that renewables provide a much greater number of jobs per unit of installed capacity than nuclear power [69]. Finally, it is not possible to credibly and definitively assure the safety of nuclear plants [70,71]. All told, a nuclear-powered future will impose a high financial cost and is vulnerable to multiple safety risks and potential impacts on public health and the environment.

Democracy too has been a casualty in the quest for nuclear energy. Construction of the Akkuyu nuclear plant has gone on despite widespread opposition to nuclear power [4,7–11]. At Akkuyu, there was no public consultation during the decision-making process or policy implementation phase for the power plant [36]. Likewise, nongovernmental organizations from Sinop were prevented from attending the Review and Evaluation Commission meeting held in Ankara for the nuclear plants to be built in Sinop [35]. What is happening in the nuclear sector is illustrative of what is happening in Turkey in general, but nuclear power plants have the "potential to adversely affect" large areas, perhaps even outside Turkey, "in the event of a disaster" [72].

Turkey's plans for nuclear power are a manifestation of authoritarian energy governance, as described by Alevgül Şorman and Ethemcan Turhan [73], and the antithesis of the "energy as commons" approach recommended by Joohee Lee, John Byrne, and Jeongseok Seo [74]. Because of the overwhelming political power of the AKP government, there is little doubt that the nuclear program will proceed despite democratic opposition, albeit nowhere near as fast as the authorities might want. In the meanwhile, we are left with the question that political theorist Langdon Winner raised decades ago: "as society adapts to the more dangerous and apparently indelible features of nuclear power, what will be the long-range toll in human freedom?" [75].

References

[1] Y. Bektur, U. Bezdegumeli, Nuclear power plant attempts in Turkey and the first licensed site, in: Proceedings of the Third Eurasian Conference "Nuclear Science and Its Application", Tashkent, 2004. https://inis.iaea.org/collection/NCLCollectionStore/_Public/38/111/38111303.pdf. (Accessed 29 September 2020).

[2] J. Jewell, S.A. Ates, Introducing nuclear power in Turkey: a historic state strategy and future prospects, Energy Res. Soc. Sci. 10 (2015) 273–282, https://doi.org/10.1016/j.erss.2015.07.011.

[3] D. Fischer, History of the International Atomic Energy Agency: The First Forty Years, International Atomic Energy Agency, Vienna, 1997.

[4] C.İ. Aydın, Nuclear energy debate in Turkey: stakeholders, policy alternatives, and governance issues, Energy Policy 136 (2020), https://doi.org/10.1016/j.enpol.2019.111041, 111041.

[5] R.F. Kartal, G. Beyhan, A. Keskinsezer, F.T. Kadirioğlu, Seismic hazard analysis of Mersin Province, Turkey using probabilistic and statistical methods, Arab. J. Geosci. 7 (2014) 4443–4459, https://doi.org/10.1007/s12517-013-1104-1.

[6] B. Aydın, N. Sharghivand, Ö. Bayazıtoğlu, Potential tsunami hazard along the southern Turkish coast, Coast. Eng. 158 (2020), https://doi.org/10.1016/j.coastaleng.2020.103696, 103696.

[7]   M. Ersoy, E. İşeri, Framing environmental debates over nuclear energy in Turkey's polarized media system, Turk. Stud. (2020) 1–27, https://doi.org/10.1080/14683849.2020.1746908.

[8]   G. Benmayor, Turkish public reluctant about going nuclear, says opinion poll, Hurriyet Daily News (2011). https://www.globalrights.info/2011/04/turkish-public-reluctant-about-going-nuclear-says-opinion-poll/. (Accessed 29 September 2020).

[9]   P. Ertör-Akyazı, F. Adaman, B. Özkaynak, Ü. Zenginobuz, Citizens' preferences on nuclear and renewable energy sources: evidence from Turkey, Energy Policy 47 (2012) 309–320, https://doi.org/10.1016/j.enpol.2012.04.072.

[10] IPSOS, Global Citizen Reaction to the Fukushima Nuclear Plant Disaster, IPSOS, 2011. http://www.ipsos-na.com.

[11] Konda, Climate Change Perception and Energy Preferences in Turkey, İklim Haber, 2018. https://www.iklimhaber.org/climatesurvey2018/. (Accessed 29 September 2020).

[12] Türkiye'de halkın üçte ikisi nükleer santrallere karşı, Euronews (2019). https://tr.euronews.com/2019/03/18/turkiye-halkin-ucte-ikisi-nukleer-santralleri-karsi-ak-partilerde-destek-orani-yuzde-50. (Accessed 29 September 2020).

[13] S. Matsuzawa, Energy democracy movements in Japan, in: J. Keahey, M. Pasqualetti, M. Nadesan (Eds.), Democratizing Energy: Imaginaries, Transitions, and Risks, Elsevier, Amsterdam, 2021.

[14] Cracks discovered during construction of Turkey's first nuclear plant, Ahval (2019). https://ahvalnews.com/akkuyu/cracks-discovered-during-construction-turkeys-first-nuclear-plant. (Accessed 29 September 2020).

[15] Roubanis, Fears emerge about the safety of Turkey's Russian-built nuclear power plant, New Europe (2019). https://www.neweurope.eu/article/fears-emerge-about-the-safety-of-turkeys-russian-built-nuclear-power-plant/. (Accessed 5 July 2020).

[16] A. Alp, Chamber to sue state over abrupt green light to Turkey's first nuclear plant, Turkey News, Hürriyet Daily News (2015). https://www.hurriyetdailynews.com/chamber-to-sue-state-over-abrupt-green-light-to-turkeys-first-nuclear-plant-76419. (Accessed 29 September 2020).

[17] BIA News Desk, Report Says Akkuyu Nuclear Plant to Cause Electricity Prices and Foreign Dependency to Increase, Bianet, 2019. https://www.bianet.org/english/environment/210091-report-says-akkuyu-nuclear-plant-to-cause-electricity-prices-and-foreign-dependency-to-increase. (Accessed 29 September 2020).

[18] S. Balkan-Sahin, Nuclear energy as a hegemonic discourse in Turkey, J. Balkan Near East. Stud. 21 (2019) 443–461, https://doi.org/10.1080/19448953.2018.1506282.

[19] E. İşeri, D. Günay, A. Almaz, Contending narratives on the sustainability of nuclear energy in Turkey, Environ. Plann. C Polit. Space (2017), https://doi.org/10.1177/2399654417704199.

[20] M. Mathai, Nuclear Power, Economic Development Discourse and the Environment: The Case of India, Routledge, Oxon, Abingdon, New York, NY, 2013. https://www.taylorfrancis.com/books/9780203100141. (Accessed 6 July 2020).

[21] M.V. Ramana, Second life or half-life? The contested future of nuclear power and its potential role in a sustainable energy transition, in: F. Kern (Ed.), The Palgrave Handbook of the International Political Economy of Energy, Part IV: Energy Transitions, Palgrave Macmillan, London, 2016.

[22] M.V. Ramana, P. Agyapong, Thinking big? Ghana, small reactors, and nuclear power, Energy Res. Soc. Sci. 21 (2016) 101–113, https://doi.org/10.1016/j.erss.2016.07.001.

[23] B.K. Sovacool, M.V. Ramana, Back to the future: small modular reactors, nuclear fantasies, and symbolic convergence, Sci. Technol. Hum. Values 40 (2015) 96–125, https://doi.org/10.1177/0162243914542350.

[24] IAEA, Market Survey for Nuclear Power in Developing Countries: Turkey, International Atomic Energy Agency, Vienna, 1973. https://inis.iaea.org/collection/NCLCollectionStore/_Public/05/149/5149007.pdf.

[25] O.B. Falls, A survey of the market for nuclear power in developing countries, Energy Policy 1 (1973) 225–242, https://doi.org/10.1016/0301-4215(73)90006-2.

[26] TEAS, Basic Facts Concerning the Proposed Nuclear Power Plant at Akkuyu in Turkey, Turkish Electricity Generation-Transmission Corporation, Ankara, 2000. https://inis.iaea.org/collection/NCLCollectionStore/_Public/31/065/31065626.pdf. (Accessed 30 September 2020).

[27] IEA, Electricity Consumption, Turkey 1990-2019, Data & Statistics, 2020. https://www.iea.org/data-and-statistics?country=TURKEY&fuel=Energy%20consumption&indicator=Electricity%20consumption. (Accessed 30 September 2020).

[28] Official Website of the President of Russia, Akkuyu Nuclear Power Plant Ground-Breaking Ceremony, President of Russia, 2018. http://en.kremlin.ru/events/president/news/57190. (Accessed 19 September 2020).

[29] DS, Turkey launches construction of first nuclear power plant, Akkuyu in Mersin, Daily Sabah (2015). http://www.dailysabah.com/energy/2015/04/14/turkey-launches-construction-of-first-nuclear-power-plant-ak-kuyu-in-mersin. (Accessed 23 September 2015).

[30] R.T. Erdoğan, No Stop, No Rest Until We Achieve Turkey's Goals for 2023, Directorate of Communications, 2019. http://wt.iletisim.gov.tr/english/haberler/detay/no-stop-no-rest-until-we-achieve-turkeys-goals-for-2023/. (Accessed 14 September 2020).

[31] O. Acar, Turkey's 2023 Vision: An Evaluation from the Energy Perspective, 2013. https://www.tepav.org.tr/upload/files/haber/1374066601-0.Ozan_Acar_Turkey_s_2023_Vision_An_Evaluation_from_the_Energy_Perspective.pdf. (Accessed 30 September 2020).

[32] Turkey's 'Mega Projects' Propel the Country Towards 2023 Targets, Investment Office of the Presidency of the Republic of Turkey, 2015. https://www.invest.gov.tr/en/news/news-from-turkey/pages/180815-turkey-me-ga-projects-2023-vision.aspx. (Accessed 30 September 2020).

[33] O. Bayulgen, E. Arbatli, S. Canbolat, Elite survival strategies and authoritarian reversal in Turkey, Polity 50 (2018) 333–365, https://doi.org/10.1086/698203.

[34] M. Schneider, A. Froggatt, The World Nuclear Industry Status Report 2020, Mycle Schneider Consulting, Paris, 2020. https://www.worldnuclearreport.org/.

[35] P. Demircan, For Sinop NPP, Turkey seeks EIA without a company for a reference reactor, Yeşil Gazete - eko-lojik, politik, katılımcı, şenlikli (2020). https://yesilgazete.org/blog/2020/07/03/for-sinop-npp-turkey-seeks-eia-without-a-company-for-a-reference-reactor/. (Accessed 1 October 2020).

[36] P. Temocin, Framing opposition to nuclear power: the case of Akkuyu in Southeast Turkey, Asian J. Peacebuild. 6 (2018) 353–377, https://doi.org/10.18588/201811.00a047.

[37] A. Bolme, A. Tanrikut, Turkey's recent decision regarding the Akkuyu NNP, in: Improving Economics and Safety of Water Cooled Reactors: Proven Means and New Approaches, International Atomic Energy Agency, Vienna, Austria, 2002, pp. 49–52. https://www-pub.iaea.org/MTCD/publications/PDF/te_1290_prn.pdf. (Accessed 29 September 2020).

[38] S. Güsten, Forging ahead on nuclear energy in Turkey, The New York Times (2011). https://www.nytimes.com/2011/03/24/world/middleeast/24iht-m24-turk-nuclear.html. (Accessed 12 September 2020).

[39] Hürriyet Daily News Turkey plans to operate 3rd nuclear power plant - Latest News, Hürriyet Daily News. (n.d.). https://www.hurriyetdailynews.com/turkey-plans-to-operate-3rd-nuclear-power-plant-47726 (Accessed 27 August 2020).

[40] IRENA, Renewable Capacity Statistics 2020, International Renewable Energy Agency, Abu Dhabi, 2020.

[41] J. Richert, Turkey: great potential, missing will, in: S. Roehrkasten, S. Thielges, R. Quitzow (Eds.), Sustainable Energy in the G20, Institute for Advanced Sustainability Studies, Potsdam, 2016, pp. 97–102. https://publications.iass-potsdam.de/rest/items/item_1906900_8/component/file_1914905/content. (Accessed 30 September 2020).

[42] B.K. Sovacool, The cultural barriers to renewable energy and energy efficiency in the United States, Technol. Soc. 31 (2009) 365–373, https://doi.org/10.1016/j.techsoc.2009.10.009.

[43] B.K. Sovacool, The intermittency of wind, solar, and renewable electricity generators: technical barrier or rhetorical excuse? Util. Policy 17 (2009) 288–296.

[44] Anadolu, Akkuyu nuclear power plant to boost employment, Daily Sabah (2017). https://www.dailysabah.com/energy/2017/12/12/akkuyu-nuclear-power-plant-to-boost-employment. (Accessed 25 September 2020).

[45] S. Uslu, Turkey to build 3rd nuclear plant with China: Erdogan, Anadolu Agency (2018). https://www.aa.com.tr/en/energy/nuclear/turkey-to-build-3rd-nuclear-plant-with-china-erdogan/20544. (Accessed 27 August 2020).

[46] Anadolu, No Need to Fear Turkey's Akkuyu Nuke Plant: Plant's CEO, Institute of Energy of South East Europe, 2015. https://www.iene.eu/no-need-to-fear-turkeys-akkuyu-nuke-plant-plants-ceo-p1742.html. (Accessed 25 September 2020).

[47] J. Harte, Building of Turkey's first nuclear plant, sited on a fault line, facing fresh questions, Reuters (2011). https://www.reuters.com/article/idUS122778134920110325. (Accessed 25 September 2020).

[48] Committee on Foreign Affairs, 2018 Commission Report on Turkey, European Parliament, Brussels, 2019. https://www.europarl.europa.eu/doceo/document/A-8-2019-0091_EN.pdf. (Accessed 25 September 2020).

[49] Associated Press, EU to apply stress tests on its nuclear plants, Deseret News (2011). https://www.deseret.com/2011/3/15/20179340/eu-to-apply-stress-tests-on-its-nuclear-plants#european-commissioner-for-en-ergy-guenther-oettinger-addresses-the-media-after-a-hastily-convened-meeting-of-energy-ministers-nucle-ar-regulators-and-industry-officials-in-brussels-tuesday-march-15-2011-the-european-union-on-tuesday-con-siders-stress-tests-t. (Accessed 25 September 2020).

[50] Turkey to expand capacity to meet energy needs with 3 nuclear power plants in action, Daily Sabah (2017). https://www.dailysabah.com/energy/2017/08/11/turkey-to-expand-capacity-to-meet-energy-needs-with-3-nuclear-power-plants-in-action-1502395900. (Accessed 25 September 2020).

[51] J. Downer, Disowning Fukushima: managing the credibility of nuclear reliability assessment in the wake of disaster, Regul. Gov. 8 (2014) 287–309, https://doi.org/10.1111/rego.12029.

[52] M.V. Ramana, Nuclear power and the public, Bull. At. Sci. 67 (2011) 43–51, https://doi.org/10.1177/0096340211413358.

[53] M.V. Ramana, A. Kumar, 'One in infinity': failing to learn from accidents and implications for nuclear safety in India, J. Risk Res. 17 (2014) 23–42, https://doi.org/10.1080/13669877.2013.822920.

[54] S. Erensü, Powering neoliberalization: energy and politics in the making of a new Turkey, Energy Res. Soc. Sci. 41 (2018) 148–157, https://doi.org/10.1016/j.erss.2018.04.037.

[55] ROSATOM Communications Dept, An EPC Contract Between AKKUYU NUCLEAR JSC and a JV Between the Russian CONCERN TITAN-2 JSC and the Turkish IC İçtaş İnşaat Sanayi ve Ticaret A.Ş. was Signed in the Framework of the Meeting, ROSATOM, 2019. https://rosatom-mena.com/press-centre/news/an-epc-contract-between-akkuyu-nuclear-jsc-and-a-jv-between-the-russian-concern-titan-2-jsc-and-the-/. (Accessed 30 September 2020).

[56] G. Bacik, S. Salur, Turkey's nuclear agenda: domestic and regional implications, Uluslararası İlişkiler (International Relations) 6 (2010) 99–116.

[57] R. Tanter, After Fukushima: a survey of corruption in the global nuclear power industry, Asian Perspect. 37 (2013) 475–500, https://doi.org/10.5555/0258-9184-37.4.475.

[58] J. Spacapan, Conventional wisdom says Turkey won't go nuclear. That might be wrong, Bull. At. Sci. (2020). https://thebulletin.org/2020/07/conventional-wisdom-says-turkey-wont-go-nuclear-that-might-be-wrong/. (Accessed 25 September 2020).

[59] H. Rühle, Is Turkey secretly working on nuclear weapons? National Interest (2015). https://nationalinterest.org/feature/turkey-secretly-working-nuclear-weapons-13898. (Accessed 25 September 2020).

[60] A. Stein, The Turkish (Nuclear) Model? 2013. http://guests.armscontrolwonk.com/archive/4115/the-turkish-nuclear-model. (Accessed 10 February 2014).

[61] E. Toksabay, Erdogan says it's unacceptable that Turkey can't have nuclear weapons, Reuters (2019). https://www.reuters.com/article/us-turkey-nuclear-erdogan/erdogan-says-its-unacceptable-that-turkey-cant-have-nuclear-weapons-idUSKCN1VP2QN. (Accessed 25 September 2020).

[62] V. Gilinsky, H. Sokolski, Taking Erdogan's critique of the nuclear non-proliferation treaty seriously, Bull. At. Sci. (2019). https://thebulletin.org/2019/11/taking-erdogans-critique-of-the-nuclear-non-proliferation-treaty-seriously/. (Accessed 30 September 2020).

[63] M. Crapo, S. Whitehouse, R.F. Ichord Jr., R. Bell, J.T. Gordon, E. Scholl, US Nuclear Energy Leadership: Innovation and the Strategic Global Challenge, Atlantic Council Global Energy Center, Washington, DC, 2019. https://www.atlanticcouncil.org/images/publications/US_Nuclear_Energy_Leadership-.pdf. (Accessed 25 May 2019).

[64] DOE, Grid Memo, Department of Energy, Washington, DC, 2018. https://www.documentcloud.org/documents/4491203-Grid-Memo.html. (Accessed 31 May 2020).

[65] A. Stirling, P. Johnstone, A Global Picture of Industrial Interdependencies Between Civil and Military Nuclear Infrastructures, Social Science Research Network, Sussex, UK, 2018, https://doi.org/10.2139/ssrn.3230021.

[66] R. Marlin, Propaganda and the Ethics of Persuasion, Broadview Press, Peterborough, ON, Canada, 2002.

[67] IRENA, Renewable Power Generation Costs in 2019, International Renewable Energy Agency, Abu Dhabi, 2020.

[68] P. Godron, M.E. Cebeci, O.B. Tör, D. Saygın, Increasing the Share of Renewables in Turkey's Power System: Options for Transmission Expansion and Flexibility, SHURA Energy Transition Center, Sabancı University, Istanbul, 2018.

[69] M. Wei, S. Patadia, D.M. Kammen, Putting renewables and energy efficiency to work: how many jobs can the clean energy industry generate in the US? Energy Policy 38 (2010) 919–931, https://doi.org/10.1016/j.enpol.2009.10.044.

[70] M.V. Ramana, Beyond Our Imagination: Fukushima and the Problem of Assessing Risk, Bulletin of the Atomic Scientists, 2011.

[71] J. Downer, M.V. Ramana, Empires built on sand: on the fundamental implausibility of reactor safety assessments and the implications for nuclear regulation, Regul. Gov. (2020), https://doi.org/10.1111/rego.12300.

[72] P. Demircan, The heartbeat of democracy heard in Turkey: reflections of an anti-nuclear activist, DiaNuke (2019). https://www.dianuke.org/the-heartbeat-of-democracy-heard-in-turkey-reflections-of-an-anti-nuclear-activist/. (Accessed 1 October 2020).

[73] A.H. Şorman, E. Turhan, The limits of authoritarian energy governance in the Global South: energy, democracy and public contestation in Turkey, in: J. Keahey, M. Pasqualetti, M. Nadesan (Eds.), Democratizing Energy: Imaginaries, Transitions, and Risks, Elsevier, Amsterdam, 2021.

[74] J. Lee, J. Byrne, J. Seo, Re-imagining energy-society relations: an interactive framework for social movement-based energy-society transformation, in: J. Keahey, M. Pasqualetti, M. Nadesan (Eds.), Democratizing Energy: Imaginaries, Transitions, and Risks, Elsevier, Amsterdam, 2021.

[75] L. Winner, The Whale and the Reactor: A Search for Limits in an Age of High Technology, University of Chicago Press, Chicago, 1986.

Fossilizing renewable energy: The case of wind power in the Isthmus of Tehuantepec, Mexico

Lourdes Alonso-Serna[a] *and Edgar Talledos-Sánchez*[b]

[a]Universidad del Mar, Huatulco, Oaxaca, México [b]CONACYT/EL Colegio de San Luis, A.C., San Luis Potosí, Mexico

1 Introduction

Energy transitions are major historical shifts that have significant implications for societies. The shift to renewables creates expectations of a more inclusive and democratic production and use of energy. Nevertheless, renewables have raised opposition among the local people stemming from the attachments and identities of local communities to a place and the concerns about procedures and distributional justice [1]. Similarly, the exclusion of the local population in the planning and decision-making processes is critical for understanding the opposition to the energy infrastructure [2, p. 247]. Thus, the challenges to renewables are eminently social, and "developers should be able to reweight project-evaluation processes to include social aspects, understand the social contexts where a project is planned and grant enough and just compensations to communities" [3, p. 220].

Social acceptance to renewables has three interrelated dimensions: sociopolitical acceptance, community acceptance, and market acceptance. Sociopolitical acceptance refers to the approval of technologies and policies by stakeholders, policymakers, and the public. Community acceptance refers to the specific endorsement of sitting decision and renewable energy projects by residents and local authorities. It depends on how the costs and benefits are shared and on the decision-making process. Finally, the third dimension of social acceptance is related to innovation and market adoption of technologies [4, pp. 2684–2685].

In this chapter, we take a different route. We argue that cases of resistance to the installation of renewable facilities [5–9] suggest that they have been placed to service "a structure of accumulation that is intensive in the use of materials and energy" [5, p. 9], constituting a continuation and not a rupture from the fossil fuel energy system. We argue that the introduction of renewable energy into an energy system that fuels the capitalist mode of production is a means of fossilizing it. The chapter builds on a historical materialist understanding of energy to foreground its central role in the production of space. The chapter presents the development of wind energy in the Isthmus of Tehuantepec in Southern Mexico as an example of what we call "the fossilization of renewables."

This chapter has three main sections. The following section presents recent proposals that underscore the central role of energy in the production of space and in capital circulation. Subsequent sections delve into the process of introducing renewables in Mexico and their use for the industrial sector. The last section looks at the production of wind energy in the Isthmus of Tehuantepec for large consumers in the industrial and retailing sectors.

2 The fossil economy: Energy and capital circulation

Recent scholarship has sought to foreground a historical materialistic approach of the role of energy in the production of space. This literature criticizes the political ecology's narrow approach toward energy as just another resource and its limited focus on the sites of extraction and the conflicts stemming from the appropriation of energy resources. This neglect is also reflected in Urban Political Ecology (UPE), which has highlighted the myriad of metabolisms that sustain urban life. Still, it has overlooked the "often invisible energy networks that make urban life possible—and spew carbon dioxide, soot, and other local pollutants in the process" [10, p. 486].

For Matt Huber, a historical-materialistic perspective to energy needs to move "from conceptions of energy as a 'thing' or a 'resource' towards a conception of energy as a 'social relation' enmeshed in dense networks of power and socioecological change" [11, p. 106]. Thus, energy as a social relationship is intrinsically historical and in a mutually constitutive relationship with society, not a transhistorical physical quality that forms the very fabric of the universe [12, p. 3].

Although the outset of the industrial revolution in England was initiated with water power [13], the shift to coal implied a historical break with various effects. First, fossil fuels allowed "the temporal *acceleration* of the pace and productivity of production. They are themselves the biotic *concentration of time*—the product of millions of years of concentrated solar energy [...] [that] produces an enormously dense energy source" [14, p. 9]. Second, they allowed the production of the spaces of production and extraction, as well as urban spaces. Third, fossil fuels also accelerated the processes of fixed capital formation allowing the absorption of large amounts of capital [15].

Coal-powered machines induced a qualitative change from the formal to the real subsumption of labor, or "the extent to which and the forms in which labor is integrated into the process of exploitation, or of production of value and surplus value" [16, p. 180]. This resulted in the acceleration of the extraction of surplus value and the worsening of control over workers. As Huber contends, "the real subsumption of labor is not only a social relation (wage-labor) but a socio-technical transformation of the labor process where machinery, etc. becomes the real master of living labor" [14, p. 14]. As Andreas Malm points out,

> The purpose of machinery – to secure absolute power over labour – was understood to necessitate a prime mover over which capital could exercise absolute power while at the same time offering capital all the power

it needed. In the powerlessness of the great powers of steam, British capital found the ideal spring of its class power. The ultimate bedrock of all that power, however, was revealed in that one little detail: the engine had to be fed with coal. *Malm [13, p. 46]*

Although the generalization of fossil fuels has been geographically uneven and a highly contradictory process, such energy resources have become "a historically specific and internally necessary aspect of the capitalist mode of production" [11, p. 105]. They have facilitated the concentration of workers and accelerated the processes of urbanization and industrialization. While the applications of fossil fuels into transportation expanded the spatial and temporal scope of goods, people, and the circulation of capital, fossil fuels, and energy more broadly, have conveyed meanings of progress, modernity, and even independence for states [17].

Fossil fuels inaugurated an era in which "growth was not episodic, evanescent or broken off after a brief efflorescence, but persistent and unremitting, a secular progression propelled by its own inner forces" [13, p. 21]. Malm calls this era of self-sustained growth the fossil economy because the ever-growing consumption of fossil fuels powers it. Drawing on this, we argue that the introduction of renewables into the electric sector does not mark a historical break to the fossil economy, but its continuation, and what is going on is a process of fossilization of the so-called renewables. Renewables are being fossilized when they are introduced into the same historical relations of production and reproduction of contemporary societies; they are used to quench the large amounts of energy that sustain life under capitalist relations. In the remaining sections, we exemplify the fossilization of renewables with the production of wind energy in Mexico, in particular in the region of the Isthmus of Tehuantepec.

3 Renewable energy in Mexico: Cheap energy for economic growth

Oil has been an abundant and strategic resource in Mexico. It is regarded as a symbol of national sovereignty since its nationalization in 1938, in addition to being a crucial element in the industrialization and urbanization of the country and a source of rents for the state. Therefore, the country's energy mix is highly dependent on fossil fuels; in 2017, 89% of primary energy came from fossil fuels, while 79% of the electricity was produced with fossil fuels and only 21% with clean sources [18, p. 5]. The transportation and industrial sectors are the largest energy consumers in Mexico, with 44% and 35% each. The industrial sector powers its productive processes with gas and electricity [19, p. 35]. In 2017 alone, 15 industrial branches accounted for 47% of the sector's total electricity consumption; they are the industries of iron and steel, cement, chemical, mining, petrochemicals, paper, glass, sugar, beer, automotive, construction, food and beverages, rubber, fertilizers, and tobacco [19, p. 39].

Renewables in Mexico have been used to meet the needs of energy-intensive industries connected with the country's growth, as stated by former President Felipe Calderón,

It is unthinkable to halt progress and economic development, especially for countries like Mexico. We cannot risk our future; we cannot stop developing our energy sector [...]. We need more energy, but we need clean energy sources; we need efficient and low carbon emission sources of energy. The solution to the dilemma between economic growth and climate change is sustainable development [...]. The answer is to produce energy without the environmental externalities that fossil fuels have for humankind. *REVE [20]*

Apart from hydropower's 12,000 MW installed capacity and some geothermal [21], renewables have a small share in the energy mix. Wind energy was introduced in the late 1990s, and it was developed in the context of the processes of neoliberalization that initiated in the 1980s. While most of the state assets were privatized in only a decade, energy remained a state monopoly due to its political and economic significance. But successive governments since the 1990s aimed to reform the legal framework to end the state monopoly in electricity. In 1992, a reform to the Law of Public Service of Electricity allowed private utilities to produce electricity as independent power producers and for self-supply. These changes enabled a parallel market of electricity between utilities and consumers in the industrial and retailing sectors; the private sector also started producing electricity for the public utility Federal Electricity Commission (Comisión Federal de Electricidad or CFE).

The parallel market of electricity increased renewables share in the electric system. In 2013, its installed capacity reached 2603.81 MW, and although wind energy accounted for 89% of the new facilities, there is also hydropower, solar, and geothermal (Table 28.1). Wind farms built in this period are all located in the Isthmus, in the state of Oaxaca, the first region with a massive energy development.

The privatization of the energy sector came about in 2013 when a reform to the law created the wholesale electricity market (Mercado Eléctrico Mayorista or MEM), where electricity and other products are traded. The MEM comprises different markets: spot energy market, power balance market, clean energy certificates and long-term auctions. In the new market, private utilities produce and trade electricity directly with their clients, or qualified users, which are companies with a minimum demand of 1 MW. In 2018, 12,000 industries could register as qualified users and chose their electricity provider [22].

The new scheme was designed to increase renewables share in the electric system. Long-term auctions (LTAs) and power purchase agreements (PPAs) are the market mechanisms that now allow renewables upscaling. The 3 auctions organized between 2015 and 2017 had wind and solar energy as the front-runners; 67 of 70 projects that won the auctions were either solar or wind [23]. Besides, the Law of Energy Transition makes it compulsory for energy-intensive industries to acquire clean energy certificates (CELs), but some industries, like steel, have opposed the measure because it would raise their production costs [24].

Lastly, the spot energy market is also designed to favor renewables. This market is based on a system of variable costs (cost of fuel, operation, and maintenance). The system operates in such a way that utilities with the lowest variable costs are the first to dispatch electricity,

TABLE 28.1   Renewables under the scheme of self-supply (1992–2013).

Source	Installed capacity (MW)	Total output allowed (GWh)	Number of facilities	Average size per facility (MW)
Wind	2317.85	8004.57	22	105.36
Hydropower	190.95	948.93	15	12.73
Geothermal	52	387	1	52
Solar	43.01	107.75	7	6.14

Own elaboration with information from Zumma, Informe: Energías Renovables en los Sectores Comerciales e Industriales de México, https://zumma.com.mx/our-work.php, n.d. (Accessed 01.10.19).

but their electricity is purchased at the price of the provider with the highest variable cost (cost of fuel, operation, and maintenance), whereas utilities with the highest variable costs—fuel oil facilities—sell electricity only during peak hours. Under this system, renewables are highly competitive because their fuel cost is null and their costs of operation are low [24].

This reform has constituted a spatial fix, or the sinking of idle capital in enduring infrastructures that allow capital's expanded circulation. The private sector's share in the electric market soared from 37% in 2013 to 50% in 2018 [25]. As Fig. 28.1 shows, wind energy has mushroomed throughout the country, with wind farms in 13 states and still a vast wind potential to tap on.

The geographical expansion of wind energy in Mexico has mainly been placed to service the need of electricity of 15 industrial sectors, such as mining (Arcelor mittal, DeAcero, Peñoles, Minera México, Cobre del Mayo); cement (Cemex, Cementos Moctezuma, Cementos Apasco); automotive industry (VW, Nissan); food and beverages (Bimbo, Nestlé, Lala, Jugos del Valle, Sabritas, Barcel, Herdez, Grupo Modelo, Cuauhtemoc Moctezuma); paper (Scribe, Kimberly Clark); glass (Vidriera Toluca, Vidriera Monterrey) [26, p. 7].

Even though wind power enters the national grid and it potentially benefits anyone who gets access to it, the electricity from this renewable source is mainly produced to satisfy the energy-intensive industries' demand. For these companies, electricity from renewable sources reduces their production costs and presumably their CO_2 emissions. Still, renewables are put to work to sustain the production processes that are akin to the fossil economy.

However, because energy is a social relation, renewables should not be abstracted from the social and political configurations that underpin their development. Thus, as was elaborated in the previous section, renewables should be seen in a dialectical relation with society. Our argument of the fossilization of renewables seeks to foreground the social relations of production and the power dynamics in which they are enmeshed. We present two ideas to make our case of the fossilization of renewables.

First, the claim that wind energy is an improvement because its primary source is ubiquitous and inexhaustible overlooks the material requirements to produce turbines and other infrastructure. Wind turbines require aluminum, copper, rare earth elements, wood epoxy, pre-stressed concrete, glass fiber reinforced plastic, carbon filament reinforced plastic, among others [27, p. 1161]. The provision of some of these materials, like rare earth elements, is being framed in terms of national security discourses that distinguish fossil fuels. Its extraction has already polluted the air, water, soil, and people's bodies. It has also triggered episodes of socio-environmental conflict and set local people against governments and corporations [28]. Thus, the assertion that wind energy is an improvement over other conventional sources is difficult to substantiate when the material requirements for new energy infrastructure and the social and environmental impacts are foregrounded.

Second, the Mexican state has subsidized wind power and other renewables to make them commercially viable. At the outset of wind energy, state officials and energy developers had to agree on the terms to connect to the electric grid; this was a crucial matter for investors because the fees would affect wind energy's economic competitiveness. Thus, state officials devised a scheme that had intermittency or fluctuation over multiple time horizons, which was an advantage for energy developers. The fees for using the electric grid were calculated on the capacity factor of wind farms, which is determined by the availability of wind. This scheme reduced the cost to connect to the grid by up to 50% [26, p. 101]. More recently, CFE's director

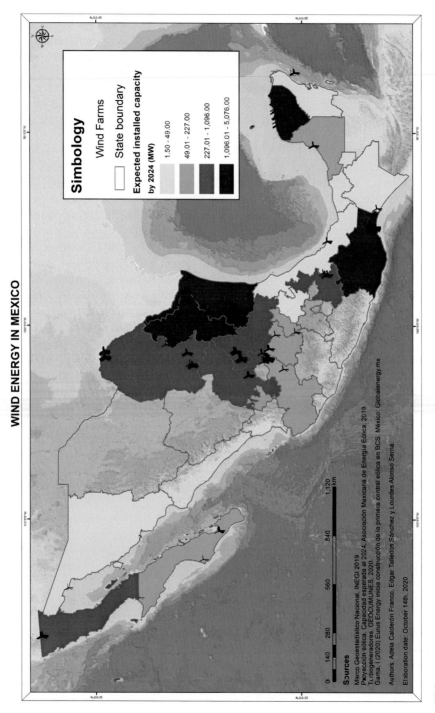

FIG. 28.1 Map of wind energy in Mexico. *Data from INEGI, Marco Geoestadístico Nacional, https://www.inegi.org.mx/temas/mg/, 2019 (Accessed 01.09.20); AMDEE, Proyección eólica. Capacidad esperada al 2024, https://amdee.org/mapas-eolicos.html, 2019 (Accessed 01.09.20); Geocomunes, Aerogeneradores, http://132.248.14.102/maps/451/download, 2020 (Accessed 02.09.20); Gama, I., Eurus energía inicia la primera central eólica en BCS, 2020, https://globalenergy.mx/noticias/alternativas/eolica/eurus-energy-inicia-construccion-de-la-primera-central-eolica-en-bcs/ (Accessed 02.09.20).*

argued that the low fees to connect to the grid and CFE's provision of baseload power are subsidies for renewables, and the financial burden is at the expense of Mexican taxpayers [29].

The ever-growing production of renewable energy is placed to keep up with production and to sustain economic growth, just as the transition toward fossil fuels did with industrial capitalism. Meanwhile, social aspects have not been addressed as the next section elaborates the case of the Isthmus of Tehuantepec.

4 The Isthmus of Tehuantepec: Cheap energy for the industrial sector

The coastal plain in the Isthmus of Tehuantepec in Oaxaca was the first region in Mexico with a utility-scale wind energy development. In 2021, there were 22 wind farms in the area with an installed capacity of 2600 megawatts (MW), representing only a fraction of the 33,000 MW of potential capacity that the region can hold [30, p. vii]. The windy area extends over various municipalities, but wind farms are for now installed in Asunción Ixtaltepec, El Espinal, Juchitán, Unión Hidalgo, and Santo Domingo Ingenio (Fig. 28.2).

The population in these municipalities is predominantly Zapotec, an Indigenous group whose culture is replete with references to resistance against the dispossession of local resources. Thus, sectors of the local population have opposed wind energy. Resistance toward wind energy is related to the lack of understanding of the cultural attachments to the land and the scant compensations from the sitting of infrastructure [31]. Grassroots movements in the region have also framed their resistance against colonialism (see Chapter 24). One campaigner argues that:

> State officials told us that the Isthmus would benefit from massive investments of millions of euros or dollars, but near 80 percent of that investment goes to buy the wind turbines produced in Germany, in Denmark, and Spain. [...] Multinationals also made the promise of employment and development; those were false promises too. [...] These companies make millions from producing electricity, and none of that remains in the region. *CGT [32]*

Just as this campaigner argues, wind energy leaves little benefit in the region. In fact, the electricity is transferred away from the region to power companies' processes elsewhere. For example, the cement company CEMEX gets 25% of its electricity from the wind farm Eurus, which is located in the locality known as La Venta [33], while the mining holdings, Grupo México and Grupo Peñoles, own a wind farm each. The former owns El Retiro in La Ventosa, and the latter owns Fuerza Eólica del Istmo in El Espinal. The baking company Bimbo gets its electricity for its 40 plants from the wind farm Piedra Larga in Unión Hidalgo [34]. There are approximately 400 companies from sectors as diverse as cement, paper, steel, beer, food, and the retailing industry (like Wal-Mart, Chedrahui, Soriana, and Suburbia) that get electricity from the Isthmus of Tehuantepec [35]. Table 28.2 shows some wind farms and their primary consumers of electricity.

Meanwhile, sectors of the population complain of the high electricity rates they have to pay and the lack of electricity in some localities in the Isthmus. Even though electricity for households is subsidized, they claim that electricity rates have increased fivefold in the last 10 years [36]. They are vocal on the social injustice of having 22 wind farms in the region while the local population pays high electricity bills. Furthermore, there is a widespread opinion

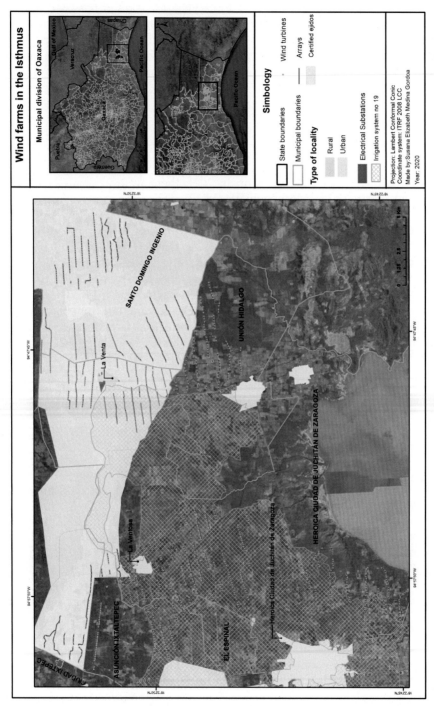

FIG. 28.2 Map of wind farms in the Isthmus of Tehuantepec. *Data from Geocomunes, http://132.248.14.102/maps/451/download, 2020 (Accessed 15.08.20); Sistema Nacional de Información de Agua, http://sina.conagua.gob.mx/sina/tema.php?tema=distritosriego, 2018 (Accessed 15.08.20); Registro Agrario Nacional, https://datos.gob.mx/busca/organization/ran, 2020 (Accessed 16.08.20); INEGI, Marco Geoestadístico Nacional, https://www.inegi.org.mx/temas/mg/, 2017 (Accessed 16.08.20); INEGI, Conjunto de Datos Vectoriales de Información Topográfica, https://www.inegi.org.mx/app/biblioteca/ficha.htm-l?upc=889463531692, 2017 (Accessed 16.08.20).*

TABLE 28.2 Wind farms in the Isthmus of Tehuantepec and main consumers of electricity.

Wind farm	Installed capacity (MW)	Main consumers	Electricity demand from main consumer (MW)	Other large consumers
Eurus	250.5	Cemex	539.35	Tec de Monterrey, Gamesa, Rotoplas, Sabritas, L'Oreal
Bii Hioxo	234	Tiendas Chedraui	165.154	Cementos Moctezuma, Bebidas mundiales, Unilever, Saint-Gobain
Eoliatec del Istmo	164	Acelor Mittal	100.82	Wal-Mart, Continental Automotive, Herdez
Eoliatec del Pacífico	160	Arcelor Mittal	160	Wal-Mart, Continental Automotive, Grupo Modelo
Demex II	137.5	Wal-Mart	137.705	Suburbia
Parques Ecológicos de México	101.9	Soriana	115.024	Cementos Apasco, Procter & Gamble, Nissan, Cuauhtémoc Moctezuma, Scribe

Own elaboration with information from Zumma, Informe: Energías Renovables en los Sectores Comerciales e Industriales de México, https:// zumma.com.mx/our-work.php, n.d. (Accessed 01.10.19).

that households should get free electricity because of the large number of wind farms in the region. The population also complains that apart from landholders who get rents, the wider population does not receive any compensation from the installation of more than 1000 turbines. While the local population in the Isthmus demands free electricity, the prices of electricity from renewable sources have dropped in the last few years. In 2015, the price per megawatt/hour was $47.48; a year later, it fell to $33.4 and to $20.57 in 2017 [37]. The electricity produced with renewable sources is cheap, but electricity in the Isthmus, even with the state subsidies, is expensive for its population.

5 Conclusions

In this chapter, we have argued that energy is a social relation enmeshed in social and power configurations. We have aimed to foreground the relations that underpin the production and consumption of wind energy in Mexico. Drawing on a historical materialist understanding of energy, we claim that in the current historical context, the introduction of renewable energy does not mark a shift from the fossil economy, but its continuation. Renewable energy is being developed to power an economic system that has the production of commodities and the circulation of capital at its core, a system that has been only possible through the abundance of fossil fuels. We present wind energy development in Mexico as a case where renewables are fossilized. Wind energy in Mexico is consumed by energy-intensive industries that have developed economic strategies to hedge their electricity needs with renewables. These companies get cheap electricity from a renewable source, continue emitting CO_2, and are responsible for severe environmental pollution, like Grupo México, a mining holding that is responsible for the spill of sulfuric acid and water sediments into various rivers and the sea. The first region in Mexico where renewables were fossilized was the Isthmus of Tehuantepec. The 22 wind farms in the region produce electricity for the industrial and retailing sectors, while the local population faces difficulties in getting access to electricity or paying their bills.

References

[1] P. Devine-Wright, Y. Howes, Disruption to place attachment and the protection of restorative environments: a wind energy case study, J. Environ. Psychol. 30 (2010) 271–280.
[2] K. Burningham, J. Barnett, G. Walker, An array of deficits: unpacking NIMBY discourses in wind energy developers' conceptualizations of their local opponents, Soc. Nat. Resour. 28 (3) (2015) 246–260.
[3] M. Pasqualetti, Social barriers to renewable energy landscapes, Geogr. Rev. 101 (2) (2011) 201–223.
[4] R. Wüstenhagen, M. Wolsink, M. Bürer, Social acceptance of renewable energy innovation: an introduction to the concept, Energy Policy 35 (2007) 2683–2691.
[5] J. Franquesa, Power Struggles: Dignity, Value, and the Renewable Energy Frontier in Spain, Indiana University Press, Indiana, 2018.
[6] J. Baka, Making space for energy: wasteland development, enclosures, and energy dispossessions, Antipode 49 (4) (2017) 977–996.
[7] A. Dunlap, Insurrection for land, sea and dignity: resistance and autonomy against wind energy in Álvaro Obregón, Mexico, J. Polit. Ecol. 25 (1) (2018) 120–143.
[8] C. Howe, Ecologics, Duke University Press, Durham, 2019.
[9] D. Boyer, Energopolitics, Duke University Press, Durham, 2019.
[10] M. Huber, Energy and social power: from political ecology to the ecology of politics, in: T. Perreault, G. Bridge, J. McCarty (Eds.), The Routledge Handbook of Political Ecology, Routledge, Oxon, 2015, pp. 481–492.
[11] M. Huber, Energizing historical materialism: fossil fuels, space and the capitalist mode of production, Geoforum 40 (1) (2009) 105–113.
[12] B.F. Towler, The Future of Energy, Elsevier, London, 2014.
[13] A. Malm, Fossil Capital: The Rise of Steam Power and the Roots of Global Warming, Verso, London, 2016.
[14] M. Huber, Lifeblood: Oil, Freedom and the Forces of Capital, University of Minnesota Press, Minneapolis, 2013.
[15] D. Harvey, Spaces of Global Capitalism: Towards a Theory of Uneven Geographical Development, Verso, New York, 2006.
[16] R.J. Das, Reconceptualizing capitalism: forms of subsumption of labor, class struggle, and uneven development, review of radial political, Economics 44 (2) (2012) 178–200.
[17] C. Raffestin, Quelques réflexions sur l'évolution des choses ou le devenir des infrastructures énergétiques, in: P. Hollmuller, B. Lachal, F. Romerio, W. Weber, J.M. Zgraggen (Eds.), Démantèlement des Infrastructures de l'énergie: Actes de la 14ème Journée du Cuepe 2003–2004, Centre universitaire d'étude des problèmes de l'énergie, 2004, pp. 3–8.
[18] SENER, Reporte de avance de energías limpias, SENER, Mexico, 2017.
[19] SENER, Balance nacional de energía, SENER, Mexico, 2018.
[20] REVE, Presidente Calderón: Hay Cambio Climático y hay Terrible Daño a la Población, 2010. https://www.evwind.com/2010/10/01/presidente-calderon-hay-cambio-climatico-y-hay-terrible-dano-a-la-poblacion/. (Accessed 19.05.19).
[21] SENER, Prospectiva de energías renovales 2016–2030, SENER, Mexico, 2016.
[22] R. Guevara, Grandes Consumidores de Energía en México: ¿Qué pasó? (1992–2013), 2020. https://energiahoy.com/2020/01/30/grandes-consumidores-de-energia-en-mexico-que-paso-1992-2013/. (Accessed 22.04.21).
[23] SENER, Anuncian SENER y CENACE Resultados Preliminares de la Tercera Subasta de Largo Plazo, 2017. https://www.gob.mx/cenace/prensa/anuncian-sener-y-cenace-resultadospreliminares-de-la-tercera-subasta-de-largo-plazo-141668. (Accessed 10.04.21).
[24] L. Alonso, Harvesting the Wind: The Political Ecology of Wind Energy in the Isthmus of Tehuantepec, Oaxaca (PhD thesis), University of Manchester, 2020.
[25] AMLO, Versión estenográfica de la conferencia de prensa matutina del Presidente Andrés Manuel López Obrador, 2019. https://lopezobrador.org.mx/2019/02/11/version-estenografica-de-la-conferencia-de-prensa-matutina-del-presidente-andres-manuel-lopez-obrador-42/. (Accessed 17.05.2019).

[26] Zumma n.d., Informe: Energías Renovables en los Sectores Comerciales e Industriales de México, https://zumma.com.mx/our-work.php (http://1.10.0.19, 10 January 2019), (Accessed 01.10.19).

[27] M.Z. Jacobson, M.A. Delucchi, Providing all global energy with wind, water, and solar power, part I: technologies, energy resources, quantities and areas of infrastructure, and materials, Energy Policy 39 (3) (2011) 1154–1169.

[28] S. Raman, Fossilizing renewable energies, Sci. Cult. 22 (2) (2013) 172–180.

[29] C. Marín, El costo de las energías limpias, Manuel Bartlett Parte I, 2020. https://www.youtube.com/watch?v=R_QLDKyQ_qI&ab_channel=MILENIO. (Accessed 19.04.21).

[30] D. Elliot, M. Schwartz, G. Scott, S. Haymes, D. Heimiller, R. George, Wind Energy Resource Atlas of Oaxaca, NREL, Colorado, 2003.

[31] M. Pasqualetti, Opposing wind energy landscapes: a search for a common cause, Ann. Assoc. Am. Geogr. 101 (4) (2011) 1–11.

[32] CGT, Rueda de Prensa, Bettina Cruz, 2018. https://www.youtube.com/watch?v=3SEs5cnkDk0. (Accessed 19.04.21).

[33] REVE, Cemex se Autoabastece de Energía Eólica, 2015. https://www.evwind.com/2015/08/06/Cemex-Se-Autoabastece-Con-Energia-Eolica/. (Accessed 04.10.20).

[34] Grupo Bimbo, Grupo Bimbo, la Conversión más Importante a Energías Renovables de la Industria Alimentaria Global, https://grupobimbo.com/es/sala-de-prensa/comunicados/grupo-bimbo-la-conversion-mas-importante-energias-renovables, n.d. (Accessed 04.10.20).

[35] R. Chacah, Beneficiadas por eólicas en el Istmo, más de 400 empresas, 2019. https://oaxaca.eluniversal.com.mx/especiales/26-02-2019/beneficiadas-por-eolicas-en-el-istmo-mas-de-400-empresas. (Accessed 12.10.20).

[36] APIIDTT, Comunicado Asamblea de Pueblos Indígenas del Istmo en Defensa de la Tierra y el Territorio, 2012. https://tierrayterritorio.wordpress.com. (Accessed 02.10.20).

[37] K. García, Subastas tiran precios de energía renovable, 2017. https://www.eleconomista.com.mx/empresas/Subastas-tiran-precios-de-energia-renovable-20171123-0032.html. (Accessed 07.10.20).

"Psychic numbing" and the environment: Is this leading to unsustainable energy outcomes in Australia?

Peta Ashworth[a] and Kathy Witt[b]

[a]School of Chemical Engineering, The University of Queensland, Brisbane, QLD, Australia
[b]Centre for Natural Gas, The University of Queensland, Brisbane, QLD, Australia

1 Introduction

Psychic numbing has been defined as a human adaptive response to problems when they become too overwhelming, threaten values and existences, or occur at scales too large to comprehend in terms of an individual's capacity to make a difference [1,2]. Gregory [2] suggests numbing has been evidenced across multiple situations where individuals, politicians, and nations' leaders avoid facing up to difficult issues. Previous examples have included nuclear wars, genocide, refugee crises, impacts of natural disasters, and pandemics.

Climate change is another issue that has invoked psychic numbing, resulting in both denial and inaction. This is despite the majority of scientists warning that the world is far beyond the tipping point of avoiding multiple climate-related impacts and disasters [3]. While the exact nature and extent of these disasters continues to be discussed, conservative Australian governments, and their voters, remain steadfast in their support for fossil fuels and related industries—a major contributor of the world's greenhouse gas emissions [4,5]. Supporting the continued use of fossil fuels seems counter to any climate-related logic, particularly given that Australia recently experienced some of the worst droughts, floods, and bushfires in its history on the back of Australia's hottest and driest year on record [6,7].

The severity and extent of human and environmental impacts from the most recent Australian bushfires have been comprehensively detailed in the Interim observations of the Royal Commission into National Natural Disaster Arrangements [8], which was established in February 2020 in response to the extreme bushfire season of 2019–20. In addition to the millions of hectares of land that burned (claimed to be between 24 and 40 million hectares), it also reports that

> Tragically, 33 people died, and smoke may well have caused many other deaths. Others suffered serious physical and emotional/psychological injuries. It is estimated that nearly 3 billion animals were killed or displaced by the bushfires, and many threatened species and other ecological communities were extensively damaged. Over 3000 homes and many other buildings were destroyed. For many people, it will take years to recover and rebuild [8, p. 5].

At the time, media coverage of the bushfires traveled swiftly around the world. The Australian Prime Minister reported to Parliament in February 2020 that 70 countries had offered assistance, and at the height of the fires, overseas resources were deployed from Canada (239), the United States (360), and New Zealand (320) [9]. Google Trends [10] also illustrate the interest the bushfires created with the normalized trend hitting 100 on January 4, 2020—both for Australia and worldwide. More poignantly, the images that circulated in the world at the time evoked strong emotional responses of both outrage and compassion, supporting arguments of Bass [11] and Slovik and colleagues [12] for the power of images to awaken peoples' senses, compared to scientific reports or statistics, at least in the short term.

While the immediate bushfire crisis sparked high levels of public attention and a heightened sense of urgency for climate action, these both appear short-lived. At the time of the bushfires, fundraising efforts went into overdrive, with multiple new charities emerging and an unprecedented amount of funds raised—more than AUD500 million within a couple of months. To illustrate this further, a review of the Australian Red Cross *Australian Bushfires Report* [13] showed that at a New Year's Eve appeal AUD11 million was raised and by the end of January, a total of AUD135 million had been donated. This total grew to AUD180 million in February, AUD203 million in March, and AUD215 million by April. This was significantly the largest amount collected by the Australian Red Cross in a single appeal but from April, only a month after the fires were declared contained, the size of donations dwindled to much smaller increments. While this effect was potentially complicated by the onset of the Coronavirus, it does suggest that the Australian public became once again numb to the impact of bushfires and other extreme weather events which signal the urgency for climate change action.

2 Psychic numbing and climate change in Australia

To further explore the phenomenon of psychic numbing to climate change, we examined the response of the Australian politicians and Australian public to climate change and their support for low carbon energy technologies as part of a climate mitigation strategy. Fielding and colleagues' [4] survey of Australian politicians in 2012 showed that respondents from the right leaning Liberal and National parties were less supportive of low carbon energy technologies, and their mean responses to questions about whether they believed climate change is happening, and the severity of its impacts were consistently below the midpoint. On the other hand, respondents from the more left leaning Labor and Greens parties held stronger supportive beliefs.

Similarly, our own surveys [5] consistently found that while the majority of the Australian public (71% in 2013 and 69% in 2017) believe that climate change is happening or that it will start happening within the next 30 years (7% in 2013 and 9% in 2017), in relation to energy technologies, participants who were supportive of the more right leaning Liberal and National parties were more likely to express support for the continued use of fossil fuels rather than low carbon technologies. Notably, both of these surveys were conducted well before the most recent bushfire events.

To further understand the current state of the Australian public's attitude toward climate change and the need for a low carbon energy transition, the results of the Australian Lowy Institute [14] poll conducted around mid-March of 2020 ($n = 2448$) are revealing. While not directly comparable, 56% of the survey population agreed that *global warming is a serious and pressing problem*, and that *we should begin taking steps now even if this involves significant costs*. This represents a 5-percentage point drop from the previous poll in 2019 [15]. Similarly, only 59% saw *climate change as a critical threat to the vital interests of Australia in the next 10 years*. While 77% identified concern about drought and water shortages, the lower figure for expressed agreement with climate change is more aligned with our earlier survey results [5]. Similar results are also echoed in the Essential Media Communications' [16] omnibus research update undertaken in January and March 2020 ($n = 1000+$). This poll shows that the Liberal/National supporters are more likely to think Australia is doing enough to address climate change compared to the Greens and Labor supporters and are much less likely to support removal of subsidies to the fossil fuel industry or setting a zero-carbon pollution target for 2030 as examples. This combination of results suggests a proportion of the Australia public still does not see climate change as a threat, even in light of the recent severe weather events [14].

While there are a number of possible explanations for these survey responses, they do point to psychic numbing, or a head-in-the-sand approach to climate change that appears to be most evident among the right leaning Australian public and politicians. In his discussions on psychic numbing in relation to genocide, Slovic [1,17] highlights the importance of affect, or feelings rather than logic, in forming our actions and responses. He and others suggest that when the reporting of statistics, such as the number of deaths, or impacts such as the number of square kilometers of bushland burned, becomes overwhelming, as was experienced with the Australian bushfires, a sense of indifference sets in. This indifference, or numbing, is linked to an individual's perceived inability to respond effectively, which in turn makes it easier for individuals to rationalize or ignore the event, as evidenced in the quote below [1, p. 79]:

> Most people are caring and will exert great effort to rescue "the one" whose needy plight comes to their attention. These same good people, however, often become numbly indifferent to the plight of "the one" who is "one of many" in a much greater problem.

Fiske and Rai [18] suggest that while this is not necessarily intended, some sectors of the public will use political cues and other social and cultural norms to rationalize responses (including inaction) which appear to be indifferent to the impacts on humans and the environment. Similarly, from research in Norway and the United States, Norgaard [19] describes how climate change facts can be easily ignored amid social pressures to fit in with others, along with the convenience of high carbon emission living and the benefits that accrue from national wealth derived from fossil fuel industries.

3 Collective storytelling and action against psychic numbing

There is no doubt that public opinion on climate change is a complex issue. If Australia is to successfully move away from its reliance on fossil fuels and transition to a low carbon energy future, it will need people, policy, and media coordination to overcome psychic numbing and move toward more nuanced approaches. These approaches will need to combine factual information and the expert voice together with images and storytelling to form targeted messages, tailored to suit different political ideologies and existing attitudes [1,11,17,20]. While there are sectors of the public and politicians who seek to maintain the status quo, the immediate and significant response to the bushfires shows how effective imagery and affect can invoke an awakening of peoples' senses. The mobilized, coordinated, and compassionate human response to the Australian bushfires is exactly the type of response required for climate change action.

The evidence of psychic numbing presented in this essay, however, points to the key challenge of finding a balance between creating enough awareness and concern to mobilize collective action in a way that does not overwhelm or lead to compassion fatigue and despondency. This suggests an opportunity for future research to detail the point at which those triggers set in across different political parties and the voting public, as well as the effects that may follow.

References

[1] P. Slovic, "If I look at the mass I will never act" Psychic numbing and genocide, Judgm. Decis. Mak. 2 (2) (2007) 79–95.

[2] R.J. Gregory, Venturing past psychic numbing: facing the issues, J. Psychoanal. Cult. Soc. 8 (2) (2003) 232–237.

[3] International Panel on Climate Change, Global Warming of 1.5°C: An IPCC Special Report on the Impacts of Global Warming of 1.5°C above Pre-Industrial Levels and Related Global Greenhouse Gas Emission Pathways, in the Context of Strengthening the Global Response to the Threat of Climate Change, Sustainable Development, and Efforts to Eradicate Poverty, World Meteorological Organization, Geneva, Switzerland, 2018.

[4] K.S. Fielding, B.W. Head, W. Laffan, M. Western, O. Hoegh-Guldberg, Australian politicians' beliefs about climate change: political partisanship and political ideology, Environ. Polit. 21 (5) (2012) 712–733.

[5] P. Ashworth, Y. Sun, M. Ferguson, K. Witt, S. She, Comparing how the public perceive CCS across Australia and China, Int. J. Greenhouse Gas Control 86 (2019) 125–133.

[6] Bureau of Meteorology, Annual Climate Statement, 2019. http://www.bom.gov.au/climate/current/annual/aus/2020. (Accessed 5 October 2020).

[7] A. Klein, Was Australia's Hottest and Driest Year on Record, 2019. https://www.newscientist.com/article/2019-2019-was-australias-hottest-and-driest-year-on-record/2020. (Accessed 5 October 2020).

[8] Commonwealth of Australia, Royal Commission into National Natural Disaster Arrangements Interim Report, 2020. https://naturaldisaster.royalcommission.gov.au/publications/interim-observations-31-august-2020. (Accessed 28 September 2020).

[9] Parliament of Australia, 2019–20 Australian Bushfires—Frequently Asked Questions: A Quick Guide, 2020. https://www.aph.gov.au/About_Parliament/Parliamentary_Departments/Parliamentary_Library/pubs/rp/rp1920/Quick_Guides/AustralianBushfires. (Accessed 28 September 2020).

[10] Google Trends, 2020. https://trends.google.com/trends/explore?date=2019-11-01percent202020-03-01&q=bushfires,donation. (Accessed 28 September 2020).

[11] R. Bass, The Book of Yaak, Houghton Mifflin, New York, 1996.

[12] P. Slovik, D. Västfjäll, A. Erlandsson, R. Gregory, Iconic photographs and the ebb and flow of empathic response to humanitarian disasters, Proc. Natl. Acad. Sci. USA 114 (4) (2017) 640–644

[13] Australian Red Cross, Australian Bushfires Report, 2020. https://www.redcross.org.au/getmedia/9fe279c7-94af-48e5-b10a-2c3b15f812e5/Report-6mth-FINAL-200708-WEB_4.pdf.aspx. (Accessed 28 September 2020).

[14] B. Strating, Australia's Shifting Mood on Climate Change, 2020. https://www.lowyinstitute.org/the-interpreter/australia-s-shifting-mood-climate-change. (Accessed 28 September 2020).

[15] N. Kassam, Lowy Institute Poll 2020: Understanding Attitudes to the World, 2020. https://poll.lowyinstitute.org/files/lowyinsitutepoll-2020.pdf. (Accessed 28 September 2020).

[16] Essential Media Communications, Essential Report, 2020. https://essentialvision.com.au/?s=climate+change&searchbutton=Search. (Accessed 3 October 2020).

[17] P. Slovic, The more who die, the less we care: confronting the deadly arithmetic of compassion, Med. Decis. Mak. 40 (4) (2020) 407–415.

[18] A.P. Fiske, T.S. Rai, Virtuous Violence, Cambridge University Press, Cambridge, 2015.

[19] K.M. Norgaard, Living in Denial: Climate Change, Emotions, and Everyday Life, The MIT Press, Cambridge, 2011.

[20] R.J. Lifton, Beyond psychic numbing: a call to awareness, Am. J. Orthopsychiatry 52 (4) (1982) 619–629.

Security

30

Deluxe energy: Newly commodified regimes of luxurious energy

Arthur Mason

Department of Social Anthropology, Norwegian University of Science and Technology, Trondheim, Norway

1 Introduction

In this chapter, I expand Christopher Berry's [1] taxonomy of "residual" expenditure to include a category of luxurious energy which I refer to as *Deluxe Energy*. Discussion of luxurious energy is largely overshadowed by idealizations of profligate, efficient, and basic energy uses [2,3]. The latter descriptions depict energy use within the embeddedness of routines of work and leisure that constitute expectations of comfort, convenience, and cleanliness [4–6]. By contrast, luxurious energy may be considered a type of luxury consumption that Werner Sombart [7, p. 87] designates as "seigniorial," by which he means display and ostentation.

For analytical purposes, I refer to this seigniorial form of energy use as Deluxe Energy. As a form of luxury, deluxe energy reflects a dual capacity for doing work: First as a method for bringing about change, and second as a form of social rivalry and ambition through display. Similar to the way Sombart understands luxury as an expression of the *nouveaux riche* who possess nothing besides their money and have no other distinctive quality other than their ability to spend the large means at their disposal on lavish living, deluxe energy communicates a materialistic and plutocratic world view. As such, the notion of deluxe energy raises several questions: For what purpose has deluxe energy become a visible expression of lavish living? How does this expression exhibit itself and who are its main agents of promotion? Finally, how did energy services transform into a type of *social good* whose chief characteristic is luxury?

2 Deluxe energy

Objects of luxury derive exceptional character and authority from their singular craftsmanship evident in design or by exorbitant price [8]. They function as a visible signature of distance from the world of practical necessity and basic need [9,10]. The social bearer of luxury thus represents an image of uniqueness and extravagance. In his well-regarded analysis, Christopher Berry [1] defines luxury as "refined" goods that pertain to four categories of universal basic needs: sustenance, shelter, clothing, and leisure. Here refinement suggests an indulgence over universal satisfaction. Thus, luxuries admit to substitution because the desire for them "lacks fervency" (p. 41). Berry's insight further suggests a paradox: On the one hand, modern advertisers employ the promise of luxury to sell as many products as possible, while on the other hand, the aim of proclaiming luxury implies exclusiveness. After all, it is in line with exclusivity that luxury goods become associated with rarity and luxury. But for this gambit to be effective, the goods in question must not only be desired but widely desired: "The image is that although only a select few now enjoy the [luxury good], many others would also like to enjoy it" (p. 5). In other words, luxury is a type of "residual" expenditure tied to a taxonomy of needs-based satisfactions [11].

The expression deluxe energy thus marks luxurious energy not only as a residual expenditure to basic energy use, but also as a mode of consumption that conspicuously exceeds basic need. Drawing on Berry's taxonomy above, deluxe energy may be considered as the fifth category of residual expenditure to a universal category of energy need. Basic energy use is a need that includes home heating, cooling, lighting, cooking, transportation, and other "essential purposes" [12, p. 3]. These energy use rights have been historically constructed from technological advances and more than a generation of inexpensive, readily available energy so that populations in the United States and Europe have come to believe in them to be obligatory and basic [13]. By contrast, deluxe energy describes a kind of conspicuous consumption whose aim is social rivalry in the form of divisions rooted in a commodity-based class system. Conspicuous consumption first appeared in the early 20th century in response to an initial stage of mass consumption. Sociologist Thorstein Veblen [14] observed social goods whose distinguishing features promoted stratification. These features emphasized exaggerated forms of superfluity that served as expressions of distance to practical necessity and absence from work.

3 Restructuring and creative destruction

In recent years, energy use has become an expression of social rivalry partly resulting from its deregulation and thus transformation from a basic need into a luxury commodity. With the restructuring of industries in electricity and natural gas, energy in the 1980s transitioned from an integrated and regulated industry into one composed of distinct but interlocking competitive segments [15]. This process reflects a departure from the industry's 60-year natural monopoly market. This doctrine imposed government ideals of growth and established relationships between utility regulators as representatives of the public and companies subject to their jurisdiction [16, p. 4]. Restructuring induced a "gale of creative destruction" [17, p. 16], the result of which is that energy can now be bought and sold through market participants [18,19].

The effects on consumption are especially subtle and take the form of increased convenience and reliability of electricity from the 1990s, creating a class of consumers described by reliance on a "high-energy civilization" [3]. From Hollywood to Silicon Valley, a higher reliance on electricity provides greater flexibility of energy uses and increased amounts and faster delivery of information. These decentralized formations draw attention to the role of commodified energy services in sustaining powerful imaginaries that shape perceptions about postindustrial progress while representing new time-space experiences. The allure of a new regime of living is intensified through the speed of internet exchange and its transfer of signals and light, a condition [20, p. 58] of "dispersed material infrastructures." For Jennifer Gabrys [20], this kind of material performativity creates an illusion of immateriality free from the chemicals, metal, and plastics of its resource requirements. In this way, industry restructuring has accompanied new forms of social differentiation, especially since the 2008 recession and austerity measures that have increased disparities between the wealthy and the middle class [21,22].

4 Public-based energy services V commodified public spaces

The residual effect of deluxe energy with its ostentatious commodified character may be contrasted with public-based energy services whose values utilize standards of basic necessity. Such standards appear in professional discourses; for example, the *Illuminating Engineering Society Handbook*—that outlines minimum illumination for public spaces and whose design specifications offer independent criteria such as esthetic preferences for indirect lighting, requirements of outdoor spaces, or evening perceptual needs, as evident in mass transit lighting and airports.

By contrast to these standards, energy users of a new era of commodified public spaces such as the American Express Centurion Lounge can, according to the website, "relax, recharge and reboot with a host of services and amenities designed to anticipate travelers' needs." Here travelers' needs are presented as being met by basic amenities, yet the services offered are luxurious compared to those provided in the surrounding non-commodified public spaces.

Situated inside the airports of America's largest cities, the Centurion Lounge is available exclusively to American Express Platinum Card Members. It offers a high-energy experience in the form of shower suites, flat-screen TVs, high-speed internet, and fresh seasonal fare. These public spaces are commodified because access requires membership through an annual fee of $400. They are public because of their location inside airports, where security and operation costs are underwritten through tax-based revenue. Their seigniorial character therefore rests upon their status as luxurious retreats of high-energy consumption that lay close to public spaces whose identical taxonomic forms of energy use maintain standards regulated through ideals of public necessity. Other examples of the seigniorial character of deluxe energy may be visible in the way automobile fuel efficiencies have been retargeted toward higher performance or in the rise of high-end chain retailers, such as Williams-Sonoma and the like, that promote trends toward professional ranges in home kitchens, with 6–8 burners and two ovens. As with the Centurion Lounge, these examples also align with Berry's definition of luxury as a residual expenditure to basic needs because of the way deluxe energy and basic use carry taxonomic form, for example, in transportation and cooking.

In his book *Bobos in Paradise*, David Brooks [23] identifies a new elite whose intertwined values of the countercultural 1960s and enterprising 1980s are expressed through luxury energy expenditures that characterize the world they occupy—these are the primary agents of deluxe energy. The esthetic commitments of Bobos (those who have wed the bourgeois world of capitalist enterprise to the hippie values of the bohemian counterculture) meld anti-establishment views with elitism while favoring individual creativity over the organizational corporate structure. Although Bobos seek to destroy the conservative beliefs of mainstream society, they also end up replacing this order with new rules to improve the performance of the new professional. Working creatively results in working continuously, which means embracing the Puritan work ethic (p. 165). As Brooks argues, the heightened visibility of this elite and the various new kinds of arenas they occupy (subject niche, demeanor, marketing, conferences, television, status-income disequilibrium) revolve around spending lavishly on deluxe energy esthetics, such as fancy kitchens and loft spaces that signify work (in its residual luxury form).

In his example of attendance at the World Economic Forum in Davos, Switzerland, Andrew Sorkin [24] identifies the deluxe energy travel requirements of this professional elite, with costs upward of $150,000. His price list includes car rental ($10,000), a helicopter trip from Zurich to Davos ($3400 each way), first-class fare from New York to Zurich ($11,000), and accommodation (up to $140,000 per week). He sums up these staggering numbers by citing the WEF annual report indicating that it brings in $185 million and spends "nearly all of it, with almost half of its costs going toward events."

5 Conclusions

To conclude, deluxe energy may be considered a type of consumptive "energy landscape" with the potential for landscape transformation [25]. Its form of luxury is dominated by electricity. Through an emphasis on social rivalry and residual expenditure, it is a subsystem of the larger basic energy environment. Finally, it is relatively dynamic, open to continuous realignment based on the social demands of material distance from basic needs.

References

[1] C. Berry, The Idea of Luxury, Cambridge University Press, Cambridge, 2004.
[2] G. Prins, On condis and cooth, Energy Build. 18 (1992) 251–258.
[3] V. Smil, Energy in the twentieth century, Annu. Rev. Energy Environ. 25 (2000) 21–51.
[4] D. Ghanem, Energy, the city and everyday life, Energy Res. Soc. Sci. 36 (2018) 36–43, https://doi.org/10.1016/j.erss.2017.11.012.
[5] E. Shove, Comfort, Cleanliness and Convenience, Berg, 2003.
[6] E. Shove, F. Trentrmann, R. Wilk, Time, Consumption and Everyday Life, Berg, 2009.
[7] W. Sombart, Luxury and Capitalism, University of Michigan Press, 1967[1913].
[8] S. Gundle, Glamour. A History, Oxford University Press, 2008.
[9] M. Featherstone, The object and art of luxury consumption, in: J. Armitage, J. Roberts (Eds.), Critical Luxury Studies, Edinburgh University Press, 2016, pp. 108–131.
[10] K. Klingeis, The power of dress in contemporary Russian society, Laboratorium 3 (1) (2011) 84–115.
[11] N. Fraser, Unruly Practices, University of Minnesota Press, 1989.
[12] NRC (National Research Council), Thinking about energy, in: P. Stern, E. Aronson (Eds.), Energy Use, Freeman & Company, 1984, pp. 14–31.
[13] T. Wilbanks, Geography and our energy heritage, Mater. Soc. 7 (3/4) (1983) 437–452.
[14] T. Veblen, The Theory of the Leisure Class, Mineola, 1912 [1994].
[15] A. Tussing, B. Tippee, The Natural Gas Industry, PennWell, 1995.

[16] R. Hirsh, Power Loss, The MIT Press, Cambridge, 1999.

[17] FERC, State of the Markets, Federal Energy Regulatory Commission, Washington, DC, 2000.

[18] J. Doucet, S. Littlechild, Negotiated settlements and the National Energy Board in Canada, Research paper, School of Business, University of Alberta, 2006.

[19] P. MacAvoy, The Natural Gas Market, Yale University Press, 2000.

[20] J. Gabry, Digital Rubbish, University of Michigan Press, Ann Arbor, 2011.

[21] J. Armitage, J. Roberts (Eds.), Critical Luxury Studies, Edinburgh University Press, 2016.

[22] P. Calefato, Luxury, Bloomsbury, London, 2014.

[23] D. Brooks, Bobos in Paradise, Simon and Schuster, 2000.

[24] A. Sorkin, A hefty price for entry to Davos, New York Times (2011). 24 January.

[25] M. Pasqualetti, S. Stremke, Energy landscapes in a crowded world, Energy Res. Soc. Sci. 36 (2018) 94–105, https://doi.org/10.1016/j.erss.2017.09.030.

31

Does security push democracy out of energy governance?

Kacper Szulecki

Climate and Energy Research Group, Norwegian Institute of International Affairs, Oslo, Norway

1 Introduction

Energy security, most often defined as the stable supplies of sufficient volumes of energy (or satisfactory energy services) at affordable prices and acceptable environmental costs, is a key policy objective for modern states. In principle, it is not only fully in line with liberal democracy but acts as a sort of necessary condition for order and democratic politics to be maintained. In the modern socio-technical imaginary, ayont lies chaos, disorder, havoc, and societal decay epitomized by the looming threat of a blackout or the historical memories of the 1973 Oil Crisis [1].

However, while securing sufficient energy supplies is an important goal that the state has in its service to society, we have to understand energy security as a concept which is deeply and intrinsically *political*. So is the definition of what constitutes energy security, what are the vulnerabilities of the energy system, what are the values that energy security seeks to protect, the acceptable costs and affordable prices, and last but not least—who is the referent object, and how are threats defined. In other words, the way insecurities are described, and the way energy provision is turned into a security issue both have important implications for energy governance and the idea of energy democratization.

What is more, security considerations can also clash with democracy. Executive steering without the need for securing social acceptance and without considering the implications of energy policy decisions beyond the aggregate state level is a temptation many expert-driven sectors face, especially in times of crisis. Think of France's formidable energy transition effort in the second half of the 1970s, known as the Messmer Plan. It involved a massive rollout of nuclear power, which was supposed to achieve the dual goal of providing energy security (in the aftermath of the 1973 Oil Crisis) and powering an industrial renaissance. However, a centralist mindset, grandiose modernist visions, and untransparent expert steering led to a

Energy Democracies for Sustainable Futures
https://doi.org/10.1016/B978-0-12-822796-1.00031-0

very visible democratic deficit, which in turn spawned protests among scientists questioning the premises of the plan, as well as different social groups.

This is why using security as justification for a move toward exceptional measures outside the eye of the public, beyond constitutional scrutiny and democratic deliberation, can indeed push democracy out of energy governance, even though both empirical and normative research shows that we need more, not less democratic legitimacy.

How is the democratization of energy governance in the times of an imminent energy transition and decarbonization constrained or empowered by talking security? This chapter seeks to explore three theoretical dichotomies. The first is the perception of security as something inherently positive or negative, which is a major division line in Critical Security Studies. I turn the reader's attention to this dichotomy to show why, how, and when security becomes an enemy of democracy. Further, the chapter looks at the change and non-change that security invocations bring about. Following observations made recently by Bruno Latour and Ole Wæver about (non-)change and (non)action, I discuss the significance of this for the transition toward a decarbonized world and the possibility of broadening the scope of democratic governance along the way. Finally, I take on the two natures of security's cession from "politics as usual" that can either take the form of Carl Schmitt's "exceptional politics," which has a clear authoritarian and anti-democratic edge, or Hannah Arendt's "politics of the extraordinary" providing a rare boost to democratic legitimacy and bringing meaning back to politics. Merged with a human security perspective, this can be a normative blueprint for thinking about the way security concerns can unlock the democratizing and emancipatory potential of the socio-technical transition toward a low carbon energy system.

2 Security as a tool of discipline or emancipation

Security is a fundamental human need, and it is only in the absence of existential threats that communities can thrive. This basic understanding of security underpins the conventional definition of *energy security*, seen as "adequate, reliable supplies of energy at reasonable prices and in ways that do not jeopardize major national values and objectives" [2]. Only if energy needs are fulfilled, and in a manner that does not put other important values under pressure, can societies and economies function without interruption. In principle, this makes energy security a basic condition for any moves toward the democratization of energy governance. But the relationship between security and democracy is not as simple as it may appear at first.

While the fundamental ontological security of a community, i.e., its ability to survive, appears as a non-negotiable and "objective" social fact, the quality of energy services that citizens take for granted, as well as the specific values and objectives that energy provision should not jeopardize, is both subjective and contingent. Even a more objectively oriented definition of energy security, emphasizing not the national economy but the equilibrium of the energy system itself, i.e., seeing energy security as the "low vulnerability of vital energy systems" [3] cannot escape the subjective element in the way different communities define what they consider to be vital.

What we as citizens expect of the energy system varies greatly, as do the values that energy supply is meant to safeguard. The threshold of basic energy security in a developing country can imply having access to a grid which can power basic appliances. Meanwhile, in highly developed societies "for most citizens, energy is available 'on tap,' it is ubiquitous and un-intrusive," which has a direct impact on how we understand energy (in)security [4].

Energy security, beyond extreme situations, thus becomes a figure of speech and a rhetor-ical device more than an objective fact [5]. It is called up in policy documents and political speeches, party programs for "keeping the lights on"—constructed more or less explicitly in opposition to a threat of a blackout, economic decline, and societal turmoil.

This is deepened by the associations of security itself. As we have already seen, security is a basic need and is "ultimately about life and death and therefore the things that ensure our continued existence" [6]. Security concerns are a top priority, but that also implies that any evocation of security can be used to catch attention, move an issue up the agenda, or even to justify exceptional measures that go beyond standard procedures. This is *securitization*, "the discursive process through which an intersubjective understanding is constructed within a political community to treat something as an existential threat [...] and to enable a call for urgent and exceptional measures to deal with the threat" [7].

This is the premise of the so-called Copenhagen School in security studies, which adopts a very strict and militaristic understanding of security, but turns our attention from what security is to what it does, i.e., the perlocutionary speech act. Through that act of calling up security, the political situation changes which can result in moving an issue outside the realm of democratic scrutiny, and allowing for *exceptional measures*—breaking with usual practices of governance—to be applied.

Securitization's emphasis on the downside of security stands in stark contrast to the way we usually think of it, i.e., in positive terms. On the surface, it also appears to contrast with the Copenhagen School's critical security sibling, known as the Welsh School. The proponents of this latter approach see survival as a basic need, unsurprisingly, and security as "survival-plus"; that is, a state in which individuals are not only able to subsist but also have the freedom to make choices. This opens the way for societal evolution toward *eman-cipation* [8].

However, the division between these two critical approaches is overstated or simply mis-placed. Securitization is not interested in security as such but rather in the way security is used and *abused* as a justification of certain actions—and inactions. Meanwhile, the emanci-patory approach of the Welsh School targets the actual content and locus of security, empha-sizing that it has to be perceived and evaluated not from the perspective of states but from that of individual people.

To illustrate how these two schools might approach the question of security in energy transitions, e.g., around increasing decentralization and share of renewables: a securitiza-tion perspective might be interested in how references to energy security are used either to block or accelerate certain elements of the transition, whereas the emancipatory perspective of the Welsh School encourages pairing considerations of security with social inclusion, e.g., through energy justice frameworks, to underline that energy transitions can make various groups differently (in)secure in distinct ways.

Taken together, these two critical approaches have a complementary normative agenda, focusing on defending the society from a disciplining use of security which suffocates dem-ocratic political contestation—and on the other hand, helping us reimagine security in a way which allows for unleashing social emancipation. These two approaches can provide very useful critical tools to analyze the politics of energy transitions and decarbonization.

3 The (non-)change of decarbonization

The imminent climate crisis increases the pressure to act and to ratchet up efforts to mitigate climate change. This pressure for action can have an anti-democratic edge too, as democratic deliberation is downplayed or even suspended with the aim of increasing decision-making effectiveness, e.g., in the form of "silver bullet" mitigation megaprojects, strong emphasis on carbon capture and storage technologies or geoengineering. Let us leave aside the debate between a democratic and a "green authoritarian" response; suffice it to say that it is neither inherently more ambitious nor effective in tackling climate change [9], but its strengths lie elsewhere and there is no reason to suspend democratizing ambitions.

However, this is not the only way security can be used as a justification in a way which is in fact countering the broader social interest. As Wæver explains, apart from being a rationale for exceptional actions, irregular behavior of the power elite and transcending normal democratic practices, securitization is also "a mode of intervention that blocks something specific and in a specific way: by defining what is not allowed to happen and can therefore be prevented by all means necessary" [10].

When the threat presented is indeed something that could potentially undermine the regular functioning of the community—for instance, a sudden failure of a single strategic infrastructure, on which an entire energy system hangs—securitization can in fact be a mobilizing impulse to move an issue up on the list of priorities, seeking a nonpartisan solution for the benefit of all. In other instances, security can be abused, i.e., a threat can be blown out of proportion to boost the executive or empower the technocrats in charge at the cost of other stakeholders, advancing a particularistic agenda [5,11].

But there is also a third securitization scenario which is gaining in prominence in the context of decarbonization. Wæver points out that securitization is "the selection of non-change" [10]. The question of what change is enabled and what is prevented becomes particularly salient in the light of decarbonization of the economy for the purpose of mitigating climate change.

Securitization is central to understanding the political (in)feasibility of necessary transition, and critical energy security studies have all the necessary tools to study this "production of non-change," identifying where, why, and by whom change which is not only possible but needed is stalled. In his recent essay "Down to Earth" [12], Bruno Latour argues that climate denial and other campaigns to block climate action are performed rhetorically around threat/defense and from attractive subject positions associated with embodying defense. Similar observations are made by the climate scientist Michael Mann, who posits that a "new climate war" is waged between the entrenched interests of the fossil industry, spreading action-skepticism toward climate mitigation, and the powers of imminent decarbonization [13]. The rhetorical commonplace of security is the most important of these subject positions.

In this mode, the referent object of security, i.e., what is to be safeguarded, is also our civilization, but it is understood more in terms of the energy-intensive and oil-consuming lifestyle. The threat, on the other hand, is identified very differently than in climate change securitization—which speaks of a "climate catastrophe," or like the increasingly impactful youth climate movement—*extinction*.

In the non-change mode, disruptive innovation and rapid change are cast as the threat. This is particularly visible in the skeptical anti-decarbonization discourse around the increased role of distributed renewable energy sources. The threats are most often located in the techno-economic realm. Renewables are portrayed as instable, unreliable, and quite possibly progenitors of energy shortages, including a total blackout [1]. Such narratives were spun in 2021 as record low temperatures hit the state of Texas and left millions without electricity for many hours or days. Despite the clear failure on the part of the grid, as well as conventional plants (gas as well as nuclear), media outlets were flooded with pictures of frozen wind turbines and the notion that renewable energy was to be blamed.

Renewables are also routinely portrayed as costly, requiring subsidies, or on the other hand, making electricity on the wholesale market too cheap, creating a money gap for investors. All these threats are based on actual features of renewable energy sources, such as intermittency, high upfront capital costs, and low operational costs (allowing renewables to bring down wholesale prices). But in the anti-decarbonization narrative, or in its next of kin, the conservative anti-decentralization narrative (sported particularly by nuclear energy enthusiasts), they are not presented as challenges to be overcome or minimized, but as dangerous departures from a "normal" functioning of the energy system, i.e., the centralized and fossil-based status quo.

This action-skepticism employing the securitization of energy can thus be deployed in ways which are not only undemocratic in terms of the governance process, participation, or franchise, but deeply anti-democratic and antisocial, as they counter the common good of climate change mitigation.

4 The anti-politics of exception and the politics of the extraordinary

If security is used to block deliberation, depoliticize energy governance, or conversely to sustain non-change, does that mean that we should object to energy security talk altogether? No—there is also an important potential in securitization, which under the right conditions can fuel democratization as well as effective decarbonization.

Without doubt, security is a prerequisite for democracy, but it can also undermine it. It can limit the space for democratic politics, when issues become framed as matters of security (securitized) and therefore removed from the sphere of deliberation and into one of a nonpolitical *state of exception*. This vision of governance is rooted in the writings of Carl Schmitt [14] who famously stated that the "sovereign is who decides on the [state of] exception" (*Ausnahmezustand*). The sovereign is the subject of politics in that (s)he knows best what direction governance should take and has the means to implement his/her will. However, these are not necessarily aligned with the needs and will of the society at large.

That was precisely the observation that underlined the activity of peace activists speaking out against Cold War securitization. Exception petrifies not only the threat/solution pair, but the broader paradigm in which values that need to be protected are defined. In energy governance, the politics of exception can easily lead to the encroachment of a state-backed governmental-industrial complex, bringing together fossil energy incumbents and conservative (in their view of energy transition, not ideological) political elites [15]. However, they can also lead to alternative energy pathways which undermine democracy and stability, e.g., the dystopian control society associated with some smart city visions, or the potential devastation of forest ecosystems though the unchecked expansion of wood-based bioenergy [16].

But this is not the only possible effect of securitization. In stark contrast to Schmitt's state of exception, the vision of politics which underpins the Securitization Theory draws on the writings of Hannah Arendt [17,18]. As Williams [19] emphasizes, securitization can lead not only to the politics of exception, but also the *politics of the extraordinary*. He suggests this can also lead to popular mobilization, self-determination, and the self-assertion of a democratic sovereign. To illustrate the implications of this extraordinary state for energy governance facing the climate crisis, we can think of the post-World War II reconstruction of Europe—a massive effort fueled by hope and enthusiasm which delivered both an infrastructural overhaul, economic recovery, and a reinvigoration of democratic politics. Barack Obama's 2008 slogan "yes, we can!" and Angela Merkel's famous quip from the 2015 migration crisis, "Wir schaffen das," are both examples of top-down impulses aimed at generating this kind of necessary enthusiasm, but they can succeed only if governments can nurture that extraordinary potential and unleash the energy of the civil society.

5 Conclusions

To sum up, while energy security is certainly desirable, we should be wary of its political uses and instances where it is used to justify actions and modes of governance by those in power. Thinking in terms of energy system security can help identify important vulnerabilities which have to be addressed to maintain the uninterrupted functioning of our societies. But following Cherp and Jewell's understanding of energy security also alerts us to the underdefined and unspecified elements which are the object of political contestation. Security is often used to advance a very particular agenda. Even if dressed up as raison d'état or national interest, it very often remains a very narrow understanding of a security to be achieved for a common good to be protected. Turning matters of energy governance into security issues (securitization) can lead to either de-politicization of things that should remain a matter of political conflict (e.g., the technological pathways toward decarbonization, the burden sharing and cost/benefit distribution of transitions, etc.), or to maintaining non-change where change should be ratcheted up.

However, security's perlocutionary and performative function should not be demonized. It is a double-edged sword, but one of the edges can help democratic politics cut through apathy and inaction to unleash civic enthusiasm and hope, enabling a transition which is not only (hopefully) leading to a carbon-free society and mitigating catastrophic climate change, but also reintroducing meaning into democracy.

For this, the emancipatory dimension of security—human security—needs to be integrated into energy security governance, and there seems to be no other way than through integrating problem-solving and normative approaches; that is, bringing more considerations of inclusiveness, energy justice and democracy into socio-technical and techno-economic blueprints for low carbon transitions. This can occur through bottom-up pressure from civil society itself, empowered by self-organization, but needs to become institutionalized both through forums for dialog (e.g., climate panels, stakeholder decarbonization commissions) and mainstreamed into public policy and the ways in which decisions are made at different levels of administration.

References

[1] K. Szulecki, J. Kusznir, Energy security and energy transition: securitisation in the electricity sector, in: K. Szulecki (Ed.), Energy Security in Europe: Divergent Perceptions and Policy Challenges, Palgrave, London, 2017, pp. 117–148, https://doi.org/10.1007/978-3-319-64964-1_5.

[2] D. Yergin, Energy Security in the 1990s, Foreign Aff. 67 (1988) 69–82.

[3] A. Cherp, J. Jewell, The concept of energy security: beyond the four As, Energy Policy 75 (2014) 415–421, https://doi.org/10.1016/j.enpol.2014.09.005.

[4] E. Commission, European Energy Security Strategy, 2014.

[5] K. Szulecki, The multiple faces of energy security: an introduction, in: K. Szulecki (Ed.), Energy Security in Europe: Divergent Perceptions and Policy Challenges, Palgrave Macmillan, London, 2017, pp. 1–29.

[6] M. Bourne, Understanding Security, Palgrave, London, 2014.

[7] B. Buzan, O. Wæver, Regions and Powers: A Guide to the Global Security Order, Cambridge University Press, Cambridge, 2003.

[8] K. Booth, Theory of World Security, Cambridge University Press, Cambridge, 2007.

[9] M. Povitkina, The limits of democracy in tackling climate change, Environ. Polit. 27 (2018) 411–432, https://doi.org/10.1080/09644016.2018.1444723.

[10] O. Wæver, What is constantly changing? Security is! in: Security Dialogue, 2019, pp. 17–18.

[11] J. Jewell, E. Brutschin, The politics of energy security, in: K.J. Hancock, J. Allison, J. Jewell, E. Brutschin (Eds.), The Oxford Handbook of Energy Politics, Oxford University Press, Oxford, 2019.

[12] B. Latour, Down to Earth. Politics in the New Climatic Regime, Polity Press, Cambridge, 2018.

[13] M.E. Mann, The New Climate War: The Fight to Take Back Our Planet, PublicAffairs, New York, 2021.

[14] C. Schmitt, Political Theology: Four Chapters on the Concept of Sovereignty, Univ. of Chicago Press, Chicago, 2008.

[15] K. Szulecki, Securitization and state encroachment on the energy sector: politics of exception in Poland's energy governance, Energy Policy 136 (2020), https://doi.org/10.1016/j.enpol.2019.111066, 111066.

[16] J. Szulecka, Towards sustainable wood-based energy: evaluation and strategies for mainstreaming sustainability in the sector, Sustainability 11 (2019) 493, https://doi.org/10.3390/su11020493.

[17] L. Hansen, Reconstructing desecuritisation: the normative-political in the Copenhagen School and directions for how to apply it, Rev. Int. Stud. 38 (2012) 525–546, https://doi.org/10.1017/S0260210511000581.

[18] O. Wæver, Politics, security, theory, Secur. Dialogue 42 (2011) 465–480, https://doi.org/10.1177/0967010611418718.

[19] M.C. Williams, Securitization as political theory: the politics of the extraordinary, Int. Relat. 29 (2015) 114–120, https://doi.org/10.1177/0047117814526606c.

32

Global energy transition risks: Evaluating the intergenerational equity of energy transition costs

Ambika Opal and Jatin Nathwani

Waterloo Institute for Sustainable Energy, University of Waterloo, Waterloo, ON, Canada

1 Introduction

Sovacool and Dworkin define energy justice as "a global energy system that fairly disseminates both the benefits and costs of energy services, and one that has representative and impartial energy decision-making" [1]. Fairly disseminating the benefits and costs of energy services means disseminating across several dimensions such as geography, economic class, age, gender, and—the focus of this chapter—time.

Climate change and its impacts will manifest over time across several generations. Mitigation and adaptation measures that are taken in the near term will affect future generations, and the severity of impacts will vary depending on the efficacy of the chosen interventions. Intergenerational equity, similar to considerations of temporal energy justice, rests on the principle that all generations, present and future, must have similar "options, quality, and access" to natural and cultural resources, and actions of one generation must not compromise future generations [2]. Applying intergenerational equity means that decisions made today must be balanced by the impacts and requirements of future generations.

As global temperatures rise, the primary effects on physical, social, and economic security will become more pronounced. Hsiang et al. estimate that for each degree of global warming, states will see a loss of 1%–20% of their annual economic output [3]. These losses will not be shared equitably across generations, with loss increasing proportionately to a rise in temperature. On the other hand, current generations will bear a larger economic burden, such as investment in mitigation and adaptation measures and addressing the inherited burden of past generations' inaction.

In this chapter, we explore and compare the economic burdens faced by different generations due to additional costs needed to finance the global energy transition and economic losses due to global warming, as well as discuss democratic policy mechanisms that could be used to generate a sufficient level of investment. In order to meet the global climate targets, current and near-term generations will shoulder a disproportionately large burden to finance the transition to a low-carbon global energy system. Future generations will not only benefit from these early investments but will also face profound and disruptive economic and social dislocations associated with the rising global temperatures [4]. We analyze the distribution of these economic impacts across generations to determine intergenerational financial equity. Three scenarios will be tested: (1) climate action beginning in 2020 sufficient to limit global warming to 1.5°C, (2) delayed climate action beginning in 2030, and (3) delayed climate action beginning in 2040. The Haudenosaunee "Seven Generations Principle" will be applied, and in the analysis we consider a span of seven generations starting from 2020, with an assumption of approximately 20 years per generation [5].

2 Analysis

Public electricity generation and heating creates approximately 41% of global carbon emissions, making it the largest single sector contributor to climate change [6]. A rapid and deep decarbonization of global energy systems is necessary to avert severely adverse impacts on social and natural systems associated with global warming.

The Intergovernmental Panel on Climate Change (IPCC) special report on global warming above 1.5°C outlines the potential global emissions scenarios, mitigation pathways, and impacts of global warming on natural and human systems. The IPCC's modeling results suggest that to stay below the 1.5°C threshold, global electricity and heating generation must come from 51% renewable sources by 2030, 77% by 2050, and almost 100% by 2100. The current share of supply generated by renewable sources in 2020 is 26% [7].

Fig. 32.1 shows that the shift to renewable energy must be rapid in the first 20–30 years, and the rate of change slows over time. This means different generations bear varying amounts of

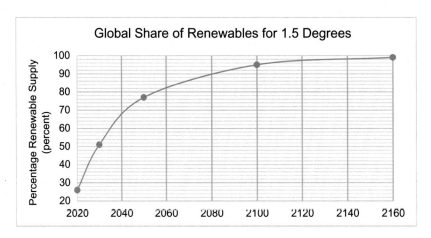

FIG. 32.1   Global renewable supply necessary for limiting warming to 1.5 degrees.

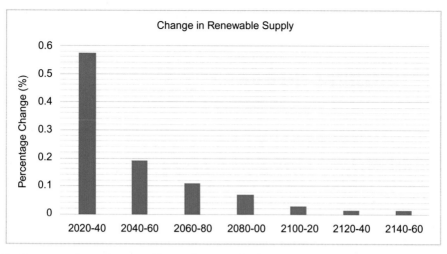

FIG. 32.2 Percentage change in renewable supply required per generation.

responsibility for replacing the existing nonrenewable energy infrastructure. In Fig. 32.1, it is assumed that the global share of renewable supply would be 95% in 2100, and 99% in 2160. Fig. 32.2 shows the relative change in renewable energy supply for which each generation will be responsible.

This distribution, however, assumes that the cost of renewable technology and its installation will remain constant over time. For a more accurate distribution of costs across generations, the change in renewable technology cost must be accounted for, as well as the discount rate. For further analysis, a 2% annual decrease in renewable technology cost and a 1% discount rate is assumed [8–10].

The IPCC estimates that an additional $830 billion 2010 US dollars (or $991.13 billion 2020 US dollars) must be mobilized globally per year between 2016 and 2050 in order to finance the energy infrastructure and technology to limit warming to 1.5°C [7]. This includes low-carbon or renewable energy technology and energy efficiency improvements. It will be assumed that the $991.13 billion annual cost extends to 2060 for the purposes of this analysis, leading to a total investment requirement of $39.645 trillion USD until 2060. The IPCC does not predict additional investment amounts beyond 2050, so it will be assumed that additional investments after 2060 are proportional to the change in percentage supply of renewable energy as shown in Fig. 32.2. Incorporating the 2% annual decline in renewable technology cost and 1% discount rate, the distribution of cost (in 2020 dollars) would be as follows. The total additional cost for financing energy transitions to limit warming to 1.5°C is calculated to be $42.492 trillion US dollars until the year 2160.

It is important to note that this cost estimate is only for one portion of the energy sector and therefore represents only a small portion of the total investment required for climate change mitigation. Cost estimates for overall climate change mitigation, adaptation, and damages range from tens to hundreds of trillions of dollars, and cost estimates for the energy sector in total are approximately $100 trillion until 2050 [11] (Table 32.1).

TABLE 32.1   Distribution of energy transition costs per generation.

Generation	Percentage of responsibility	Additional cost (billion $/year)	Total additional cost (billion $)
2020–2040	78.8	1674.21	33,484.17
2040–2060	14.5	308.84	6160.83
2060–2080	4.6	97.24	1944.84
2080–2100	1.6	33.65	673.01
2100–2120	0.35	7.45	149.05
2120–2140	0.10	2.06	41.26
2140–2160	0.03	1.14	22.85

The analysis thus far shows that earlier generations will face a disproportionately large cost burden to finance the global energy transition. On the other hand, generations further in the future will face larger climate risk and economic cost due to impacts of global warming. Fig. 32.3 shows the IPCC's global surface warming projections until 2300. As this chapter assumes that humanity is successful in limiting global warming to 1.5°C by 2160, approximately the bottom edge of scenario B1 is used for modeling purposes.

Hsiang et al. estimates that the United States will see a GDP loss of 1.2% per degree of warming per year associated with global warming [3]. For developing countries, Hsiang et al. estimates that the GDP loss will be 2%–20% per degree per year [3]. Kompas et al. estimate GDP losses between 0.144% and 7.98% per year in the long term for 1–2 degrees of warming [13]. For this chapter, a global estimate of 2% GDP loss per degree per year will be assumed. Since electricity and heating systems produce 41% of global emissions, $2 \times 0.41 = 0.82\%$ will be the assumed GDP loss due to electricity and heating system global warming contributions. It

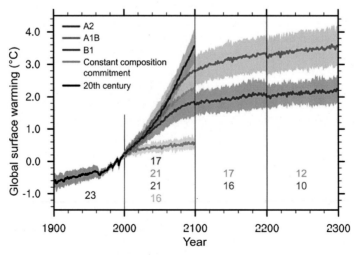

FIG. 32.3   IPCC climate projections to 2300 [12].

TABLE 32.2 Total GDP and GDP losses due to energy transition and global warming per generation.

Generation	GDP per year ($ trillion USD)	GDP loss from energy transition (%)	GDP loss from global warming (%)	Total GDP loss (%)
2020–2040	142	1.18	0.30	1.48
2040–2060	173.27	0.178	0.46	0.633
2060–2080	211.42	0.0460	0.70	0.746
2080–2100	257.97	0.0130	0.88	0.888
2100–2120	314.77	0.00237	0.98	0.982
2120–2140	384.08	0.000537	1.02	1.02
2140–2160	468.65	0.000244	1.05	1.05

is assumed that the total proportion of emissions due to electricity and heat remains constant at 41% over time. Historical proportional emissions can be found in Ref. [14].

It is also assumed that the global GDP growth will be 1% per year between 2020 and 2160. While this assumption is likely not accurate in the short-term (global GDP growth was 2.48% in 2019), a 1% assumption will be used due to declining rates of GDP growth and the uncertain sustainability of long-term economic growth due to climate change [15]. Table 32.2 shows the estimated GDP at the beginning of each generation and the percent of total GDP loss due to the electricity and heating component of global warming, which is calculated from the following equation:

$$\text{Percentage GDP loss from global warming per year} =$$
$$\left(\frac{\text{GDP} \times \text{degrees warmed} \times 2\% \text{ GDP loss per degree} \times}{41\% \text{ emissions from electricity and heat}} \right) / \text{GDP}$$

Table 32.2 also converts the total additional investment cost for the energy transition (as shown in Table 32.1) to a proportion of GDP, and the total GDP percentage loss per year for each generation.

3 Evaluation

Fig. 32.4 shows the distribution of the intergenerational financial burdens associated with the electricity and heating sector, measured in GDP loss per year.

This analysis shows that the current generation (2020–2040) faces the largest financial burden regarding energy transitions out of the next 7 generations, almost 1.5 times higher than the next highest generation in 2140–2160. The following generation, from 2040 to 2060, faces the lowest financial burden, and subsequent generations face an increasing burden per generation. Thus, in order to limit global warming to below 1.5 degrees, the intergenerational financial burden will not be equal.

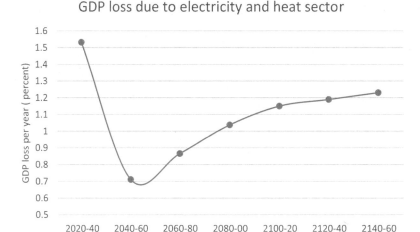

FIG. 32.4   Total GDP loss per generation due to the electricity and heating sector.

The high financial burden on the current generation is in part due to the historical inaction by previous generations. Sanderson and O'Neill calculate that delayed action on climate change mitigation increases costs by $0.6 trillion USD per year in 2020 [16]. It is tempting to delay mitigation further since over time GDP increases, the cost of renewable technology decreases, and the principle of discounting makes future investment more attractive than present investment. However, delayed action will contribute to higher cumulative costs over seven generations.

To evaluate the intergenerational cost of delayed action, the model was used to calculate the GDP losses per year if action was delayed until 2030 or 2040. Global warming values due to delayed action were taken from Schaeffer et al. [17]. It was assumed that 50% of the 2020–2040 infrastructure costs would be shifted to the 2040–2060 generation in the 2030 start scenario, and 100% in the 2040 start scenario. The following graph shows the intergenerational financial burdens due to the infrastructure and climate effects of the heating and electricity sector if significant action starts in 2020, 2030, and 2040 (Fig. 32.5).

Two main insights can be gathered from this figure.

(1) As action is delayed, the cumulative financial impacts of climate change increase, a fact supported by many including [18]. The cumulative financial burden can be represented as the area under each of the curves, which is 6.36 for action started in 2020, 8.53 for 2030, and 9.56 for 2040. This means that over seven generations, if action is delayed by 10 years, the world will cumulatively have to spend 25.4% more to mitigate and adapt to climate change in the electricity and heating sector, and if action is delayed by 20 years, cumulative spending will be 33.5% higher.

(2) The generation that benefits the most from delayed action is the current generation, as the GDP per year losses decrease drastically for the 2020–2040 generation as action is delayed further. On the other hand, the generation that would be adversely impacted most by delayed action is the following generation—our generation's children.

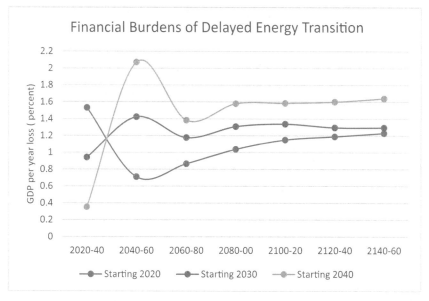

FIG. 32.5   Intergenerational trends in energy transition financial burdens, current and delayed action.

The purpose of this analysis was to determine the trend in intergenerational financial burdens associated with energy transitions. The specific values calculated are highly uncertain and sensitive to assumed values, though the model was tested to ensure it was not highly sensitive to changes in assumed values. The model's assumptions and limitations are listed here.

(1) The model only looks at costs and impacts in the public electricity and heat sector, but does not include other elements of energy such as transportation and industrial processes.

(2) The model assumes a GDP growth of 1% per year for the period of 2020–2160.
  (a) This is an assumption of the average GDP per year between 2020 and 2160, based on long-term scenarios modeled by Leimbach et al. [19].

(3) The model calculates the total additional costs regarding the global electricity and heat sector transition to be $42.492 trillion USD (in 2020 dollars) between 2020 and 2160.
  (a) This value is roughly equivalent to or lower than other estimates of additional costs for the global energy transition (for example, $23 trillion to 2050 in IRENA [9] and $73 trillion total in Jacobson et al.) [20].

(4) The model assumes a decrease in renewable technology cost by 2% per year and a 1% discount rate [8,10].

(5) The model assumes that the electricity and heating sector will retain a 41% share in global emissions between 2020 and 2160 [6].

(6) The model does not account for other major costs in the electricity and heat sector, such as stranded assets in the fossil fuel sector, or the social cost of carbon.

(7) The IPCC values and predictions that were used in this report have inherent uncertainties.

4 Recommendation—Cap-and-invest

The analysis above shows that significant investment is needed between 2020 and 2040 in order to meet the electricity and heating system changes required to limit global warming to 1.5°C. Future generations have a significantly smaller investment burden but face increasing GDP losses due to effects of global warming. This analysis answers the question of when investments should occur, but not the question of how. Current financing strategies and policies in place are not on track to meet the scale of investment required, at least for the period of 2020–2040.

Nathwani and Ng propose a "cap-and-invest" strategy that would cap emissions at the maximum levels allowed under decarbonization goals necessary to remain within 1.5-degree warming pathways and invest in mitigation and innovation initiatives to adhere to these goals [21]. The cap-and-invest strategy is based on a General Environmental Tax (GET) that would "[create] a large 'pool of capital' to de-risk investment in emerging low-carbon solutions in support of an energy infrastructure resilient to the threat of climate change." The general environmental tax (GET) would be an economy-wide tax on consumption, whose revenues would flow into an Environmental Trust Fund (ETF), a fund intended to reduce the intergenerational effects of climate change. The ETF would be governed at arm's length from government entities to ensure disengagement from politics and short time horizons. ETF funds would be allocated to private, public, and community climate mitigation initiatives that drive innovation in the clean technology industry and/or support global decarbonization goals. The concept of public trust funds intended to benefit future generations is similar in principle to state-managed pension funds such as the Canada Pension Plan or the Government Pension Fund of Norway.

To be successful and equitable, the GET and ETF instruments would need to support the democratization of energy systems. We recommend five principles on which to base the development and implementation of the GET and ETF [22].

(1) Citizens are politically engaged and are provided with democratic mechanisms for participation in and engagement with the GET and ETF.
(2) Principles of social justice and inclusion are used to develop the GET and ETF.
(3) The GET and ETF are publicly controlled and disengaged from politics and government.
(4) The ETF is used primarily to support local and community initiatives.
(5) The use of ETF funds is decided through participatory and community planning.

In order to limit warming to 1.5 degrees and reduce cumulative impacts, action must be taken now. The concept of an economy-wide consumption tax and arm's-length governance for a public trust fund is one example of an equitable and democratic mechanism for distributing benefits and costs of climate change over generations.

5 Conclusions

This chapter shows that to limit warming to 1.5°C, financial burdens regarding the electricity and heating sector cannot be perfectly intergenerationally equal and energy just. Significant and disproportionate investment is needed in the short term to finance the transition to renewable energy, and as the planet warms, future generations will see a continually increasing financial burden to adapt to climate change and global warming. One potential mechanism to amass the necessary short-term investment and long-term impacts is a cap-and-invest strategy that includes a General Environmental Tax on economy-wide consumption, as well as an Environmental Trust Fund that governs these funds to improve intergenerational climate equity.

The model presented in this chapter predicted distributions in intergenerational financial burdens for three scenarios: climate action beginning in 2020, and delayed action beginning in 2030 or 2040. It was found that if action is taken now, the current 2020–2040 generation would face the highest financial burden, and the 2040–2060 generation the lowest. If action is delayed, an increasing burden is placed on the next generation from 2040 to 2060, and cumulative costs over seven generations increase significantly. In short, if we do not take significant action now to catalyze the energy transition, we are harming at least the next six generations, most significantly the children of today. In order to promote energy justice and energy democracy, the intergenerational and temporal distributions of climate-related energy costs and damage must be accounted for, and powerful momentum for decarbonizing global energy systems must begin now.

References

[1] B.K. Sovacool, M.H. Dworkin, Energy justice: conceptual insights and practical applications, Appl. Energy 142 (2015) 435–444.
[2] E. Brown Weiss, Climate change, intergenerational equity, and international law, Vt. J. Environ. Law 9 (2007) 615.
[3] S. Hsiang, R. Kopp, A. Jina, J. Rising, M. Delgado, S. Mohan, K. Larsen, Estimating economic damage from climate change in the United States, Science 356 (6345) (2017) 1362–1369.
[4] F. Ackerman, E.A. Stanton, C. Hope, S. Alberth, J. Fisher, B. Biewald, What We'll Pay if Global Warming Continues Unchecked, Natural Resource Defence Council, 2008.
[5] R. Walker, D. Natcher, T. Jojola, Reclaiming Indigenous Planning, vol. 70, McGill-Queen's Press-MQUP, 2013.
[6] International Energy Agency (IEA), CO2 Emissions From Fuel Combustion, 2020.
[7] J. Rogelj, D. Shindell, K. Jiang, S. Fifita, P. Forster, V. Ginzburg, C. Handa, H. Kheshgi, S. Kobayashi, E. Kriegler, L. Mundaca, R. Séférian, M.V. Vilariño, Mitigation pathways compatible with 1.5°C in the context of sustainable development, in: V. Masson-Delmotte, P. Zhai, H.-O. Pörtner, D. Roberts, J. Skea, P.R. Shukla, T. Waterfield (Eds.), Global Warming of 1.5°C. An IPCC Special Report on the Impacts of Global Warming of 1.5°C Above Pre-Industrial Levels and Related Global Greenhouse Gas Emission Pathways, in the Context of Strengthening the Global Response to the Threat of Climate Change, Sustainable Development, and Efforts to Eradicate Poverty, Intergovernmental Panel on Climate Change, 2018.
[8] M. Taylor, Renewable Power Generation Costs in 2019: Latest Trends and Drivers, International Renewable Energy Agency, 2019.
[9] IRENA, Global Energy Transformation: A Roadmap to 2050, International Renewable Energy Agency, Abu Dhabi, 2018.
[10] F. Egli, B. Steffen, T.S. Schmidt, A dynamic analysis of financing conditions for renewable energy technologies, Nat. Energy 3 (2018) 1084–1092.
[11] IRENA, Global Energy Transformation: The REmap Transition Pathway, 2019 Edition—Background Report, International Renewable Energy Agency, Abu Dhabi, 2019.
[12] G.A. Meehl, T.F. Stocker, W.D. Collins, P. Friedlingstein, A.T. Gaye, J.M. Gregory, A. Kitoh, R. Knutti, J.M. Murphy, A. Noda, S.C.B. Raper, I.G. Watterson, A.J. Weaver, Z.-C. Zhao, Global climate projections, in: Climate Change 2007: The Physical Science Basis. Contribution of Working Group I to the Fourth Assessment Report of the Intergovernmental Panel on Climate Change, Cambridge University Press, Cambridge, United Kingdom and New York, NY, USA, 2007, p. 762. Figure 10.4.

[13] T. Kompas, V.H. Pham, T.N. Che, The effects of climate change on GDP by country and the global economic gains from complying with the Paris climate accord, Earth's Future 6 (8) (2018) 1153–1173.

[14] H. Ritchie, M. Roser, Emissions by Sector. Our World in Data, 2020.

[15] Macrotrends, World GDP Growth Rate 1961–2020, 2020.

[16] B.M. Sanderson, B.C. O'Neill, Assessing the costs of historical inaction on climate change, Sci. Rep. 10 (1) (2020) 1–12.

[17] M. Schaeffer, L. Gohar, E. Kriegler, J. Lowe, K. Riahi, D. van Vuuren, Mid-and long-term climate projections for fragmented and delayed-action scenarios, Technol. Forecast. Soc. Chang. 90 (2015) 257–268.

[18] N. Glanemann, S.N. Willner, A. Levermann, Paris climate agreement passes the cost-benefit test, Nat. Commun. 11 (1) (2020).

[19] M. Leimbach, E. Kriegler, N. Roming, J. Schwanitz, Future growth patterns of world regions–A GDP scenario approach, Glob. Environ. Chang. 42 (2017) 215–225.

[20] M.Z. Jacobson, M.A. Delucchi, M.A. Cameron, S.J. Coughlin, C.A. Hay, I.P. Manogaran, A.K. von Krauland, Impacts of Green New Deal energy plans on grid stability, costs, jobs, health, and climate in 143 countries, One Earth 1 (4) (2019) 449–463.

[21] J.S. Nathwani, A.W. Ng, A "Cap and Invest" Strategy for Managing the Intergenerational Burden of Financing Energy Transitions (No. 869), ADBI Working Paper Series, 2018.

[22] M.J. Burke, J.C. Stephens, Energy democracy: goals and policy instruments for sociotechnical transitions, Energy Res. Soc. Sci. 33 (2017) 35–48.

Democratizing energy through smart grids? Discourses of empowerment vs practices of marginalization

Ekaterina Tarasova[a,b] *and Harald Rohracher*[a]

[a]Linköping University, Linköping, Sweden [b]Södertörn University, Huddinge, Sweden

1 Smart grids and the empowered consumer

Smart grids—the digitalization of electricity grids allowing two-way communication between electricity suppliers and users—are generally portrayed as a precondition to accommodate a high share of intermittent renewable energies and an important step toward a sustainable energy system. Consumers and households are supposed to play a much more active and empowered role in smart energy systems [1], as is explicitly stressed in many policy documents [2–5] where the empowered user plays a central role in legitimizing new regulation and public investments. An example is the latest legislation to re-design the European electricity market ("Clean Energy for all Europeans package"). As a press release of the European Commission states: "The new rules are designed to empower energy consumers to play an active role in driving the energy transition and to fully benefit from a less centralized, and more digitalized and sustainable energy system" [6]. Analyzing Swedish policy documents, Wallsten has found that users are presented as "economically motivated individuals" as well as "portrayed as being empowered through enhanced market choices, through enriched information about electricity consumption and as prosumers who produce their own electricity" [4, p. 216].

Strengers summarizes visions of an active smart grid user as "an efficient micro-resource manager" who is interested in and capable of engaging with energy consumption [7]. The association of smart grids with energy democratization is also reflected in public discourses, as Michram et al. demonstrated in the case of the Netherlands and the United Kingdom [8]. The transformation of user roles has been discussed as a transition from a role of electricity consumer to a role of energy citizen, a user with an increased sense of individual responsibility toward energy consumption [9–11]. Szulecki [12] considers users developing into responsible energy citizens and prosumers as a core element of an ideal model of energy democracy, thus establishing a conceptual connection between changing roles of users and democratization in the energy system. Rhetorics of user empowerment and increased participation in the energy system may be a sign of increased importance of democratization in the energy sphere.

However, to what extent such ideas of empowerment, justice, and democracy materialize in the practice of smart grid implementation is yet to be discovered. Research on energy transitions suggests that relations between innovations for a more sustainable energy system and democratization are far from straightforward [13]. Several empirical studies of smart grid users, investigating what user empowerment and active users actually mean (e.g., Ref. [4]) and under what conditions users will become engaged energy citizens [11], come to a conclusion that ideas about active and engaged users remain fuzzy. The conflation of "citizenship" with users, consumers, or beneficiaries individualizes engagement and tends to hide issues of injustice, inequality, and marginalization [14]. The extent to which energy system configurations associated with smart grids actually create spaces for the empowerment of users and social inclusivity remains an open question.

While visions of active and empowered smart grid users may be representative of some social groups, they can hardly give an idea of how users, in general, will engage with the smart grid. Users may have different conditions and opportunities (e.g., Ref. [15]) and not least interest to engage with energy technologies [16]. Specific empirical investigations, for instance, of the use of smart meters by elderly people [17] have not found raised awareness of energy use and more control of their energy consumption. In a similar vein, Sovacool et al. [15] raise concerns about the negative impacts of the smart meter roll-out in the United Kingdom on potentially vulnerable groups such as energy poor or less educated consumers. Potential consequences for vulnerable users would be confused about the use of smart meters or problems with paying the additional costs incurred by smart meters. Similar concerns are voiced on social interactions within smart homes [18] when, for example, less technically interested household members increasingly lose control over their home compared to more technically enthusiastic family members.

It is thus important not only to highlight processes that may lead to a democratization of the energy system and consumer empowerment but also to analyze and unpack processes of smart grid implementation that may have reverse effects and intensify inequalities and injustices. As McCauley et al. warn, new injustices may develop in the course of sustainability transitions [19], and we need to be attentive not only to what happens with old inequalities and injustices but also to the potential rise of new ones when the smart grid is being developed. In the remainder of this chapter, we briefly discuss the social inclusivity of smart grid roll-out in Sweden as an element of a democratic energy system. We draw on a research project on the potential marginalization or even exclusion of certain groups of smart grid users from the expected benefits of smart grid development in Sweden that both authors are part of and focus on how interests and needs of some users may be disregarded and even "designed out" in the implementation process of smart grids.

2 From social groups to conditions and practices

The marginalization of smart grid users can be understood in different ways. In contrast to the conventional electricity grid, smart grid services, such as favorable tariffs for those shifting their electricity use to off-peak times, often require additional equipment (e.g., monitors, control devices) which has to be bought by the user, or they require the use of certain apps or webservices. We may speak of "hard" forms of exclusion when certain users do not have the resources to access smart grid services-whether economically ("energy poor") because they have special needs, or because they lack the competencies to navigate a digitalized environment ("IT literacy"). Despite the focus on "active users" and public funding of a smart grid transition, the techno-economic configurations of the grid which currently emerge may benefit already resourceful consumers. This imbalance may become further pronounced by "weaker" forms of exclusion such as the marketing and adaptation of smart grid applications to particular and economically interesting customer segments. Advertisements of smart homes often only show middle-class families steering their smart single-family houses and family cars. Conditions that may lead to the marginalization of users in smart grids thus overlap to some extent with conditions of digital exclusion since a digitalized electricity grid is part of a general digitalization of society. Underlying conditions reducing participation in a digital society are lack of knowledge, skills, and motivation as well as insufficient access to the Internet and computer appliances [20].

Interviews with Swedish organizations representing potentially marginalized user groups, that the first author conducted in a research project on marginalized households in smart grids, however, revealed that it is not meaningful to make general statements about specific social groups and their use of digital and/or energy products and services. There is a great diversity in groups such as elderly people, energy poor, or inexperienced IT users. Similar to the suggestion of Bouzarovksi and Petrova to think about energy poverty in terms of conditions that lead to energy vulnerabilities instead of socio-demographic approaches [21], we shift the focus from identifying social groups that may be negatively affected by smart grids to specifying conditions and practices that may leave some users more vulnerable and marginalized in smart grid development than others. For instance, it is not the age of a user that makes him or her less likely to make use of a smart energy product or service, but rather a habit, interest, and/or skill to use digital products and services. It is well-known that usage of digital products and services does correlate with age, but age does not lie at the center of vulnerability. The same applies to other groups such as recent immigrants from countries with a low digitalization rate who can be very heterogeneous in their actual IT competencies and moreover often quickly adapt to the new situation [20].

Our findings suggest that conditions and practices that may shape the marginalization of users in smart grids can be found on individual and structural levels. On the individual level, conditions, interests, and needs of individual users play a role. Limited knowledge of the energy system, reduced IT skills, inability to afford a smart (energy) product or service are examples of such individual conditions. Inflexible routines of some users may make them more sensitive in a situation when users are asked to adjust their energy consumption in line with electricity price fluctuation. Moreover, users may not be interested in reducing their electricity consumption due to negligible effects on electricity costs or they generally reject smart grid innovations for reasons such as privacy issues [16]. Designing a more inclusive smart grid would mean being more sensitive to such individual circumstances and interests and actively involving such groups in the design of smart grids or developing services that particularly cater to these groups. Making design processes of smart grids more inclusive of different interests and needs would require strategic efforts. This could happen through collaborations with interest organizations that represent consumers that are identified as potentially marginalized in smart grids or recruiting participants of test projects from diverse social groups.

On a structural level, the marginalization of users in smart grids is shaped by the state of infrastructure, legal provisions, and regulations of the energy sector, channels for public participation, design, and development of smart energy products and services. Infrastructure for smart grids includes the Internet and electricity networks. The respondent from an organization working with rural areas points out that there are still places that are not covered by the high-speed Internet in Sweden, which is essential for smooth access to smart grid services. Participation in smart grids thus becomes also a question of spatial "peripheralization" where IT and energy peripheries overlap [22].

3 Configuring users in smart grid projects

The extent to which smart grids create conditions for the involvement of a broad diversity of users is not necessarily decided and visible from the outset, but materializes in the course of implementations when users become "configured" as part of the new smart grid constellations [23]. At the core of the Swedish smart grid implementation strategy have been large demonstration projects, such as on the island of Gotland (see Ref. [4]) or in the cities of Stockholm and Malmö (see Ref. [24]). Such projects are not just about the testing of new technologies, but about testing and learning from a whole set of socio-technical arrangements, such as the integration of households in smart grids and the adaptation of their social practices, the formation of new visions and expectations, or new business models redefining relations between energy suppliers and customers. The value of such experiments is based on the possibility of creating and convincing certain publics about the usefulness of the arrangements put to test, and of their potential for replication and upscaling [25]. The creation of publics through the demonstrative character of experiments and the relationship to democratic processes through the voluntary or involuntary participation of a wide range of actors is the main theme of the sociological analysis of experiments [26]. Brice Laurent analyzes pilot projects thus as "technologies of democracy" [25], socio-material practices that organize the participation of different publics in the definition and treatment of collective problems.

The large-scale smart grid demonstration project on Gotland is instructive in this respect. While the inclusive character of this field trial and the focus on a diversity of households was widely communicated, this inclusive perspective was successively narrowed down during the implementation process. Not only was preference given to households with enough electricity loads to shift (i.e., are high-consuming), but micro-producers who were interested in selling electricity generated by their PV-panels back to the grid (and thus are typical examples of prosumers) were denied participation in the pilot project with the argument that they would make the trial too complex. Not least, the trial saw a shift from an active role of households in managing their daily energy use to automatic control systems and steering through utilities [4]. Such narrowing of ideas about smart grid users may shape the design of smart grids in a way that makes it unattractive for households similar to the ones excluded from the smart grid trial, for example, households wanting to cover a large part of their electricity use through own production from solar energy. The practice of implementing smart grids through pilot projects thus configures the new socio-technical smart grid arrangements in particular ways and shapes whose interests and needs are inscribed in smart grid development. As respondents in our project suggest, engaging households that are hardly interested in energy products and services requires considerably more effort than involving resourceful users. Our analysis of smart grid pilot projects shows that despite rhetoric of inclusion and participation, the new socio-technical configurations rather serve to keep incumbent actors and hierarchical relations between suppliers and users in place.

4 Concluding points

In our brief discussion we encounter the paradoxical situation that novel energy system configurations are legitimized by imaginaries of a more democratic, empowering, and inclusive energy system, while the socio-material assemblages that emerge and eventually congeal around these technologies still seem largely driven by pre-existing power relations and the interest of incumbent energy system actors.

It seems that in practice little attention is paid to the diversity of user perspectives in smart grid development and to the potential of alternative smart grid configurations. Without embracing diversity and the engagement of users in development processes to a higher extent, it is hardly possible to imagine that smart grid development will democratize the energy system. We see that injustices in smart grid development can potentially materialize with respect to the recognition and incorporation of diverse user perspectives consisting of various experiences, interests, lifestyles, and needs. Social science research can contribute to a more inclusive smart grid development by unpacking the socio-material practices of implementing smart grids and by drawing attention to the marginalization of different social groups in this process.

Funding

This work was supported by the Swedish Energy Agency [grant number 44341-1] and the Göteborg Energi Research Foundation.

References

[1] G.P.J. Verbong, S. Beemsterboer, F. Sengers, Smart grids or smart users? Involving users in developing a low carbon electricity economy, Energy Policy 52 (2013) 117–125.

[2] F. Gangale, A. Mengolini, I. Onyeji, Consumer engagement: an insight from smart grid projects in Europe, Energy Policy 60 (2013) 621–628.

[3] D. Geelen, A. Reinders, D. Keyson, Empowering the end-user in smart grids: recommendations for the design of products and services, Energy Policy 61 (2013) 151–161.

[4] A. Wallsten, Assembling the Smart Grid: On the Mobilization of Imaginaries, Users and Materialities in a Swedish Demonstration Project, Linköping University, Tema Teknik och social förändring, Linköping 730, 2017.

[5] Y. Strengers, Smart Energy Technologies in Everyday Life: Smart Utopia?, Palgrave Macmillan, New York, 2013.

[6] European Commission, Clean Energy for All Europeans: Commission Welcomes European Parliament's Adoption of New Electricity Market Design Proposals, 2019.

[7]   Y. Strengers, Smart energy in everyday life, Interactions 21 (4) (2014) 24–31.

[8]   C. Milchram, R. Hillerbrand, G. van de Kaa, N. Doorn, R. Künneke, Energy justice and smart grid systems: evidence from the Netherlands and the United Kingdom, Appl. Energy 229 (2018) 1244–1259.

[9]   M. Goulden, B. Bedwell, S. Rennick-Egglestone, T. Rodden, A. Spence, Smart grids, smart users ? The role of the user in demand side management, Energy Res. Soc. Sci. 2 (2014) 21–29.

[10] E. Heiskanen, K. Matschoss, P. Repo, Engaging consumers and citizens in the creation of low-carbon energy markets, in: ECEEE Summer Study Proceedings, 2015, pp. 2123–2132.

[11] M. Ryghaug, T.M. Skjølsvold, S. Heidenreich, Creating energy citizenship through material participation, Soc. Stud. Sci. 48 (2) (2018) 283–303.

[12] K. Szulecki, Conceptualizing energy democracy, Environ. Polit. 27 (1) (2018) 21–41.

[13] M.J. Burke, J.C. Stephens, Political power and renewable energy futures: a critical review, Energy Res. Soc. Sci. 35 (2018) 78–93.

[14] B. Lennon, N. Dunphy, C. Gaffney, A. Revez, G. Mullally, P. O'Connor, Citizen or consumer? Reconsidering energy citizenship, J. Environ. Policy Plan. 22 (2) (2020) 184–197.

[15] B.K. Sovacool, P. Kivimaa, S. Hielscher, K. Jenkins, Vulnerability and resistance in the United Kingdom's smart meter transition, Energy Policy 109 (2017) 767–781.

[16] N. Kahma, K. Matschoss, The rejection of innovations? Rethinking technology diffusion and the non-use of smart energy services in Finland, Energy Res. Soc. Sci. 34 (2017) 27–36.

[17] G. Barnicoat, M. Danson, The ageing population and smart metering : a field study of householders' attitudes and behaviours towards energy use in Scotland, Energy Res. Soc. Sci. 9 (2015) 107–115.

[18] L. Nicholls, Y. Strengers, J. Sadowski, Social impacts and control in the smart home, Nat. Energy 5 (2020) 180–182.

[19] D. McCauley, V. Ramasar, R.J. Heffron, B.K. Sovacool, D. Mebratu, L. Mundaca, Energy justice in the transition to low carbon energy systems: exploring key themes in interdisciplinary research, Appl. Energy 233–234 (2018) 916–921.

[20] T. Hornliden, F. Sprung, En Studie om Digitalt Utanförskap, Umecon, Svenska Stadsnätsföreningen, 2016.

[21] S. Bouzarovski, S. Petrova, A global perspective on domestic energy deprivation: overcoming the energy poverty-fuel poverty binary, Energy Res. Soc. Sci. 10 (2015) 31–40.

[22] K. O'Sullivan, O. Golubchikov, A. Mehmood, Uneven energy transitions: understanding continued energy peripheralization in rural communities, Energy Policy 138 (2020) 111288.

[23] S. Woolgar, Configuring the user: the case of usability trials, in: J. Law (Ed.), A Sociology of Monsters: Essays on Power, Technology and Domination, Routledge, London/New York, 1991, pp. 58–99.

[24] D. Parks, Energy efficiency left behind? Policy assemblages in Sweden's most climate-smart city, Eur. Plan. Stud. 27 (2) (2019) 318–335.

[25] B. Laurent, Political experiments that matter: ordering democracy from experimental sites, Soc. Stud. Sci. 46 (5) (2016) 773–794.

[26] N. Marres, D. Stark, Put to the test: for a new sociology of testing, Br. J. Sociol. (2020) 423–443.

Contested scales of democratic decision-making and procedural justice in energy transitions

Chad Walker, Stacia Ryder, Jean-Pierre Roux, Zoé Chateau, and Patrick Devine-Wright

Department of Geography, College of Life and Environmental Sciences, University of Exeter, Exeter, United Kingdom

1 Introduction

In parallel with recent community-led action, published research within energy democracy [1,2] and energy justice [3,4] has proliferated over the past decade. Though nascent, advancements are being made rapidly in both sets of literature, which include critical reviews offering extensive discussions on these emergent bodies of literature [5–7]. In this chapter, we respond to some of the noted shortcomings in the literature by (a) linking energy democracy to a wider theory of democracy, and (b) providing empirical evidence to ground energy democracy-related analyses. We use contrasting case studies from Canada and the United States to contribute to the conceptual debate on different understandings of energy democracy and how these manifest in diverse democratic contexts. Moreover, we showcase the importance of thinking pragmatically about the challenges of employing the concept of energy democracy in relation to regional (or non-local) energy policy.

Our aim with these case studies is to demonstrate how emergent social movements' actions to resist, reclaim, and restructure facets of a wider energy system [2]can politicize the deployment of energy infrastructure. Unlike the majority of the literature, our case studies also draw attention to other dimensions of democracy beyond direct citizen involvement to demonstrate democracy in practice across multiple governance scales and in different energy infrastructure and national contexts. In some instances, these processes involve the use of established democratic institutions (e.g., provincial elections in Ontario, Canada) to further group interests across multiple governance scales. Further, our case studies illustrate how different types of infrastructure [onshore wind energy and unconventional oil and gas (UOG)] can shape democratic politics and how these facets may interact in different ways over space and time.

2 Energy democracy, liberal democracy, and justice

Given that our case studies are drawn from Canada and the United States, we nest our understanding of energy democracy within the wider democratic theory of agonistics [8] to offer a more general explanation for how energy becomes politicized within liberal-democracies. Modern liberal-democracy is characterized through the dynamic relationship between contradictory commitments to liberalism and democracy:

> On the one side we have the liberal tradition constituted by the rule of law, and the defence of human rights and the respect of individual liberty; on the other the democratic tradition whose main ideas are those of equality, identity between governing and governed and popular sovereignty. There is no necessary relation between those two distinct traditions but only a contingent historical articulation. *Mouffe [9, pp. 2–3]*

Neither tradition's values can be fully realized without risking the collapse of liberal-democracy because the exercise of democratic rights always entails exclusionary identity formation and the exercise of power in contexts of pluralism of values. In a liberal-democracy, limits are always placed on the exercise of sovereignty by the people, but these limits depend on contingent interpretations of what human rights mean at a given moment in a given context, and are thereby only the expression of a prevailing hegemony constituted by the exercise of power. These limits are therefore always contestable and set through pragmatic, contingent negotiations which may offer temporary stabilizations between contesting forces through the establishment of the hegemony of one over the other.

We conceive energy democracy as a bundle of processes through which individuals, groups, and incumbents attempt to exercise power over how and where energy infrastructure is deployed within a liberal-democracy. This suggests a related interest in procedural justice. Yet, accepting an agonistic characterization of the wider democratic context seems to preclude the possibility of a value-neutral conception of procedural justice or a theoretical solution to the paradox of liberal-democracy [8]. This conflict may be even more evident in cases where the idea of democracy is applied in one area (e.g., an entire country) and procedural justice across another (e.g., a city, town, or village). This scalar mismatch can reveal a tension, reorienting our understanding of what is fair, just, or even democratic—and to whom. Empirically, there are widely established metrics or terms through which both researchers and affected publics/communities evaluate whether a decision-making process was fair [10–15]. These metrics have also been widely operationalized in environmental decision-making best practice guidance [16]. Yet case studies also produce ample evidence of the fundamentally contested nature of what counts as the fulfillment of these procedural metrics for a particular decision-making process in a complex context [13,14,17,18].

We, therefore, do not evaluate our case studies against established generic metrics and from a limited set of perspectives. This would essentially amount to showing how certain processes fall short of a procedural ideal from the perspective of some actors. Instead, in this chapter, we compare how claims of injustice emerge from directly affected communities (i.e., communities proximate to wind farms and UOG), and the actions taken to contest the experienced injustice. We argue that relevant actors (e.g., politicians) need to think pragmatically about energy issues that span multiple scales (e.g., global, sub-national, local) and discuss how alternative (hypothetical) actions by provincial/state governments may have better served procedural justice and liberal democratic ideals.

3 Wind energy development in Ontario, Canada

This chapter's lead author (Walker) conducted graduate research in Ontario and Nova Scotia, Canada between 2011 and 2017. His focus was on better understanding the factors associated with local support and opposition to wind energy development [19,20]. Through a combination of interview and survey work with people living within 2 km of a wind turbine, his work contributed to the idea that procedural justice, and in particular the ability for local residents to impact a project, was the most powerful force in shaping local support [15].

His entire research program was shaped by rising local opposition to wind turbines in Ontario during the mid-to-late 2000s. In 2009 and just prior to beginning to pursue his master's degree, the provincial government passed the Green Energy and Green Economy Act (i.e., the Green Energy Act or GEA). This law severely limited local planning authority in the development of wind turbines, leaving municipal governments and local residents without the ability to say no [21]. Then-Premier of Ontario Dalton McGuinty was proud of the GEA, stating that the new law would not allow municipalities to object to wind turbines simply "because they don't like them" [22]. In practice, this meant that any renewable energy approval could only be heard by the provincial Environmental Review Tribunal (ERT), and under objections related to either serious harm to human health or the environment. McRobert et al. [21] write that such narrow statutory grounds made successful appeals very unlikely given such "evidentiary burden" (p. 13).

McGuinty and his Liberal government were able to pass the GEA into law because first, in 2007, they were elected to a majority government. With this power, and under the start of a global economic recession, they decided that a move toward more renewable energy generation was needed. The government wanted to build a "green energy" economy, and because of aging infrastructure and promises to eliminate coal-fired electricity, felt it had to invest in new kinds of clean energy projects [21, p. 1]. Indeed, public opinion polls from across Ontario also suggested the move toward building more wind (and solar) energy projects would be popular [23]. Yet, because less than 10% of Ontario's population was (and is) rural, it is not surprising that there was a high level of support for the idea of wind energy. That is because due to the practicalities of large-scale wind energy development, it is only rural communities that can possibly host turbines.

Unlike province-wide survey work, Walker's research looked at public opinion on the ground in host communities. In some of this work [15], he found that for a variety of reasons (notably procedural injustice), 79% of those living close to turbines (within 2 km) opposed them. His other work has shown how much anti-wind turbine sentiment has crept into policy discourse. One person interviewed even said of a local politician who later won his riding (i.e., electoral district), "his main platform is 'stop the turbines'" [24, p. 670]. That anti-wind energy political messaging was effective at the local level while being largely ignored at the provincial level suggests a mismatch between a kind of democracy that is dismissed versus one that matters.

Beginning only 2 years after the GEA was passed, people have written about how the Liberal approach to bulldozing rural autonomy or local decision-making ability in the context of wind energy contributed to recent losses in the two subsequent provincial elections [25]. Research has suggested that this lack of procedural justice [15], alongside the perception of negative health effects from turbines [26,27] and issues like property value loss to a lesser degree [28,29] was driving much of the anti-wind turbine sentiment. The focus of such opposition varied by community, though calls for larger setback distances (i.e., from 550 m) and outright moratoriums were the most common [30,31].

In combination with other factors [32], this opposition toward wind energy contributed to a loss of a Liberal minority government in 2018, when populist Doug Ford [33] led a Progressive Conservative Party to a resounding, majority win. Some of his earliest moves in office included removing both provincial carbon pricing [34] and the Green Energy Act [35]. This is in spite of the fact that such moves have since been unpopular [32,36] and have resulted in a myriad of extra costs and lawsuits [37]. Yet even today, and more than a decade after the GEA was passed, its destructive legacy still lives on. This is highlighted by the fact that 90 townships and counties in Ontario have rejected the environmental and socio-economic benefits of wind energy in passing largely symbolic resolutions stating they are not a willing host for turbines. Despite their lack of legal standing, doing so may have allowed for like-minded communities to come together, signal their discontent, and aggregate some kind of political momentum.

From our Ontario case, the take-home message may be that in attempting to drive renewable energy development via the power that was earned through successful provincial democratic elections, the Liberal government of Ontario failed those rural communities who actually have to live with the daily-life reality of wind turbines. Eventually, this caught up to them at the ballot box. If instead, the government had approached the wind energy file via local energy democracy (i.e., with elements of procedural justice), they could have realized that the strong local opposition was not irrational, but a sign of [energy] injustice. Addressing this problem could have been done in several ways—the most obvious of course is to not go forward with development at all. Though in the context of the climate crisis and a need to lower emissions from electricity generation, a better option may have been to develop policies and programs that promote community ownership and the ability to regain the decision-making power which comes alongside it. It would be difficult to imagine provincial-level support for wind energy would have been negatively impacted by such a move. In fact, evidence from across Europe suggests the opposite—that wider, mostly urban populations are more likely to support renewable energy development when done in ways that give local areas a voice and keep benefits within communities.

4 Unconventional oil and gas development, Colorado, United States

This chapter's second author (Ryder) studied issues of power and procedural justice in multiscale governance processes for regulating unconventional oil and gas (UOG) taking place near communities in Colorado. Taking place between 2016 and 2018, this consisted of 57 interviews with residents, as well as engaging in participant observation and critical policy analysis. In this research, she explored regulatory tensions between the state of Colorado and local communities, wherein historically the latter have had little say over how and where energy production occurs.

As UOG production has grown in Colorado (and the United States more broadly), so too have concerns over impacts to people, communities, and the environment [38–41]. Oil and gas have primarily been governed as a state-level issue, which has led to state pre-emption of local efforts to regulate the practice, particularly efforts at subjecting site proposals to local zoning regulations [42,43]. This is true even in states like Colorado, where Home Rule laws grant increased leeway to cities and counties to self-govern, and all other types of development are subjected to local zoning regulations [44,45]. In this context, justifications for local and state-level regulation are presented as normative. Proponents of local regulation draw on the creation of municipal zoning laws to serve the purpose of determining what type of developments take place where. Those in favor of state-level regulations point to the significant state-wide benefits (i.e., tax revenues).

As Colorado has continually denied local governments the right to regulate where proposed developments can be located, residents along the Northern Colorado Front Range (such as within the communities of Boulder and Fort Collins) formed organizations to push back on proposed fracking projects in their communities. They have asked for more stringent regulation and local autonomy, appealing to their local councils and state legislators. They have also filed lawsuits and successfully created ballot measures both locally—to create moratoriums and at the state level—to push for more strict regulations, particularly in terms of well setback distances. While short-term moratoriums (6 months) established by local councils have gone unchallenged, longer-term moratoriums (2 + years) passed by local ballot measures and councils (i.e., to allow for studies on health impacts) have been challenged by the state and the Colorado Oil and Gas Associations as they argue these amount to de-facto bans. Community activist efforts have been multi-faceted and unrelenting, as one participant indicated:

> We're trying all angles, right? We're trying the courts…we're lobbying at the state house and voting and doing a ballot measure. We're supporting Colorado Rising [an organization against 'neighbourhood fracking']…and we're working on the public, like ad campaign.

While a host of concerns have shaped proposed regulations, of utmost importance has been the push to expand the distance required between well sites and occupied buildings (i.e., homes, schools, and hospitals). Here, we focus on this driver to illustrate issues at the intersection of procedural justice and democracy in the context of energy decision-making.

Perhaps the best proxy for democracy in the United States is the public capability to vote directly on an issue at hand. As such, ballot measures are a crucial component of democracy and have been an avenue for pursuing democratic consensus on energy issues in Colorado. Over the last decade, organizations for and against more regulations of UOG have proposed ballot measures for the people of Colorado to vote on, both at the state and local levels. These initiatives have been met with varying success, and offer insight into the importance of scalar thinking in the context of procedural justice and energy democracy.

On local levels, community members have created city or county initiatives where residents vote on placing moratoriums on fracking within their boundaries. In nearly all communities where moratoriums were placed on the ballot, residents showed a higher turnout and voted to approve these moratoriums. In several cases, these actions drew lawsuits from industry and were struck down as they operationally conflicted with state law. Yet there is at least a sense that the democratic practice of voting on the local ballot measures coincided with ideals of procedural justice—that is, those who would likely be most impacted by the decision had the capacity to influence the decision-making process.

Yet, if we focus on state-level ballot initiatives, we find that the overall will of Coloradans has not necessarily aligned with the desires of communities who face existing or potential impacts of UOG. Since 2014 in Colorado, at least nine ballot measure initiatives related to fracking have been put forward. Three of these initiatives aimed to increase setback distances, one was directed toward enhancing local control over UOG, and one was to ban fracking in the state. Of these five, only one (Proposition 112, a 2018 vote to increase setback distances to 2500 ft) made it to the ballot. Despite some speculation to the contrary, three of the four that did not make it on the ballot lacked sufficient voter signatures. Participants interviewed explained that various barriers—such as limited access to economic resources—can make it more difficult for organizations to put a measure on the ballot:

> You have a [state] constitution now that is becoming amendable if you've got enough money to pay for the signature gatherers. But if you're a grassroots effort, it's almost impossible…we've concluded that the system isn't broken, the system is fixed.

Further, efforts to stymie these local efforts represent potential threats to democracy. In 2016, a UOG organization donated over $1 million to Amendment 71, which changed the requirements needed for citizen-initiated ballots. It now requires petitioners to collect 2% of voter signatures from all 35 state senate districts. The Amendment was ultimately passed, and despite challenges to the constitutionality of the measure, it has been ruled constitutional. Even with this change, community organizations eventually succeeded in getting a measure on the ballot, which, one respondent frames as a win in itself:

> I think the ballot measure, I think running a statutory ballot measure is a big deal because that's citizen driven, citizen voice.

Finally, and despite making it to the state-wide ballot, 55% of voters rejected Colorado Proposition 112. Here, we can see a case of a democratically decided energy decision. Yet it also appears to constitute an issue of procedural justice, where those residents likely to be the most impacted by UOG development have been unable to meaningfully influence the decision. That is, more distant and state-wide voters, who might only experience indirect benefits of drilling via state tax revenues, had equal standing in the regulatory decision as people who lived nearby. Essentially, we see that what is democratically decided at a state level creates potential procedural injustices for localities facing drilling in both rural and urban areas of the state. If the decision to make this state-level change to the regulations were left to local communities meaningfully impacted by UOG, would it constitute a democratic process? This example demonstrates the need to understand what bounds might be placed around the concept of democracy to ensure that democratic processes do not supersede the need for those most impacted by a particular decision to take a lead role in meaningfully influencing the corresponding decision-making process and outcome.

5 Discussion and conclusions

Set toward advancing an understanding within the bourgeoning field of energy democracy, our case studies reveal what we see as a scalar tension between the use of provincial/state democratic structures, and the procedural injustice experienced by local communities playing host to energy development. That is, when we look closely enough, we reveal strong anti-democratic actions toward people living closest to and most impacted by these energy projects. This chapter thus adds empirical evidence to the work of Sovacool [7] who argues that improving procedural justice at the local level in the deployment of renewable energy can produce "co-benefits" including improved democracy.

In this chapter, we employed a definition of energy democracy as a bundle of democratic processes aimed at exercising power over how [energy] infrastructure is deployed. In some instances, these are established processes specific to particular jurisdictions, such as provincial elections (Ontario) or ballot measures (Colorado). When these are seen to create undemocratic outcomes or injustices *at the local level*, we see the rise of more informal social movements, or self-identified reluctant activists (e.g., coalition of "unwilling hosts" in Ontario and local-level organizations in Colorado) that attempt to establish new ways of expressing group interests. In this way, we see the inherent tensions that exist within liberal-democracies [9]. Decisions to develop energy infrastructure must navigate this complex system where individual liberty is valued alongside popular sovereignty across spatial and jurisdictional scales. How each should be weighed and how we balance them in energy futures that are organized and governed across sometimes large (i.e., state/provincial, national, or international) scales [46] is a question that deserves more attention going forward. The Ontario case study in particular also adds further evidence that linking social science and geographic energy research [47,48] with rural studies may be a fruitful way to understand the urban/rural divide in energy transitions [49].

Perhaps our chapter's most significant contribution is that it demonstrates the usefulness of using an agonistic view of liberal democracy when considering multiple dimensions of democratic action in practice; including macro-scale processes (i.e., province or state voting mechanisms) alongside more deliberative, direct, and participatory actions (often at the local scale) [2]. This draws attention to the complex scalar interactions that constitute group identity formation and mobilization of diverse democratic mechanisms to further interests. In doing so we link the concept of energy democracy to a broader theory of democracy by showing how established and diverse democratic processes in two different liberal-democracies can be mobilized by opposing groups of citizens, elected officials, and incumbents to exercise control over energy infrastructure deployment. Furthermore, these same rules of the game can also be used to decrease democratic participation in order to further either renewable energy or fossil fuel interests. Contending groups do not just play within these fixed rules but also seek to use democratic mechanisms to limit the democratic processes available to their opponents.

Ultimately, our case studies identify clear tensions across local and sub-national scales that lead us to recommend a twofold understanding of energy democracy. First as a bundle of processes through which individuals, groups, and incumbents attempt to exercise power over how and where energy infrastructure is deployed within a liberal-democracy. There is great value in researching both conventional democratic processes of participation (voting) as well as more deliberative mechanisms of citizen input to decisions. Secondly, our case studies recommend an understanding of energy democracy as an outcome of new socio-technical configurations affecting communities through the unique spatial distribution of different types of energy infrastructure. We hope that by introducing these concepts into academic and policy discourse, we can stimulate more critical investigations looking at the underlying assumptions of energy democracy, including whose voices actually matter, and whose do not.

References

[1] M.J. Burke, J.C. Stephens, Energy democracy: goals and policy instruments for sociotechnical transitions, Energy Res. Soc. Sci. 33 (2017) 35–48.
[2] K. Szulecki, Conceptualizing energy democracy, Environ. Polit. 27 (1) (2018) 21–41.
[3] K. Jenkins, D. McCauley, R. Heffron, H. Stephan, R. Rehner, Energy justice: a conceptual review, Energy Res. Soc. Sci. 11 (2016) 174–182.
[4] B.K. Sovacool, M.H. Dworkin, Energy justice: conceptual insights and practical applications, Appl. Energy 142 (2015) 435–444.
[5] B. Van Veelen, Negotiating energy democracy in practice: governance processes in community energy projects, Environ. Polit. 27 (4) (2018) 644–665.
[6] K. Szulecki, I. Overland, Energy democracy as a process, an outcome and a goal: a conceptual review, Energy Res. Soc. Sci. 69 (2020), 101768.
[7] B.K. Sovacool, M. Martiskainen, A. Hook, L. Baker, Decarbonization and its discontents: a critical energy justice perspective on four low-carbon transitions, Clim. Chang. 155 (4) (2019) 581–619.
[8] C. Mouffe, Agonistics: Thinking the World Politically, Verso Books, London, 2013.
[9] C. Mouffe, The Democratic Paradox, Verso Books, London, 2000.
[10] P. Devine-Wright, Environment, democracy, and public participation, in: D. Richardson, N. Castree, M. Goodchild, A. Kobayashi, W. Liu, R. Marston (Eds.), International Encyclopedia of Geography: People, the Earth, Environment and Technology, Wiley-Blackwell, New Jersey, 2017, pp. 1–10.
[11] J. Dwyer, D. Bidwell, Chains of trust: energy justice, public engagement, and the first offshore wind farm in the United States, Energy Res. Soc. Sci. 47 (2019) 166–176.
[12] J. Firestone, B. Hoen, J. Rand, D. Elliott, G. Hübner, J. Pohl, Reconsidering barriers to wind power projects: community engagement, developer transparency and place, J. Environ. Policy Plan. 20 (3) (2018) 370–386.
[13] C. Gross, Community perspectives of wind energy in Australia: the application of a justice and community fairness framework to increase social acceptance, Energy Policy 35 (5) (2007) 2727–2736.
[14] N. Simcock, Procedural justice and the implementation of community wind energy projects: a case study from South Yorkshire, UK, Land Use Policy 59 (2016) 467–477.
[15] C. Walker, J. Baxter, Procedural justice in Canadian wind energy development: a comparison of community-based and technocratic siting processes, Energy Res. Soc. Sci. 29 (2017) 160–169.
[16] S. Dietz, N. Stern, Why economic analysis supports strong action on climate change: a response to the Stern Review's critics, Rev. Environ. Econ. Policy 2 (1) (2008) 94–113.
[17] J. Firestone, C. Hirt, D. Bidwell, M. Gardner, J. Dwyer, Faring well in offshore wind power siting? Trust, engagement and process fairness in the United States, Energy Res. Soc. Sci. 62 (2020), 101393.
[18] D. Van der Horst, NIMBY or not? Exploring the relevance of location and the politics of voiced opinions in renewable energy siting controversies, Energy Policy 35 (5) (2007) 2705–2714.
[19] C. Walker, "Winds of Change": Explaining Support for Wind Energy Developments in Ontario, Canada (Masters thesis), Western University, 2012.

[20] C. Walker, Wind Energy Policy, Development, and Justice in Ontario and Nova Scotia, Canada: A Comparison of Technocratic and Community-Based Siting Processes (PhD thesis), Western University, 2017.

[21] D. McRobert, J. Tennent-Riddell, C. Walker, Ontario's green economy and green energy act: why a well-intentioned law is mired in controversy and opposed by rural communities, Renew. Energy Law Pol. Rev. 7 (2016) 91–112.

[22] Canadian Press, New law will keep NIMBY-ism from stopping green projects: Ont. Premier, CBC News (2009). Retrieved October 11 2020, Accessed from: https://www.cbc.ca/news/technology/ new-law-will-keep-nimby-ism-from-stopping-green-projects-ont-premier-1.805978.

[23] Ipsos Reid, Wind Energy in Ontario, 2010, Retrieved September 11 2020. Accessed from: https://www.ipsos. com/sites/default/files/publication/2010-07/4868.pdf.

[24] C. Walker, L. Stephenson, J. Baxter, "His main platform is 'stop the turbines'": political discourse, partisanship and local responses to wind energy in Canada, Energy Policy 123 (2018) 670–681.

[25] J.M. McGrath, Why So Many Rural Ontarians Can't Stand Wind Power, TVO, 2015. Retrieved October 11 2020. Accessed from https://www.tvo.org/article/why-so-many-rural-ontarians-cant-stand-wind-power.

[26] B. Deignan, E. Harvey, L. Hoffman-Goetz, Fright factors about wind turbines and health in Ontario newspapers before and after the Green Energy Act, Health Risk Soc. 15 (3) (2013) 234–250.

[27] C. Walker, J. Baxter, D. Ouellette, Adding insult to injury: the development of psychosocial stress in Ontario Wind Turbine communities, Soc. Sci. Med. 133 (2015) 358–365.

[28] R.J. Vyn, R.M. McCullough, The effects of wind turbines on property values in Ontario: does public perception match empirical evidence? Can. J. Agric. Econ. 62 (3) (2014) 365–392.

[29] C. Walker, J. Baxter, S. Mason, I. Luginaah, D. Ouellette, Wind energy development and perceived real estate values in Ontario, Canada, AIMS Energy 2 (4) (2014) 424–442.

[30] S.D. Hill, J.D. Knott, Too close for comfort: social controversies surrounding wind farm noise setback policies in Ontario, Renew. Energy Law Pol. Rev. 1 (2) (2010) 153–168.

[31] C. Walker, J. Baxter, D. Ouellette, Beyond rhetoric to understanding determinants of wind turbine support and conflict in two Ontario, Canada communities, Environ Plan A 46 (3) (2014) 730–745.

[32] E. Lachapelle, S. Kiss, Opposition to carbon pricing and right-wing populism: Ontario's 2018 general election, Environ. Polit. 28 (5) (2019) 970–976.

[33] B. Budd, The People's champ: doug ford and neoliberal right-wing populism in the 2018 Ontario provincial election, Polit. Gov. 8 (1) (2020) 171–181.

[34] H. Millar, E. Bourgeois, S. Bernstein, M. Hoffmann, Self-reinforcing and self-undermining feedbacks in subnational climate policy implementation, Environ. Polit. (2020) 1–20.

[35] C. Mang-Benza, C. Hunsberger, Wandering identities in energy transition discourses: political leaders' use of the "we" pronoun in Ontario, 2009–2019, Can. Geogr. 64 (3) (2020) 516–529.

[36] C. Walker, Bill 4 and the removal of cap and trade: a case study of carbon pricing, climate change law and public participation in Ontario, Canada, J. Environ. Law Pract. 33 (1) (2020) 35–72.

[37] E. McIntosh, Doug Ford facing second lawsuit over environmental assessment changes, Natl. Obs. (2020). Retrieved October 11 2020, Accessed from: https://www.nationalobserver.com/2020/08/31/news/ doug-ford-facing-second-lawsuit-over-environmental-assessment-changes.

[38] J.L. Adgate, B.D. Goldstein, L.M. McKenzie, Potential public health hazards, exposures and health effects from unconventional natural gas development, Environ. Sci. Technol. 48 (15) (2014) 8307–8320.

[39] M.L. Finkel (Ed.), The Human and Environmental Impact of Fracking: How Fracturing Shale for Gas Affects Us and Our World, ABC-CLIO, Santa Barbara, CA, 2015.

[40] R.W. Howarth, R. Santoro, A. Ingraffea, Methane and the greenhouse-gas footprint of natural gas from shale formations, Clim. Chang. 106 (4) (2011) 679–690.

[41] A.E. Ladd, Motivational frame disputes surrounding natural gas fracking in the Haynesville Shale, in: A.E. Ladd (Ed.), Fractured Communities: Risk, Impacts, and Protest against Hydraulic Fracking in U.S. Shale Regions, Rutgers University Press, New Jersey, 2018, pp. 149–172.

[42] A. Shaffer, S. Zilliox, J. Smith, Memoranda of understanding and the social licence to operate in Colorado's unconventional energy industry: a study of citizen complaints, J. Energy Nat. Resour. Law 35 (1) (2017) 69–85.

[43] T. Silvy, Extraction announces plan for East Greeley operations, will drill mostly outside school hours, Greeley Tribune (2018). Retrieved October 7 2020, Accessed from: https://www.greeleytribune.com/news/local/ extraction-announces-plan-for-east-greeley-operations-will-drill-mostly-outside-school-hours/.

[44] S.S. Ryder, Developing an intersectionally-informed, multi-sited, critical policy ethnography to examine power and procedural justice in multiscalar energy and climate change decision making processes, Energy Res. Soc. Sci. 45 (2018) 266–275.

[45] S.S. Ryder, Unconventional regulation for unconventional energy in Northern Colorado? Municipalities as strategic actors and innovators in the United States, Energy Res. Soc. Sci. 26 (2017) 23–33.

[46] R. Cowell, G. Ellis, F. Sherry-Brennan, P.A. Strachan, D. Toke, Rescaling the governance of renewable energy: lessons from the UK devolution experience, J. Environ. Policy Plan. 19 (5) (2017) 480–502.

[47] M.J. Pasqualetti, Opposing energy landscapes: a search for common cause, Ann. Am. Assoc. Geogr. 101 (4) (2011) 907–917.

[48] M.J. Pasqualetti, Social barriers to renewable energy landscapes, Geogr. Rev. 101 (2) (2011) 201–223.

[49] M. Naumann, D. Rudolph, Conceptualizing rural energy transitions: energizing rural studies, ruralizing energy research, J. Rural. Stud. 73 (2020) 97–104.

35

Mind the gap: Citizens, consumers, and unequal participation in global energy transitions

Breffní Lennon[a,b] *and Niall P. Dunphy*[a,b]

[a]Cleaner Production Promotion Unit, School of Engineering and Architecture, University College Cork, Cork, Ireland [b]Environmental Research Institute, University College Cork, Cork, Ireland

1 Introduction

The current energy transition, with its rush to exploit renewable energy sources (RES), has raised significant questions with regard to the roles and expectations of citizens in the energy system. Historically, the experience for citizens has been characterized by narrow commodity and consumer choice narratives that have ignored the environmental and social consequences of energy-related production and consumption practices. This transition too is not without its discontents [1], including deeply ingrained ambiguities on what it actually means for citizens. Issues concerning procedural, distributional, and substantive justice are often ignored in favor of a notional public acceptance [2–6]. While popular and policy-related discourses have made tentative links to the transformative potential of citizen participation in driving the current energy transition [7–9], an emerging social gap between support for renewable energies in principle and local opposition to renewable energy projects is already challenging collective efforts to transition. This has been compounded by "the business as usual" approach taken by energy incumbents and their challenges to public consultation, which is very often informed by neoliberal framing of citizens as potential threats rather than as collaborative partners [10,11]. Therefore, current variances in accepted notions of what is energy citizenship must be adequately addressed if we are to realize a just and equitable transition to more sustainable modes of energy consumption and production.

2 Deconstructing neoliberal energy citizenship

In a recent paper [12], we highlighted how the concept of energy citizen has been co-opted into the neoliberal trope of "citizen-as-consumer," which consequently removes any transformative potential from new and emergent forms of citizen participation in the transition. We also highlighted how current debates on energy democracy have opened up space for examining alternative roles for citizens that could potentially challenge the inherent contradictions existing in neoliberal logic with its championing of carefully stratified forms of participation.

However, as Szulecki and Overland [13] subsequently pointed out, understandings of energy democracy in the literature remain fragmented at best, with limited adherence to wider theorizations on democracy. Therefore, it is essential we close this gap in understanding and continue to identify those factors inhibiting meaningful participation in the energy transition. Also, it is important that we continue to unpack the many (mis)understandings and (mis) representations of what an energy citizen should be, as presented by numerous competing polities engaged in the energy sphere.

Given the numerous social inequalities already intersecting the organizational and social structures affecting our individual lived experiences [14–18], it is imperative we incorporate the perspectives and understandings of those sections of the citizenry who have traditionally been sidelined or unheard. Ironically, it has often been the commonly accepted interpretations of how energy citizens are expected to participate that have in fact limited their ability to engage in a meaningful way.

Energy citizenship has become increasingly popular as a nebulous term used to capture the many ways citizens are seen and expected to participate in the energy transition. Energy citizenship is also invoked in the formalized language of national governments and supranational organizations, such as the European Union. So much so that it is routinely used as a standard bearer of sorts for citizen participation and applied interchangeably with that of the active consumer in a growing number of policy documents [12,19–21]. In order to unpack the term, we must first acknowledge how certain inequalities are locked-in by the very framing of particular conceptualizations of energy citizenship. For instance, *inclusivity* (e.g., who belongs? Who is considered to be an energy citizen and in what context?), *distribution of power* (e.g., what are the decision-making processes and who is involved? How are citizens' interests protected?), and the *apportioning and utilizing of certain rights* (e.g., are citizens conceived solely through a market lens, limited to consumer rights?), etc.

Also, situating participation within democratic governance structures will not in itself translate into fairer, more equitable experiences for citizens, but rather will most likely reinforce existing inequalities. If we examine the spectrum of experiences in extant democracies, many have failed to mitigate against social inequality and have in fact overseen an acceleration and widening of these gaps [22]. These inequalities are further compounded when we consider the citizen-as-consumer trope with its linking of consumption and affordability to the rights and privileges of citizens when accessing energy. Energy citizens are expected to play an "active role in the transition to a low carbon energy future" [23], yet the legal and structural frameworks to facilitate and support such actions can still act as barriers to participation, even when expressly designed to achieve that goal [24]. To paraphrase van Waas and Jaghai [25], all energy citizens are created equal, but some remain more equal than others.

TABLE 35.1 Examples of citizen/consumer participation in the energy system.

(1) The passive consumer, the traditional passive receiver of energy;
(2) the active consumer, encouraged to use their purchasing power to influence the market in certain directions;
(3) the good citizen, who just needs to be informed and they will "do the right thing";
(4) the constitutionalist, for whom everything is about understanding and enforcing your legal rights;
(5) the producer, who either individually or collectively with others is involved in the production of energy; and
(6) the challenger, who engages in public debate, organizes and protests to promote alternative perspectives on energy and the energy system.

The kinds of roles citizens are being permitted to play or even want to play within the energy system, vary from country to country and are underscored by the ideological positions of decision-makers in each jurisdiction. These roles can be seen to occupy a continuum ranging from individualist to collectivist engagements. All have very real implications for citizens' interpretations of inclusivity, democracy, and ultimately the acceptability of renewable energy technologies. In previous work, we identified six modes of citizen participation currently in practice within the energy domain [26,27]. These are presented in Table 35.1.

Unfortunately, for many citizens the role assigned to them is very often that of the passive consumer, or if they have a certain degree of economic leverage, an active consumer. Both feed into paternalist representations of what constitutes a good citizen, with its positivist emphasis on linear-decision-making and maintaining existing (unequal) socio-economic structures [24,26,28,29]. While we acknowledge that this may suit individual citizens, whose priorities and attentions lie elsewhere, the absence of choice points to a democratic deficit for the majority of citizens. This is especially true for those citizens wishing to meaningfully participate in change, whether as energy producers or as challengers to the status quo. A deep irony of the current transition is that those tasked with maintaining the status quo (keeping the lights on, so to speak) are also those expected to deliver the transformative change this transition, in particular, requires.

In addition, perceptions and expectations of what constitutes participation vary, arising to tensions among stakeholders even when sharing a common goal [30]. Where decision-makers approach citizens in good faith (partaking in substantive participative engagement either individually or collectively at the community level), if the multiple historic and socially situated contexts within that community are not taken into consideration the potential for conflict and misunderstanding remains high [31]. Sensitivity to citizens' situated knowledge must become integral to the approaches of leaders and decision-makers if we are to leverage the full potential of collective action [32].

3 Conclusions

As we can see with the democratic project more generally, participation must be decoupled from the destructive impulses of consumption if we are to have a realistic chance at achieving the type of transformation that is required to reduce vulnerability to climate change. We must broaden our perspective to embrace more inclusive ecologies of participation that take into consideration the diversity of publics engaged in socio-technical systems like the energy domain, with Chilvers et al.'s relational co-productionist framework offering a useful point of departure for understanding the diversities and inequalities locked into participation [33]. Ultimately, the transition must move beyond simplified technical perspectives and embrace those socio-political concerns that prioritize both the material security of citizens and the ecological integrity of our resource bases, while also planning for the inevitable rebound effects [34–37]. Otherwise, we have little chance of meeting the challenges of the climate crisis and by extension the energy transition.

References

[1]   B.K. Sovacool, M. Martiskainen, A. Hook, L. Baker, Decarbonization and its discontents: a critical energy justice perspective on four low-carbon transitions, Clim. Chang. (2019), https://doi.org/10.1007/s10584-019-02521-7.

[2]   M. Wolsink, Social acceptance revisited: gaps, questionable trends, and an auspicious perspective, energy res, Soc. Sci. 46 (2018) 287–295, https://doi.org/10.1016/j.erss.2018.07.034.

[3]   P. Upham, C. Oltra, À. Boso, Towards a cross-paradigmatic framework of the social acceptance of energy systems, Energy Res. Soc. Sci. 8 (2015) 100–112, https://doi.org/10.1016/j.erss.2015.05.003.

[4]   K. Jenkins, D. McCauley, R. Heffron, H. Stephan, R. Rehner, Energy justice: a conceptual review, Energy Res. Soc. Sci. 11 (2016) 174–182, https://doi.org/10.1016/j.erss.2015.10.004.

[5]   S. Fuller, D. McCauley, Framing energy justice: perspectives from activism and advocacy, Energy Res. Soc. Sci. (2016) 1–8.

[6]   F. Bartiaux, M. Maretti, A. Cartone, P. Biermann, V. Krasteva, Sustainable energy transitions and social inequalities in energy access: a relational comparison of capabilities in three European countries, Glob. Transit. 1 (2019) 226–240, https://doi.org/10.1016/j.glt.2019.11.002.

[7]   D. Soares da Silva, L.G. Horlings, The role of local energy initiatives in co-producing sustainable places, Sustain. Sci. 15 (2020) 363–377, https://doi.org/10.1007/s11625-019-00762-0.

[8]   O. Escobar, Pluralism and democratic participation: what kind of citizen are citizens invited to be? Contemp. Pragmatism. 14 (2017) 416–438, https://doi.org/10.1163/18758185-01404002.

[9]   J.C. Stephens, Energy democracy: redistributing power to the people through renewable transformation, Environment 61 (2019) 4–13, https://doi.org/10.1080/00139157.2019.1564212.

[10] J. Clarke, J. Newman, N. Smith, E. Vidler, L. Westmarland (Eds.), Creating Citizen-Consumers Changing Publics and Changing Public Services, SAGE Publications Ltd, 2007.

[11] I.A. Fontenelle, M. Pozzebon, A dialectical reflection on the emergence of the 'citizen as consumer' as neoliberal citizenship: the 2013 Brazilian protests, J. Consum. Cult. (2018), https://doi.org/10.1177/1469540518806939.

[12] B. Lennon, N.P. Dunphy, C. Gaffney, A. Revez, G.M. Mullally, P. O'Connor, Citizen or consumer? Reconsidering energy citizenship, J. Environ. Policy Plan. 22 (2020) 184–197, https://doi.org/10.1080/1523908X.2019.1680277.

[13] K. Szulecki, I. Overland, Energy democracy as a process, an outcome and a goal: a conceptual review, Energy Res. Soc. Sci. 69 (2020), https://doi.org/10.1016/j.erss.2020.101768, 101768.

[14] M.L. Hoffman, Bike Lanes Are White Lanes: Bicycle Advocacy and Urban Planning, University of Nebraska Press, Lincoln & London, 2016, https://doi.org/10.2307/j.ctt1d4v13q.6.

[15] S. Bouzarovski, N. Simcock, Spatializing energy justice, Energy Policy 107 (2017), https://doi.org/10.1016/j.enpol.2017.03.064.

[16] R. Listo, Gender myths in energy poverty literature: a critical discourse analysis, Energy Res. Soc. Sci. 38 (2018) 9–18, https://doi.org/10.1016/j.erss.2018.01.010.

[17] E. Royce, Poverty and Power: The Problem of Structural Inequality, Rowman and Littlefield, Lanham, 2015.

[18] S. Buzar, Energy Poverty in Eastern Europe: Hidden Geographies of Deprivation, Ashgate, 2007.

[19] P. Devine-Wright, Energy citizenship: psychological aspects of evolution in sustainable energy technologies, in: J. Murphy (Ed.), Governing Technology for Sustainability, Routledge, London, 2007, p. 240.

[20] A. Titus, D. Kuch, I. Director, Emerging dimensions of networked energy citizenship: the case of coal seam gas mobilisation in Australia, Commun. Polit. Cult. 47 (2014) 35.

[21] M. Ryghaug, T.M. Skjølsvold, S. Heidenreich, Creating energy citizenship through material participation, Soc. Stud. Sci. 48 (2018) 283–303, https://doi.org/10.1177/0306312718770286.

[22] F. Alvaredo, L. Chancel, T. Piketty, E. Saez, G. Zucman, World Inequality Report 2018, 2018.

[23] Department of Communications Climate Action and Environment, Ireland's Transition to a Low Carbon Energy Future 2015–2030, Gov. White Pap. Energy, 2015, p. 122. https://www.dccae.gov.ie/documents/EnergyWhite Paper-Dec 2015.pdf.

[24] M.M. Sokołowski, Renewable and citizen energy communities in the European Union: how (not) to regulate community energy in national laws and policies, J. Energy Nat. Resour. Law. 38 (2020) 289–304, https://doi.or g/10.1080/02646811.2020.1759247.

[25] L. van Waas, S. Jaghai, All citizens are created equal, but some are more equal than others, Netherlands, Int. Law Rev. 65 (2018) 413–430, https://doi.org/10.1007/s40802-018-0123-8.

[26] G. Mullally, N.P. Dunphy, P. O'Connor, Participative environmental policy integration in the Irish energy sector, Environ. Sci. Pol. 83 (2018) 71–78, https://doi.org/10.1016/j.envsci.2018.02.007.

[27] N.P. Dunphy, B. Lennon, Citizen Participation in the Energy System at the Macro Level. A Deliverable of the Energy Polities Project, University College Cork, 2020.

[28] S. Barr, Beyond behavior change: social practice theory and the search for sustainable mobility, in: E.H. Kennedy, M.J. Cohen, N. Krogman (Eds.), Putt. Sustain. into Pract. Appl. Adv. Res. Sustain. Consum, Edward Elgar, Cheltenham, 2015, pp. 91–108, https://doi.org/10.4337/9781784710606.

[29] J. Pykett, M. Saward, A. Schaefer, Framing the good citizen, Br. J. Polit. Int. Rel. 12 (2010) 523–538, https://doi.org/10.1111/j.1467-856X.2010.00424.x.

[30] F. Ruef, M. Stauffacher, O. Ejderyan, Blind spots of participation: how differently do geothermal energy managers and residents understand participation? Energy Rep. 6 (2020) 1950–1962, https://doi.org/10.1016/j.egyr.2020.07.003.

[31] H. Kim, S.H. Cho, S. Song, Wind, power, and the situatedness of community engagement, Public Underst. Sci. 28 (2019) 38–52, https://doi.org/10.1177/0963662518772508.

[32] J.C. Stephens, Diversifying Power: Why We Need Antiracist, Feminist Leadership on Climate and Energy, Island Press, Washington, DC, 2020.

[33] J. Chilvers, H. Pallett, T. Hargreaves, Ecologies of participation in socio-technical change: the case of energy system transitions, energy res, Soc. Sci. 42 (2018) 199–210, https://doi.org/10.1016/j.erss.2018.03.020.

[34] T. Pricen, The Logic of Sufficiency, MIT Press, Cambridge, MA, 2005.

[35] G. Wallenborn, Rebounds are structural effects of infrastructures and markets, Front. Energy Res. 6 (2018) 1 13, https://doi.org/10.3389/fenrg.2018.00099.

[36] P. Casal, Why sufficiency is not enough, Ethics 117 (2007) 296–326, https://doi.org/10.1086/510692.

[37] S. Sorrell, B. Gatersleben, A. Druckman, The limits of energy sufficiency: a review of the evidence for rebound effects and negative spillovers from behavioural change, Energy Res. Soc. Sci. 64 (2020), https://doi.org/10.1016/j.erss.2020.101439, 101439.

Worse than its reputation? Shortcomings of "energy democracy"

Veith Selk and Jörg Kemmerzell

Institute of Political Science, Technical University of Darmstadt, Darmstadt, Germany

1 Introduction: The promising notion of post-carbon energy democracy

The normative idea of energy democracy is multilayered, envisioning that green energy transition not only stands for a low-carbon transition but also can lead to both strengthening and expanding democracy. First, it is associated with an inclusive and open form of civic participation that entails civic-expert (co)decision-making [1]. Second, it relates to the economic dimension by transforming the ownership of means of energy production, for instance, through energy cooperatives and small businesses, which contrast with the big business, big government imaginary associated with fossil industrialism. Flat hierarchies of small-unit energy production and their anchoring in civic communities seem to enforce overall democratization [2]. Third, participatory procedures and small-scale cooperative units promise a more broadly shared and inclusively produced form of knowledge, as is the case in the iconic figure of energy democracy, the *prosumer*. The prosumer is not only an ecologically conscious consumer but also a smart operator of an eco-business as well as a well-informed and community-oriented citizen [3].

In a seminal contribution, Kacper Szulecki has captured these dimensions and given the concept of "energy democracy" a concise definition. The first dimension in Szulecki's concept is "civic ownership" [2, p. 36]. This entails the aforementioned material participation of citizens as owners or producers of energy. His second dimension, "participatory governance" [2, p. 36], encompasses inclusive governance instruments of co-decision making, planning, and implementation. "Popular sovereignty" [2, p. 36], the third dimension, means that in energy democracy stakeholders as well as account holders do have their say and cooperatively foster political autonomy. In the following, we will build upon Szulecki's useful three-dimensional concept and assess whether these normative promises of energy democracy are viable.

Energy Democracies for Sustainable Futures
https://doi.org/10.1016/B978-0-12-822796-1.00036-X

2 Asymmetric ownership

An apparent result of contemporary green energy transitions is the spread of ownership. The empirical significance of this dimension is indeed striking. As new eco-business models emerge, we can observe a resurgence of community-based enterprises as well as manifold actors becoming entrepreneurs in the energy sector. Around renewable energy installations, a new network of plant engineering and service providers is flourishing. This process has a progressive appeal because it challenges the entrenched business models of the fossil giants. It sometimes even breathes the spirit of counter-culture as "the early process of innovation and diffusion of the renewable energies" in particular was driven by the "ecological and alternative movement" [4, p. 115].

However, the pushback against old economic power players and their vested interests should neither be conflated with the anti-corporate protests since the 1970s, e.g., against the nuclear industry nor equated with the emergence of a more democratic energy regime, as some authors assume [5]. Changes in ownership patterns are exactly that: a restructuring of forms of energy production and, for the most part, corporate ones. Without challenging the general property structures of the economy, the restructuring or "the process of creative destruction" of business models and the transformation of ownership do not lead to more inclusive and material democratization of the means of production [6, p. 81]. On the contrary, energy transitions can result in an unequal and elitist form of energy production. One reason for this is the problem of asymmetric ownership. Material ownership in energy infrastructure presupposes material resources: capital to invest in renewable energy or landed property to set up energy plants, or political influence, foremost bargaining and justificatory power to win subsidies.

The problem of asymmetric ownership counteracts the progressive-inclusive rhetoric in which the well-off proponents of green energy transition sometimes indulge. Survey data on the course of the German *Energiewende*, one of the most discussed instances of the green energy transition, reveal that a relative majority of respondents evaluate the *Energiewende* both as unfair and elitist [7]. This stands in stark contrast to an understanding of energy transitions as a means of democratization that fosters a more widespread distribution of resources and opportunities. In short, regarding civic ownership, energy democracy fails to live up to its promise. An analysis of the other dimensions, participatory governance, and popular sovereignty, yields a similar result.

3 Unequal participation

In the wake of the "participatory revolution," civic participation has become an important feature of Western democracies [8]. This is not specific to the energy sector, but it is a general means of governing in modern post-industrial societies because especially well-educated, affluent, and highly individualized members of the citizenry are not satisfied with representation alone; they want to have their own say [9,10]. The expectations of earlier, more egalitarian proponents of participation [11], however, have been frustrated, since the increase in participatory innovations went hand-in-hand with a severe "participation gap" [12]. In particular, we observe a general pattern in all Western societies: the less educated, less moneyed citizens abstain not only from conventional participation, e.g., voting, at disproportionately high levels but even more from participatory governance [13]. This participation gap also exists in participatory arrangements in energy transition due to insufficient procedures and selective participation patterns. Resource-rich citizens are largely overrepresented in these arrangements [14]. Borrowing a term of Niklas Luhmann's systems theory, at this point a "re-entry" of problems occurs, which we already have observed in the ownership dimension: social selectivity, elitism, and political inequality [15, p. 150].

4 Diminishing popular sovereignty

Neoliberal modernization caused a shift from state-centered to networked governance at least since the 1970s. This has led to an increase in participation, albeit an asymmetric one [16]. At the same time, it has weakened the economic dimension of democracy. This advent of "post-democracy" not only triggers the regression of representative party-democracy but also involves setbacks in democratizing the economy [17]. This process is at work in the energy transition in a particularly effective way.

To understand this, it is important to recall the form of modern democracy that took shape over the course of the 20th century in the Western world. It was fundamentally intertwined, certainly to a stronger extent in continental Europe than in the English-speaking world, with an interventionist welfare state adding a material dimension to formal political rights and liberties. The social-democratic theorists of this development had recognized that, in order to achieve popular sovereignty, material autonomy at the workplace must underpin formal autonomy at the ballot box [18]. Without industrial democracy, they argued, elites alone will dominate electoral democracy. It is noteworthy that in some instances this industrial democracy had progressed the furthest within the energy and heavy industries, as was the case, for instance, in the German *Montanindustrie* (i.e., coal, iron, and steel). Sectors such as this have come especially under pressure in the wake of the green energy transition.

5 (Old-time) industrial democracy

It is telling that, in the case of West Germany, the most egalitarian model of co-determination (i.e., between management and labor representatives) was introduced in the mining, iron, and steel industries. Shortly after the foundation of the Federal Republic of Germany, through the threat of a general strike, German unions brought forth far-reaching co-determination structures. In May 1951, the federal parliament passed the law on co-determination in mining companies and the iron and steel producing industry. It provides for equal representation of employee representatives on the supervisory boards of all large corporations in the sector. According to this law, one member of the board of directors, the labor director, cannot be appointed or dismissed against the majority of the employee representatives. Co-determination therefore not only refers to supervisory rights but also includes participation in management decisions [19].

Furthermore, the realization of collective participation in old-time industrial democracy was not limited to the institutional structures of corporate democracy. It was also at work on the mobilization side. In his book on *Carbon Democracy*, Timothy Mitchell describes the particular working conditions in the mining sector of Western societies as an incubator for the emergence of democratic solidarity among the common workers and their capacity for political action on their own, not least because of the nature of coal as a labor-intensive resource [20]. Against this backdrop, and in a more moralistic vein [21], Jedediah Purdy has recently evoked the Miners for Democracy revolt in 1969 as a model for an eco-friendly and democratic form of mobilization in the energy sector [22, p. 134]. However, this historic experience, described by Mitchell and prescribed by Purdy, stands in stark contrast to the de facto democratic trade-offs of a post-carbon post-democracy. Thus, with regard to collective sovereignty, which was at least partially realized within the old-time industrial democracy paradigm, the current discussion about energy democracy has a blind spot: diminishing popular sovereignty.

6 Conclusions

In analyzing energy transitions, it is necessary to distinguish between effects on ecology on the one hand and consequences for democracy on the other. With regard to the currently dominant forms of an environmentally friendly low-carbon transition, we cannot observe a general democratization effect. In many ways, particularly because of diminishing popular sovereignty, the opposite comes true. To indicate these shortcomings does not mean denouncing the green transition. Rather, this may help avoid moralization, understand the legitimate conflicts that emerge from it, and detect the blind spots of nascent post-carbon energy democracy talk.

References

[1] I. Blühdorn, M. Deflorian, The collaborative management of sustained unsustainability: on the performance of participatory forms of environmental governance, Sustainability 11 (4) (2019) 1189.
[2] K. Szulecki, Conceptualizing energy democracy, Environ. Polit. 27 (1) (2018) 21–41.
[3] W.-P. Schill, A. Zerrahn, F. Kunz, Prosumage of solar electricity: pros, cons, and the system perspective, Econ. Energy Environ. Policy 6 (1) (2017) 7–31.
[4] R. Mautz, The expansion of renewable energies in Germany between niche dynamics and system integration, Sci. Technol. Innov. Stud. 3 (2) (2007) 113–131.
[5] C. Morris, A. Jungjohann, Energy Democracy: Germany's Energiewende to Renewables, Palgrave Macmillan, London, 2016.
[6] J.A. Schumpeter, Capitalism, Socialism, and Democracy, third ed., Harper, New York, 2008 (1950).
[7] D. Setton, Soziales Nachhaltigkeitsbarometer der Energiewende 2018, Institute for Advanced Sustainability Studies, Potsdam, 2019.
[8] M. Kaase, The challenge of the "participatory revolution" in pluralist democracies, Int. Polit. Sci. Rev. 5 (4) (1984) 299–318.
[9] S. Tormey, The End of Representative Politics, Polity Press, Cambridge, 2015.
[10] U. Beck, The Reinvention of Politics: Rethinking Modernity in the Global Social Order, Polity Press, Cambridge, 1997.
[11] C. Pateman, Participation and Democratic Theory, Cambridge University Press, Cambridge, 1970.
[12] R.J. Dalton, The Participation Gap. Social Status and Political Inequality, Oxford University Press, Oxford, 2017.
[13] K. Armingeon, L. Schädel, Social inequality in political participation: the dark side of individualisation, West Eur. Polit. 38 (1) (2015) 1–27.

[14] V. Selk, J. Kemmerzell, J. Radtke, In der Demokratiefalle? Probleme der Energiewende zwischen Expertokratie, partizipativer Governance und populistischer Reaktion, in: J. Radtke, W. Canzler, M. Schreurs, S. Wurster (Eds.), Energiewende in Zeiten des Populismus, Springer, Wiesbaden, 2019, pp. 31–66.

[15] N. Luhmann, Die Politik der Gesellschaft, Suhrkamp, Frankfurt, 2002.

[16] R.A. Rhodes, Understanding governance: ten years on, Organ. Stud. 28 (8) (2007) 1243–1264.

[17] C. Crouch, Post-democracy, Polity Press, Cambridge, 2004.

[18] J. Dewey, Liberalism and social action, in: J.A. Boydston (Ed.), The Later Works, vol. 11, Southern Illinois University Press, Carbondale, IL, 1987, pp. 1–65 (Original work published 1935).

[19] W. Abelshauser, Deutsche Wirtschaftsgeschichte. Von 1945 bis zur Gegenwart, second ed., C. H. Beck, Munich, 2011.

[20] T. Mitchell, Carbon Democracy. Political Power in the Age of Oil, Verso, London, 2011.

[21] A. Scerri, Review of Jedediah Purdy, this land is our land: the struggle for a new commonwealth, Social. Democr. (2021), https://doi.org/10.1080/08854300.2021.1889873.

[22] J. Purdy, This Land Is our Land: The Struggle for a New Commonwealth, Princeton University Press, Princeton, 2019.

III. Risks

Conclusion: A call to action, toward an energy research insurrection

Alexander Dunlap

The Centre for Development and the Environment, University of Oslo, Oslo, Norway

1 Introduction

Energy Democracies for Sustainable Futures has 36 chapters discussing the challenges, proposals, and hopes for cultivating energy transition and democracy. The book confronts the realities of "top-down" socio-technical design, but also the possibilities for reconfiguring these relationships by democratizing energy development. The book's introduction frames the present socio-ecological situation within the history of hydrocarbon and nuclear development, before discussing the onset of renewable energy. This includes recognizing the expansive concerns related to participation, the possibilities for creating global energy transitions as well as the limitations of local politics.

The book covers a wide breadth of topics. Part I examines the possibilities created by imagining energy systems differently. Karen Hudlet-Vazquez and colleagues remind us that community energy systems "can reproduce neoliberal values in relation to the roles of the state, the individual and participation," noting that "utopian" futures "can therefore have unexpected consequences." Karen Hudlet-Vazquez and colleagues, however, do not deny an endless amount of alternative energy system possibilities. Rudy Khsar and Ry Brennan demonstrate how decentralization is crucial to forming energy democracies. Brennan applying the bioregional concept to energy infrastructure shows how localizing energy production and consumption into "technoregions" can decrease dependence on decaying wildfire-prone high-tension power lines, meanwhile improving energy efficiency and decreasing consumption. This includes authors reimagining energy systems through social movements, expanded democratic and community governance practices. Part II explores solar transition in India, community adaptations of micro grids in Puerto Rico, and, as Caroline Wright shows, the difficult, but important energy justice possibilities held by energy cooperatives. Demonstrating the transparency, judicial and procedural challenges for offshore wind energy development in Brazil, Thomas Xavier and colleagues outline comprehensive procedural, participatory, and monitoring pathways to improve democratic processes and work toward energy justice.

Taking another step toward critical engagement, Part III looks at the "insecurities" and "constraints" of energy democracy. Bidtah Becker and Dana Powell reveal how resource extractivism and energy development collide with CoVID-19 in Navajo territory, with the latter exaggerating existing "disconnection, intergenerational trauma, and infrastructural precarity." Following this is a look into the Argentinian side of the "lithium triangle," examining the impact of lithium mining necessary for the production of lithium-ion batteries instrumental for facilitating the so-called "green" energy transition. The chapters here turn to examine authoritarian energy governance systems, the politics of nuclear energy, wind energy development, and even, as Peta Ashworth and Kathy Witt show, the "psychic numbing" underlining political apathy, inadequate environmental policy and, consequently, prolongs the onset of climate catastrophe. More still, energy security framings, high-consumption practices, and energy democracy are further unpacked. The findings from Ekaterina Tarasova and Harald Rohracher investigating "smart grids" in Sweden convey a reoccurring message across many of the chapters: the current "course of sustainability transitions" are intensifying old "inequalities and injustice," meanwhile potentially creating new ones.

Energy Democracies remains an impressive collection of thoughtful contributions that will prepare readers to confront the organizational, operational, and environmental justice challenges inherent with the energy transition, decarbonization, and, of course, energy democracy. In the tradition of *Energy Democracies* critical outlook, this conclusion seeks to widen further this critical discussion. Highlighting areas for investigation, the conclusion briefly discusses five areas in need of greater attention and care. This includes challenging further: (1) The "fossil fuels versus renewable energy dichotomy" and related supply-webs; (2) quantitative data collection and energy models; (3) the normative language in energy research; (4) greater engagement with degrowth literature; and (5), in line with the book, further unpacking and questioning democracy. By further unraveling these areas, the conclusion seeks to indicate doorways toward an insurrection of energy research, challenging—if not overthrowing—existing conceptions, methodologies, terminology, economic and political forms of organization, and, consequently, research.

2 Toward an energy research insurrection

Democratizing energy systems will make social and, potentially, ecological improvements, becoming indispensable for creating *real energy transitions*. On the other hand, taking into account the assemblage of production and enmeshed socio-political factors, the book demonstrates well that democratization is "not enough." The roots of techno-capitalism, its epistemology, concepts, forms of organization, and organizational structure have normalized various types of violence, making the infrastructural harms emanating from them appear invisible to people who are not directly and "quickly"—as opposed to "slowly" [1]—impacted by them. Locating reoccurring gaps in the literature, this conclusion seeks to increase critical pressure and precision to begin a process of *real energy transition*. Firstly, an insurrection in energy research situates itself in recognizing that "energy transition," as it currently stands, is a myth or hope and not a historical fact [2,3]. Unfortunately, as the introduction to this book outlined, the only energy transition in the process—operating on a planetary scale—started over 200 years ago and it is a transition in the wrong direction: from renewable and low-energy technologies to industrialized systems dependent on increasing energy and material intensive computational technologies.

This conclusion wants to widen the research agenda, pushing readers to go deeper with their assessment of energy infrastructural development and transition. This necessitates questioning all the little Latourian "black boxes" that operate within energy research. The themes discussed below, we can call, in popular parlance, research pathways toward decolonizing energy transition [4] or, thinking of Foucault's [5] "insurrection of subjugated knowledges," providing projectiles for an "insurrection in energy research." Thinking of the history of "social acceptance," outlined by Susana Batel [6], which traditionally conceptualized "local opposition as deviant and something to understand only in order to be overcome." Insurrectionary energy research responds by privileging dissenting perspectives within research areas—stakeholders opting to question the ideologies of economic growth, technological progress, and reject project development entirely—to create space for often under representative recalcitrant and so-called "insurgent" voices.[a] This opens space for experiences, knowledge, and perspectives about the socio-ecological realities of energy infrastructures to come to the foreground within public and policy discourses. Aside from privileging dissenting voices, the remainder of this section will present five areas in need of greater consideration, in hopes to ferment an insurrection in energy research. This entails widening the focus and taking more holistic—and embedding a (non-modeling) life cycle-oriented—approaches within the energy and infrastructural research. Implicitly, this contends that by recognizing the depth of the "energy problem" people can create stronger foundations for institutional change, meanwhile charting creative pathways for potential solutions—making rebellious and ecologically sustainable dreams lived realities. While not complete in any way, the following offers five noticeable blind spots in need of greater uptake, acknowledgment, and research.

2.1 Fossil fuels vs renewables: Participatory inclusion for invisibilizing supply-webs

The resource extractive reality behind so-called "renewable energy" and "lower-carbon" supply-webs is underestimated and largely ignored. In the book's collection, the introduction references the issues of rare earth mining, meanwhile, Melisa Escosteguy and colleagues reveal the realities behind securing lithium-ion batteries for electric vehicles and other lower-carbon infrastructures. This also includes Lourdes Alonso-Serna and Edgar Talledos-Sánchez noting the multi-dimensional extractive reality behind wind turbines. The issue of energy supply-webs remains largely neglected in the discipline, gaining popularity only recently in reports and social science research [7–11]. The severity—even calamity—related to mining, processing, and manufacturing for "energy democracy" remains generally underacknowledged. The socio-ecological issues relating to the impacts of energy infrastructure supply-webs makes or break the concepts of "transition," "clean," "renewable," and "green" energy. This dimension deserves further attention and inclusion within research, remembering that behind every operational site of energy infrastructure is extensive mining and manufacturing supply-webs.

Consider these conservative estimates. The World Bank [10], based on ambitious global temperatures scenarios, explains: "[D]emand for aluminum, indium, and silver are expected to *increase by more than 300% by 2050* from the [2018] base scenario, while the demand for copper, iron, lead, neodymium, and zinc is expected to increase *by more than 200percent*" (emphasis added). The situation, however, is radically underestimated. In the EU alone,

---

[a] These voices and concerns are, in fact, normal and reasonable, yet are implicitly understood "deviant," "wrong" and, when taking direct action, "insurgent" from statist and capitalist developmentalist perspectives.

demand for lithium, dysprosium, cobalt, neodymium, and nickel increase by up to 600% in 2030 and up to 1500% in 2050. Batteries for electric vehicles and low-carbon technologies will drive the 2030 demand for lithium up by 1800% and cobalt by 500%, and in 2050 demand will increase by almost 6000% for lithium and 1500% for cobalt [12]. These approximations, however, still do not take into account distributional, or "secondary," infrastructures (e.g., transformers, power lines), e-bikes, scooters, and "smart grid" technologies. Nor does this anticipate unexpected increases in electric vehicle demand currently in process in Norway.

The extractive statistics above are limited, having missing data and reductive by methodological nature, which is discussed more below. Yet importantly, these numbers do not discuss mineral processing (leaching, etc.), smelting, transportation, manufacturing, and corresponding socio-ecological conditions. A recent study by Andrea Brock and colleagues [13] is insightful, revealing the complications and problems related to solar manufacturing. The article reminds us once again that so-called renewables are deeply integrated with and dependent on the global economy and other toxic industries, which—of course—necessitate fossil fuels. Supply-webs reveal the harsh reality that fossil fuels and renewable energy are deeply intertwined and dependent on each other [7,14]. Every machine—electric, digitalized or not—in mining sites, processing plants, manufacturing facilities, and the transportation sector relies on hydrocarbons. So-called renewable energy infrastructures are dependent on hydrocarbons in every single phase of their existence, from their conception to decommissioning into landfills [9]. This is why fossil fuel+ is a more accurate term for "renewable energy," because every aspect of wind, solar, and hydrological infrastructures are dependent on extensive—under and unaccounted for—uses of hydrocarbons. Meanwhile, the "+" or "2.0" (in Spanish and Italian) indicates the energy harnessing, or green extractivist character, that absorbs wind, solar, and hydrological kinetic energy into the energy grid and capitalist industries [7,14]. This "plus" or "2.0," critically, however, does not take into account disrupting solar-landscape cycles or wind "velocity deficits" created by wind turbines that are empirically proven to have "statistically significant impacts on near-surface air temperatures and humidity as well as surface sensible and latent heat fluxes" [15]. This results in land drying, stressing water resources, and climatic temperature increases [15], which resonates with ethnographic works [16–18]. The marketed claim of "renewable inexhaustibility" requires sensitivity and greater critical reflection. Furthermore, it should be remembered, as others have pointed out [4,19,20], raw material extractivism and processing remains an issue for community-based energy systems, which intersects with the various issues of environmental justice issues covered within *Energy Democracies*. This contends, as I have elsewhere [14], that environmental justice and "social acceptance" needs to take place at every point along with a supply web, which stresses the importance of Myles Lennon [4] advocacy for "supply chain solidarity" and other important steps toward trying to make wind, solar, and water infrastructures actually—and really—renewable.

2.2 Epistemology/ontology of quantitative and modeling studies

The extractive data referenced above from the World Bank and European Commission (EC) on raw material required for lower-carbon energy infrastructures, like most data of this type, is quantitative and based on models. This, as most researchers know, means there are serious limitations, or reductions, in terms of acknowledging ecosystem toxification, political violence, discrimination practices, and the overall cumulative effect of multiple industries working together: timber, mining, infrastructure, conservation, smelting, manufacturing, transportation, and so on. What constitutes "data," "how it is collected," the ideological assumptions underlining the research (e.g., liberal governance), and, finally, the models' data is plugged into that rely on making "assumptions." This research is reductive and remains particularly influential to public policy. Working on raw material extraction for lower-carbon infrastructures, a modeling researcher explains to me, "What we are able to do is *to* characterize the context around that mining project." Moreover, the referenced World Bank and EC studies are lacking or do not have data on numerous minerals. This extends to faulty or outdated (by decades) environmental impact assessments and does not extend to many so-called "smart" or "clean" energy devices relying on digital platforms.

The culture, methods, and authority granted to quantitative data is a research-policy issue that pervades all energy research. Democratic decision-making does not extend to questioning the dominance of epistemological cultures, which democratic reform can potentially improve by widening or creating spaces of participation and debate, yet there is no guarantee. Antti Silvast and colleagues [21] have begun to look into the psychology and practice of energy modelers, which deserves continued research to understand the socio-disciplinary blind spots normalized into public policy and knowledge production in general. Moreover, this issue of quantitative data modeling speaks to the issue of epistemic discrimination and/or racism within universities [22]. This does not only play out with modeling, but how environmental impact assessments [23] and other assessments ignore local knowledge or Indigenous sciences.

Epistemic discrimination emerges when challenging or looking to other explanations outside Western science or biomedicine to understand the impacts of energy infrastructure. The issue of electromagnetic fields and illness are reoccurring issues across energy infrastructure conflicts [18,24]. The materialist focus inherent in biomedical approaches ignores energy flows or vitality within the body, which—inversely—is central to traditional Chinese medicine with the concept of Qi [25]. Medical anthropology is clear about the necessity of building "asymmetry" and "medical pluralism" [26], or simply mutual respect, between biomedicine and other established sciences—pejoratively and indiscriminately lumping different sciences and practices together as "alternative medicines"—to understand different phenomenon or ailments within people. Recognizing other medicines and sciences to understand the reality of energy infrastructure, demonstrates an area necessitating an insurrection of subjugated knowledge subordinated by modern science [27,28]. The insurrection or decolonizing of energy research necessitates not only questioning—and deconstructing—the culture of quantitative data modeling, but also creating openings and spaces for different epistemological sciences such as traditional Chinese and Ayurveda medicines to contribute with revealing the extent of issues related to energy infrastructural development, operation, and decommissioning.

2.3 The poverty of energy research language

The terms employed in energy research are inaccurate, expressing ideological bias, misleading hopes, and (unintentionally) weave complicity with expanding and renewing capitalist socio-ecological destruction in the "Global South" and "North." Whether an issue of academic rigor or distancing from brutal—even genocidal [29]—extractivism, this conclusion asks that we call energy infrastructure what it is, which is not a "farm," "park," "clean," "green," "renewable," or even "low-carbon."

There is, as already mentioned, no such thing as renewable energy as we presently know it [14]. This is not to say it is impossible to create renewable energy, because nothing is impossible. This, however, will require establishing careful reciprocal relationships with habitats and creating serious efforts at supply-chain solidarity that are actively working to heal and create *real restoration*, as opposed to the "restorative" manipulations of market-based conservation [30,31]. Creating community renewable energy ecologies (CREE), as developed by Christina Siamanta [20], demand great effort to close and remediate the socio-ecologically degrading extractive processes related to the production/consumption energy generation systems. The first step to creating CREEs is to change the discourse—the language—and move toward accurate portrayals of energy infrastructures. This entails, as the supply-web subsection above contends, that the "renewable energy versus fossil fuels dichotomy" is false [7,14], the two energy systems are deeply intertwined to the point that research unraveling these connections, to put it lightly, is strenuous. Even textbooks on the matter admit: "there can be no such thing as a 'sustainable supply chain'" [32]. Furthermore, wind and solar generation infrastructure are not "parks," "farms" and, in most instances, are not even "public utilities" because of private-sector domination and profiteering. In reality, these infrastructures are "factories" [33], elaborate technological assemblages that industrialize and colonize rural environments with the values of utility, production, profiteering, and efficiency. Meanwhile, wind and solar factories require maintenance and upkeep, emerging as 21st-century factories relying on robots, algorithms, and centralized digitalized controls. So let us start calling so-called renewable energy infrastructure what they really are: power plants, factories, or fossil fuel + infrastructures, not a "farm," "park," or "renewable."

Friendly infrastructural terminology is misleading. Eyebrows can also raise at the term "lower-carbon" or "decarbonization." The term "lower-carbon" infrastructure is safer, but even this term—taking into account the extensive supply web—could be potentially misleading. While life-cycle assessments are comparing conventional and fossil fuel + power plants [34], because of the epistemological culture and data oversights mentioned above, even the term "lower-carbon" infrastructure is debatable when employed within a techno-capitalist system organized around infrastructural expansion and economic growth. This highlights the importance of the idea and literature on "degrowth," which the next section turns.

2.4 Degrowth now!

Contributors within *Energy Democracies* implicitly recognize the necessity to reduce energy consumption. Discussing "convivial assemblages of energy" or Arthur Mason's critique of "Deluxe Energy" indicate this, yet the book had little to no direct engagement with the degrowth literature. Degrowth advocates for the planned reduction of energy and material throughput to restore balance with the planet, meanwhile reducing inequality and improving human well-being [35]. While *Energy Democracies* remains implicitly ground in this tension, it could not be overstated enough that renewable community energy ecologies are pointless without a degrowth ethic, challenging—if not hostile—to the (capitalist) objective of economic, material, and energy growth. Moreover, considering the popularity of degrowth and, with the exception [36], the relative lack of specific engagement with energy infrastructures makes it an important site of co-development between research schools. Expanding and/or developing degrowth technologies for energy transition remains an important area to develop and expand [37] because medium-scale wind and solar power plants do not necessarily challenge growth imperatives [20]. In Germany, Rommel and colleagues [36] remind us community energy projects "are dominated by the middle class" and participants are not "critical of technology, and as such, few members explicitly criticize excessive consumption and energy use. In fact, CRE [community renewable energy] is dominated by technophile eco-modernists." Lacey-Barnacle and Nicholls, in Chapter 14, remind us how "municipal energy models allow for little democratic involvement in decision-making procedures, which despite being publically funded, leave democratic processes to local government executives, offering no opportunities for wider public input into decision making by local government tax payers." While Lacey-Barnacle and Nicholls highlight important procedural issues, it is important to remember that energy democracy does not necessarily ensure degrowth [4,20,36,38]. The book extensively documents on-site energy infrastructural procedural and governance issues, yet stronger and more explicit links with supply-webs and degrowth require further development.

2.5 The limitations of democracy

Offering a genealogy of democracy, Ferit Güven's [39] demonstrates a reading of its origins as a socio-political control mechanism designed to adapt to the growing Ancient Greek City state. Güven reveals how this model of political control was an important export of colonialism, consequently arguing for the necessity of "decolonizing democracy." This perspective complements and mirrors various anarchist critiques of democracy, some celebrating notions of "direct democracy" [40,41] or others refuting it [42–44]. There is always a risk of democracy reproducing the same colonial-statist dynamics [45,46]. *Energy Democracies* demonstrates across numerous countries, theories, and field sites how democratizing energy offers important procedural improvements, yet it still has to contend with the fact—as other contributors in energy research reveal [4,20,36,38]—that "direct democracy is just representative democracy on a smaller scale" [44].

Whether municipal governance or tribal chiefs, who actually adequately represents the "community" and allows their participation in designing their energy futures. As mentioned above, there are middle-class, eco-modernist, and consumerist tendencies that pervade community renewable energy systems [36], and, as the majority of works within this volume indicate, there are significant deficiencies in capacity, participation, and distribution of costs and benefits. Then there is the question, what if some people do not want more energy infrastructure? Does the majority rule? Does this majority silence the voice of nonhumans, deciding what flora and fauna are allowed to live or let die? Furthermore, direct democracies then have to confront supply-webs, grid contracting issues (e.g., feed-in-tariff), ideological and consumer tendencies [4,14,38]. The issue of social engineering, or manipulating, representative democracy [47], local governments [48], or big assemblies [44] is an area with room for further exploration. It remains a challenge, as book contributors demonstrate, how people will choose to develop infrastructure, and who they will do it with (municipalities, worker coopts, state initiatives, corporations, NGOs, nonprofits, etc.). At this point, as *Energy Democracies* reveals, the situation is rather constrained on the ground, even if this is changing as we speak.

Drawing on anarchist critique [44,45], energy democracies require special attention to avoid reproducing specialists that control political, economic, and technical processes. This includes reinforcing centralized political and economic systems via definitions of decentralization, which can lead to exclusions or, more accurately, minimizing the voice critical and/or rejecting claims of technological progress, anticipated ecological degradation, and profiteering. While energy development might always require some level of technical expertise, *Energy Democracies* has been exceptional in stressing the duel need of direct democracy and profound decentralization together. Encouraging smaller-scales, microgrids, direct democracy, and creating close proximity between production and consumption are crucial ways forward to creating real community renewable energy ecologies [20]. The concern to consider moving forward is whether democracy is overemphasizing the means over the ends, creating bureaucratic controls unresponsive to local needs, and together creating a system that always discriminates against the nonhuman and specific humans racialized and classed within techno-capitalist society. While an important avenue to develop, direct democracy requires caution as it can potentially reinforce and reproduce the failures of representative democracy, we know too well. Rejecting authoritarianism, and allowing thinking past the confines of democracy to other forms of autonomist, technoregionalist, and decentralized forms conscious of supply-web violence and national grid domination remain pathways for an insurrection against energy transition as we have it today.

3 Conclusions

This concluding chapter has located areas to reconsider. An insurrection within energy research can begin by examining the political ecology of supply-web violence invisibilized by the marketing of the "renewability," "green," and "clean" claims of fossil fuel + power plants. This entails also questioning and further deconstructing the epistemological and quantitative culture providing inadequate and incomplete data that fashions claims of "green" and "renewable" infrastructures. Complementing this is changing our language around energy infrastructure to communicate their socio-ecological realities more accurately. This continues to encourage linking degrowth with technology and energy research, as well as advancing critical inquiry—as this book does—on the reality and complications associated with direct energy democracy. These areas in need of further connection, development, and research amount to a call for decolonizing energy research and development and encouraging researchers to engage in an individual or collective insurrection of energy transition as we know it. Getting to the roots of socio-ecological catastrophe, an insurrection of energy transition begins with researching and supporting post-developmental and degrowth processes that not only deconstruct the costly myths of "renewable energy," "transition" and "democracy" that mask trends of authoritarianism and epistemic discrimination in the name of "inclusion," "participation," and "sustainability." But this also includes working toward a praxis designed to create community renewable energy ecologies, in the most honest sense of the term, to create a future without discrimination and developing habitats with healthier rivers, plants, soils, trees, and relationships imbued with care, ready to defend against destructive development in everybody's backyard, whether mine, factory, or wind turbine.

References

[1]   R. Nixon, Slow Violence and the Environmentalism of the Poor, Harvard University Press, Cambridge, 2011.

[2]   C. Bonneuil, J.-B. Fressoz, The Shock of the Anthropocene: The Earth, History and Us, Verso Books, New York, 2016.

[3]   V. Smil, Energy Transitions: Global and National Perspectives, Praeger, Santa Barbara, 2016.

[4]   M. Lennon, Energy transitions in a time of intersecting precarities: from reductive environmentalism to antiracist praxis, Energy Res. Soc. Sci. 73 (2021) 1–10.

[5]   M. Foucault, "Society Must Be Defended:" Lectures at the College De France 1975–1976, Picador, New York, 2003 (1997).

[6]   S. Batel, Research on the social acceptance of renewable energy technologies: past, present and future, Energy Res. Soc. Sci. 68 (2020) 1–5.

[7]   End the "Green" Delusions: Industrial-Scale Renewable Energy Is Fossil Fuel+, Verso Blog, 2018. https://www.versobooks.com/blogs/3797-end-the-green-delusions-industrial-scale-renewable-energy-is-fossil-fuel. Accessed 10-05-2018.

[8]   A Just(Ice) Transition Is a Post-Extractive Transition: Centering the Extractive Frontier in Climate Justice, War on Want & London Mining Network, 2019. https://catapa.be/wp-content/uploads/2019/09/Post-Extractivist_Transition.pdf. (Accessed 20 February 2019).

[9]   B.K. Sovacool, A. Hook, M. Martiskainen, A. Brock, B. Turnheim, The decarbonisation divide: contextualizing landscapes of low-carbon exploitation and toxicity in Africa, Glob. Environ. Chang. 60 (2020) 1–19.

[10]   Minerals for Climate Action: The Mineral Intensity of the Clean Energy Transition, The World Bank Group, 2020. http://pubdocs.worldbank.org/en/961711588875536384/Minerals-for-Climate-Action-The-Mineral-Intensity-of-the-Clean-Energy-Transition.pdf. Accessed 13-05-2020.

[11]   A Material Transition: Exploring Supply and Demand Solutions for Renewable Energy Minerals, War on Want, 2021. https://waronwant.org/sites/default/files/2021-03/A%20Material%20Transition_report_War%20on%20Want.pdf. Accessed 20-03-2021.

[12]   Critical Raw Materials for Strategic Technologies and Sectors in the Eu: A Foresight Study, European Commission, 2020. https://ec.europa.eu/docsroom/documents/42881. Percentage break down by Meadhbh Bolger from Friends of the Earth Europe in a forthcoming report, Accessed 20-11-2020.

[13]   A. Brock, B.K. Sovacool, A. Hook, Volatile photovoltaics: green industrialization, sacrifice zones, and the political ecology of solar energy in Germany, Ann. Am. Assoc. Geogr. (2021) 1–23.

[14]   A. Dunlap, Does renewable energy exist? Fossil fuel+ technologies and the search for renewable energy, in: S. Batel, D.P. Rudolph (Eds.), A Critical Approach to the Social Acceptance of Renewable Energy Infrastructures—Going beyond Green Growth and Sustainability, Palgrave, London, 2021, pp. 1–12.

[15]   S.A. Abbasi, Tabassum-Abbasi, Tasneem-Abbasi, Impact of wind-energy generation on climate: a rising spectre, Renew. Sust. Energ. Rev. 59 (2016) 1591–1598. 1594.

[16]   C.F. Lucio, Conflictos Socioambientales, Derechos Humanos Y Movimiento Indígena En El Istmo De Tehuantepec, Universidad Autónoma de Zacatecas, Zacatecas, 2016.

[17]   J. Franquesa, Power Struggles: Dignity, Value, and the Renewable Energy Frontier in Spain, Indiana University Press, Bloomington, 2018.

[18]   A. Dunlap, Renewing Destruction: Wind Energy Development, Conflict and Resistance in a Latin American Context, Rowman & Littlefield, London, 2019.

[19]   M.J. Burke, J.C. Stephens, Political power and renewable energy futures: a critical review, Energy Res. Soc. Sci. 35 (2018) 78–93.

[20]   Z.C. Siamanta, Conceptualizing alternatives to contemporary renewable energy development: community renewable energy ecologies (Cree), J. Polit. Ecol. 28 (1) (2021) 258–276.

[21]   A. Silvast, E. Laes, S. Abram, G. Bombaerts, What do energy modellers know? An ethnography of epistemic values and knowledge models, Energy Res. Soc. Sci. 66 (2020) 1–8.

[22]   R. Grosfoguel, Del extractivismo económico al extractivismo epistémico y ontológico, Rev. Int. Comun. Desarrollo 1 (24) (2016) 123–143.

[23]   R. Lawrence, R.K. Larsen, The politics of planning: assessing the impacts of mining on sami lands, Third World Q. 38 (5) (2017) 1164–1180.

[24]   A. Dunlap, Bureaucratic land grabbing for infrastructural colonization: renewable energy, l'amassada and resistance in Southern France, Hum. Geogr. 13 (2) (2020) 109–126.

[25]   T.J. Kaptchuk, Chinese Medicine: The Web That Has No Weaver, Random House, 2000.

[26] H.A. Baer, Medical pluralism: an evolving and contested concept in medical anthropology, in: M. Singer, P.I. Erickson (Eds.), A Companion to Medical Anthropology, Wiley-Blackwell, Malden, 2011, pp. 405–423.

[27] V. Shiva, Staying Alive: Women, Ecology and Development, Zed Books, London, 2002 (1989).

[28] Q.G. Eichbaum, L.V. Adams, J. Evert, M.-J. Ho, I.A. Semali, S.C. van Schalkwyk, Decolonizing global health education: rethinking institutional partnerships and approaches, Acad. Med. 96 (3) (2021) 329–335.

[29] A. Dunlap, The politics of ecocide, genocide and megaprojects: interrogating natural resource extraction, identity and the normalization of erasure, J. Genocide Res. (2020) 1–26.

[30] Antipode Foundation, Intervention—"Accumulation by Restoration: Degradation Neutrality and the Faustian Bargain of Conservation Finance", Antipode Foundation, 2017, pp. 1–4. https://antipodefoundation. org/2017/11/06/accumulation-by-restoration/.

[31] A. Dunlap, S. Sullivan, A faultline in neoliberal environmental governance scholarship? Or, why accumulation-by-alienation matters, Environ. Plan. E Nat. Space 3 (2) (2020) 552–579.

[32] Y. Bouchery, C.J. Corbett, J.C. Fransoo, T. Tan, Sustainable Supply Chains: A Research-Based Textbook on Operations and Strategy, vol. 4, Springer, New York, 2017, p. 2.

[33] A. Dunlap, M. Correa-Arce, 'Murderous energy' in Oaxaca, Mexico: wind factories, territorial struggle and social warfare, J. Peasant Stud. (2021) 1–27.

[34] H. Li, H.-D. Jiang, K.-Y. Dong, Y.-M. Wei, H. Liao, A comparative analysis of the life cycle environmental emissions from wind and coal power: evidence from China, J. Clean. Prod. 248 (2020) 119–192.

[35] J. Hickel, Less Is More: How Degrowth Will Save the World, Random House, 2020.

[36] J. Rommel, J. Radtke, G. Von Jorck, F. Mey, Ö. Yildiz, Community renewable energy at a crossroads: a think piece on degrowth, technology, and the democratization of the German energy system, J. Clean. Prod. 197 (2018) 1746–1753. 1751.

[37] C. Kerschner, P. Wächter, L. Nierling, M.-H. Ehlers, Degrowth and technology: towards feasible, viable, appropriate and convivial imaginaries, J. Clean. Prod. 197 (2018) 1619–1636.

[38] A.A. Smith, D.N. Scott, 'Energy without injustice'? Indigenous ownership of renewable energy generation, in: S.A. Atapattu, C.G. Gonzalez, S.L. Seck (Eds.), The Cambridge Handbook of Environmental Justice and Sustainable Development, Cambridge University Press, Cambridge, 2021, pp. 383–398.

[39] F. Güven, Decolonizing Democracy: Intersections of Philosophy and Postcolonial Theory, Lexington Books, London, 2015.

[40] M. Bookchin, Libertarian municipalism: an overview, Green Perspect. 24 (1991) 1–6.

[41] D. Graeber, Direct Action: An Ethnography, AK press, Oakland, 2009.

[42] This Is What Democracy Looks Like, Elephant Editions, 2006. https://theanarchistlibrary.org/library/ various-authors-this-is-what-democracy-looks-like.

[43] P. Gelderloos, The Failure of Nonviolence: From Arab Spring to Occupy, Left Bank Books, Seattle, 2013.

[44] Fire Extinguishers and Fire Starters, Crimethinc, 2011. https://crimethinc.com/2011/06/08/fire-extinguishers-and-fire-starters-anarchist-interventions-in-the-spanish-revolution-an-account-from-barcelona. Accessed 20-04-2021.

[45] P. Gelderloos, Worshiping Power: An Anarchist View of Early State Formation, AK Press, Oakland, 2017.

[46] Reconsidering the logistics of autonomy: ecological autonomy, self-defense and the polícia comunitaria in álvaro obregón, mexico, Conference Paper 8, Emancipatory Rural Politics Initative (ERPI), 2018. https://www. tni.org/files/article-downloads/erpi_cp_8_dunlap.pdf. Accessed 20-03-2018.

[47] E.S. Herman, N. Chomsky, Manufacturing Consent: The Political Economy of the Mass Media, Random House, New York, 2010 (1989).

[48] J. Verweijen, A. Dunlap, The evolving techniques of social engineering, land control and managing protest against extractivism: introducing political (re)actions 'from above', Polit. Geogr. 83 (2021) 1–9.

Alexander Dunlap is a postdoctoral fellow at the Centre for Development and the Environment, University of Oslo. His work has critically examined police-military transformations, market-based conservation, wind energy development, and extractive projects more generally in Latin America and Europe. He has published two books: *Renewing Destruction: Wind Energy Development, Conflict and Resistance in an American Context* (Rowman & Littlefield, 2019) and, the co-authored, *The Violent Technologies of Extraction* (Palgrave, 2020).

Index

Note: Page numbers followed by *f* indicate figures and *t* indicate tables.